CHIEF

THE EVOLUTION, DEVELOPMENT AND ROLE OF THE CHIEF PETTY OFFICER IN THE UNITED STATES NAVAL SERVICE

TURNER PUBLISHING COMPANY

TURNER PUBLISHING COMPANY

"Chiefs" Staff:
Joe B. Havens, Contributing Editor

Turner Publishing Company Staff:
Chief Editor: Robert J. Martin

Library of Congress
Catalog Card Number: 96-61997
ISBN: 978-1-56311-248-5

Additional copies may be purchased from
Turner Publishing Company.

This publication was compiled using available information.
The publisher regrets it cannot assume liability for errors
or omissions.

Endsheets: USS *Chief* (MCM-14)

Photo this page: USS *Estes* (AGC-12)

TABLE OF CONTENTS

Acknowledgements

"CHIEF: The evolution, development and role of the chief petty officer in the U.S. Naval Service" never could have been completed without the great help and support of persons whose names appear below. First and foremost, my son, Emmett A. Havens, a graphic design artist, whose vision and creative imagination inspired the cover design. Certainly others, to name a select few, made significant contributions, namely, Lee Miller Hughes; Leigh Taylor Hughes; Shipmate Harold Mull, who literally scanned the Naval Historical Center and kept sending messages of encouragement and a passionate plea for patience on my part; staff of the naval institute press; Phyliss Hansen, the accomplished Public Affairs Officer at Cnett, NAS, Memphis; then there was Dr. Dean C. Allard and staff at the Naval Historic Center; Master Chief Petty Officer of the Navy, John Hagan and his very capable Chief Journalist, Neil R. Gillibeau. My heartfelt appreciation to all! Godspeed!

Dedication

This book is dedicated to a special breed of people – chief petty officers, past, present and future, of the United States Navy and to all who aspire to attain this coveted status. Also, to the United States Navy itself, an entity wherein this status rightfully belongs. For all the men and women who have aspired to and gained the status of chief petty officer bound by the common thread – love of country and its heritage that fuels the passionate opportunity to enhance the lives of all we touch during our service. For the elite group privileged to be called "chief."

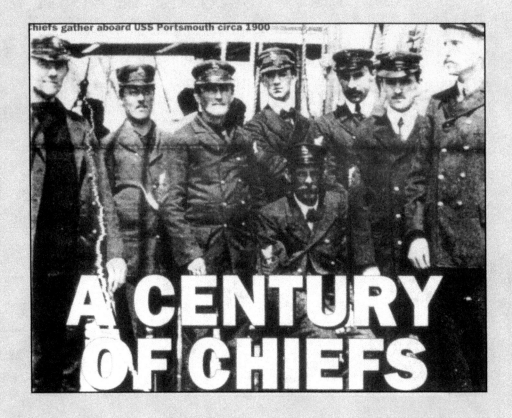

Chiefs gather aboard USS Portsmouth circa 1900

A CENTURY OF CHIEFS

Joe B. Havens, USN (Retired)

This book is proudly presented

to the

United States of America

by

Joe B. Havens

Senior Chief Hospital Corpsman

U.S. Navy, Retired

and

All those shipmates whose

names and/or memories

appear within

Chief of Naval Operations
11 December 92
The Chief Petty Officer

Admiral Frank B. Kelso, II

Achieving the rank of chief petty officer is the most significant rite of passage enlisted men and women in the United States Navy experience. We expect – we demand – more of them, not because they have been advanced on pay grade, but because they have become chief petty officers.

In the United States Navy, the title of chief petty officer carries with it responsibilities and privileges no other armed force in the world grants enlisted people. These responsibilities and privileges exist because, for more than 100 years, chiefs have freely accepted duty beyond call, and, by their actions and performance, commanded the respect of their seniors as well as their juniors.

The example these chiefs set inspires our young men and women to this day. Certainly what Americans see in our impressive young sailors is the tradition of devotion and dedication the first chiefs set with their sacrifices and their valor.

Their successors, today's chief petty officers, are no less dedicated. They prove their worth every day, and they continue to meet great challenges and endure adversity to protect our nation's interests.

My challenge to chief petty officers of the 21st century is to reaffirm the commitment to the faith and fellowship that has allowed their comrades in arms before them to wear "the hat" with tremendous pride for a century.

It is my great privilege to salute the chief petty officers of the past, the present and the future.

Frank B. Kelso, II
Admiral, U.S. Navy

The Secretary of the Navy
Washington

Sean O'Keefe

10 December 1992

It is said that chief petty officers form the backbone of the United States Navy. I believe that chiefs run the division, teach the young sailors, repair the equipment, and train the junior officers. They are at the heart of all that we do in the Navy today.

I am very happy that a retired senior chief petty officer with an outstanding career in the Navy behind him has chosen to undertake the important task of writing this centennial study of Navy chief petty officers.

As all who have served in the Navy will agree, there is no higher honor within the Navy than to be addressed by that hundred-year-old question, "What do we do now, chief?" This book will provide the reader with a solid understanding of why our Navy chiefs are always prepared with a good answer to that often difficult query – and why they are so important and vital to the continued success of the United States Navy.

Cordially,

Sean O'Keefe
Acting

Rear Admiral Albert M. Sackett, USN (Retired)

Rear Admiral Albert M. Sackett, USN (Retired)

Rear Admiral Albert M. Sackett was born in Victor, Iowa, on June 24, 1920. He joined the Navy in 1937 and completed recruit training at Naval Training Center, Great Lakes. His initial assignments included the USS *Northampton*, Surface Diesel School at the Submarine Base, New London, Connecticut, and the USS *Jamestown*. He attained the rate of chief petty officer prior to receiving his commission.

After being commissioned an ensign in 1944, he served as engineering officer and executive officer of the USS LCI (R) 7a4 and then as commanding officer of USS LCI (R) 34, and commanding officer of USS LST-990 and USS LSM-109.

His subsequent assignments included USS *Joseph P. Kennedy* (DD-850) as engineering officer; squadron engineering officer with Destroyer Squadron 16; Staff of Commander Destroyer *Flotilla Four* as Material and Operations Officer; Operations, Plans and Security Officer, Naval Training Center, Great Lakes; and commanding officer of USS *Preston* (DD-795).

After a course in Naval Warfare at the Naval War College, he became the first commanding officer of the guided missile destroyer USS *John King* (DDG-3).

Rear Admiral Sackett then served as head, Missile Systems Training Branch, Bureau of Naval Personnel and special assistant to the Under Secretary of the Navy before assuming command of the guided missile cruiser USS *Gridley* (CG-21).

He became commanding officer of the Naval Destroyer School, Newport, RI, in June 1969. In August 1970 he returned to the Bureau of Naval Personnel for duty as director of the Officer Distribution Division, where he was responsible for the duty assignments of the 72,000 officers in the Navy.

Upon being selected to Flag Rank in 1972, he was assigned duty as chief of Naval Technical Training with headquarters at Naval Air Station, Memphis, Tennessee. His duties encompassed the responsibility for almost all enlisted training in the Navy – recruit, submarine, surface, and aviation, conducted at 53 subordinate commands throughout the country. The average on-board staff and student population exceeded 60,000.

He next assumed duties as Commandant, Ninth Naval District and Commander Naval Base, Great Lakes.

Admiral Sackett retired from the United States Navy on July 1, 1977, and resides in Memphis, Tennessee, where he serves as director, Community Relations, Memphis Publishing Company.

Included among the many awards Rear Admiral Sackett has received are the American Defense Service Medal; Asiatic-Pacific Campaign Medal with Seven Stars; Navy Occupation Service Medal; China Service Medal; National Defense Service Medal with Bronze Star; Armed Forces Expeditionary Medal; the Philippine Liberation Ribbon with One Star; the Navy Commendation Medal with Gold Star; the Meritorious Service Medal with Gold Star; the Legion of Merit with Gold Star; and the Distinguished Service Medal.

Rear Admiral Sackett holds a BA degree in social science from George Washington University. He is married to the former Patricia Soer, also of Victor, and they have two daughters and four sons.

Admiral Elmo Russell Zumwalt, Jr., USN (Retired)

Elmo Russell Zumwalt Jr., was born in San Francisco, CA, on Nov. 29, 1920, son of Dr. E.R. Zumwalt and Dr. Frances Zumwalt. He attended Tulare (California) Union High School, where he was an Eagle Scout and Class Valedictorian and the Rutherford Preparatory School, at Long Beach, CA, before his appointment to the United States Naval Academy, Annapolis, MD, from his native state in 1939. As a midshipman he was president of the Trident Society, vice-president of the Quarterback Society, twice winner of the June Week Public Speaking Contest, (1940, 1941), company commander in 1941 and regimental three striper in 1942, and participated in intercollegiate debating. Graduated with distinction and commissioned ensign on June 19, 1942, with the Class of 1943, he subsequently progressed to the rank of admiral, to date from July 1, 1970.

Following graduation from the Naval Academy in June 1942, he joined the destroyer USS *Phelps*, and in August 1943 was detached for instruction in the Operational Training Command, Pacific, at the USS *Robinson*, and for "heroic service as evaluator in the Combat Information Center...(of that destroyer), in action against enemy Japanese battleships during the Battle of Leyte Gulf, 25 October 1944..." he was awarded the Bronze Star with Combat "V." The citation further states: "During a torpedo attack on enemy battleships, Lieutenant Zumwalt furnished information indispensable to the success of the attack...".

After the cessation of hostilities in August 1945, until December 8 of that year, he commanded (as prize crew officer) *HIMJS Ataka*, a 1200-ton Japanese river gunboat with 200 officers and men. In that capacity he took the first shop since the outbreak of World War II, flying the United States flag, up the Whangpoo River to Shanghai. There they helped to restore order and assisted in disarming the Japanese.

He next served as executive officer of the destroyer USS *Saufley*; and in March 1946 was transferred to the destroyer USS *Zellars*, as executive officer and navigator. In January 1948 he was assigned to the Naval Reserve Officers Training Corps Unit of the University of North Carolina at Chapel Hill, where he remained until June 1950. That month he assumed command of the USS *Tills* in commission in reserve status. That destroyer escort was placed in full active commission at Charleston Naval Shipyard on Nov. 21, 1950, and he continued to command her until March 1951, when he joined the battleship USS *Wisconsin* as navigator.

"For meritorious service as navigator of USS *Wisconsin* during combat operations against enemy North Korean and Chinese Communist forces in the Korean Theater from 21 November 1951—30 March 1952..." he received a Letter of Commendation, with Ribbon and Combat "V," from Commander *Seventh Fleet*. The letter continues: "As navigator his competence and untiring diligence in assuring safe navigation of the ship enabled the commanding officer to devote the greater part of his attention to planning and gunfire operations. His performance of duty was consistently superior in bringing the ship through dangerously mined and restricted waters, frequently under adverse conditions and poor visibility. He assisted in the planning of the combat operations...(and) piloted *Wisconsin* into the closest possible inshore positions in which maximum effect could be obtained by gunfire...".

Detached from USS *Wisconsin* in June 1952, he attended the Naval War College, Newport, Rhode Island, and in June 1953 reported as head of the shore and overseas bases section, Bureau of Naval Personnel, Navy Department, Washington, DC. He also served as officer and enlisted requirements officer and as action officer on medicare legislation. Completing that tour of duty in July 1955, he assumed command of the destroyer USS *Arnold J. Isbell*, participating in two deployments to the *Seventh Fleet*. In this assignment he was commended by the Commander, Cruiser-Destroyer Forces, United States Pacific Fleet for winning the battle efficiency competition for his ship and for winning excellence awards in engineering, gunnery, antisubmarine warfare, and operations. In July 1957 he returned to the Bureau of Naval Personnel for further duty. In December 1957 he was transferred to the office of the Assistant Secretary of Navy (Personnel and Reserve Forces), and served as special assistant for naval personnel until November 1958, then as special assistant and naval aide until August 1959.

Admiral Elmo Russell Zumwalt, Jr., USN (Retired)

Ordered to the first ship built from h keel up as a guided-missile ship, USS *Dewey* (DLG-14), building at the Bath (Maine) Iron Works, he assumed command of that guided missile frigate at her commissioning in December 1959, and commanded her until June 1961. During this period of his command, *Dewey* earned the excellence award in engineering, supply, weapons, and was runner-up in the battle efficiency competition. He was a student at the National War College, Washington, DC, during the 1961-62 class year. In June he was assigned to the office of the assistant secretary of defense (International Security Affairs), Washington, DC, where he served first as desk officer for France, Spain, and Portugal, then as director of arms control and contingency planning for Cuba. From December 1963 until June 21, 1965 he served as executive assistant and senior aide to the Honorable Paul H. Nitze, Secretary of Defense and the Secretary of the Navy, he was awarded the Legion of Merit.

After his selection for the rank of rear admiral, he assumed command in July 1965 of Cruiser-Destroyer Flotilla *Seven*. "For exceptionally meritorious service..." in that capacity, he was awarded a Gold Star in lieu of a Second Legion of Merit. In August 1966 he became director of the Chief of Naval Operations Systems Analysis Group, Washington, DC, and for "exceptionally meritorious service...as director, Systems Analysis Division, Office of the chief of Naval Operations, deputy scientific officer to the Center for Naval Analyses, during the period from August 1966—August 1968..." he was awarded the Distinguished Service Medal. The citation further states in part:

"Rear Admiral Zumwalt, by direction of the Chief of Naval Operations, established the Systems Analysis Division and rapidly developed it into a highly effective, responsive organization. Under his leadership, the division has assisted in generating within the navy a better understanding of requirements, problems and a more effective presentation of those requirements in major program areas which will strongly influence the combat capabilities of United States Naval Forces through the next generation. He has displayed exceptional acumen, integrity, tact and diplomacy as personal representative of the chief of Naval Operations, not only in dealings within the Department of Defense, but also in testifying before congressional committees. Among the major analyses completed under his direct supervision were the major Fleet Escort, antisubmarine warfare force level. tactical air, surface-to-surface missile, and war-at-sea studies. Additionally, under Rear Admiral Zumwalt's guidance, the Center of Naval Analyses has been restructured, and its methodologies clearly defined with such precision as to ensure that completed studies will reflect thoroughness, comprehensiveness, and accuracy when subjected to closes scrutiny...".

In September 1968 he became Commander Naval Forces, Vietnam and chief of the Naval Advisory Group, United States Military Assistance Command, Vietnam. President Richard M. Nixon nominated him as Chief of Naval Operations on April 14, 1970. Upon being relieved as Commander Naval Forces, Vietnam, on May 15, 1970, he was awarded a Gold Star in lieu of a second Distinguished Service Medal for exceptionally meritorious service. He assumed command as chief of Naval Operations on July 1, 1970 and retired from that position on July 1, 1974.

Since his retirement from naval service in July 1974, he has written a book entitled *On Watch* (1976), based on his four years as chief of Naval Operations. He and his late son, Elmo Russell Zumwalt III, Esq., co-authored a book, *My Father, My Son* (1986), which was later produced as a movie, describing their battle to deal with the son's Agent Orange induced cancer. He ran as the democratic candidate for the United States Senate in Virginia in 1976. He is currently serving as president of Admiral Zumwalt & Consultants and as chairman of the following organizations: The National Marrow Donor Program, Phelps-Stokes Fund, Ethics and Public Policy Center, the Marrow Foundation, The Agent Orange Coordinating Council, and Fleet Aerospace, Inc.

His official home address is Arlington, VA. He is married to the former Mouza Coutelais-du-Roche of Harbin, Manchuria, and they have one son, James G. Zumwalt and two daughters, Ann Zumwalt Coppola and Mouzetta Zumwalt Weathers. Their eldest son, Elmo Russell Zumwalt III, died of cancer in August 1988.

Admiral Jeremy Michael Boorda

Former Chief of Naval Operations

Commmissioning of USS *Chief*

Thank you Captain Garrold, thank you chief petty officers here today, and welcome to the crew, the families and the friends of our newest United States Navy ship, the USS *Chief*. It is truly wonderful to be here with all of you on this very special day.

What a great day for Navy people and for our Navy today! For the second time in our history, we will have a warship named for a very special group of Navy professionals, the chief petty officers of the finest, strongest Navy in the world.

Chief - what a great name for a Navy warship. *Chief* - the very name says it all—the best, number 1, a winner in every way - what a great name!

There are two very proud Secretaries of the Navy today. Our present secretary, John Dalton, asked me to tell you of his pride in this new ship, of those who will sail in her, and of the outstanding men and women, the chief petty officers, whose service for over 100 years has been a shining example for every shipmate, officer and enlisted.

The other proud secretary is one who served not long ago in that position, Lawrence Garrett, who during his tour decided to name MCM-14 the second USS *Chief* in honor of the chief petty officer community on their 100th anniversary. Secretary Garrett is here today and I know he made a wonderful decision, a decision he will remember and we will remember for many years as this ship, our newest chief, serves and serves well in our great Navy.

There are so many special people here today that I hesitate to single out any one of you for, in fact, you are all special. I must, however, mention the former and present Master Chief Petty Officers of the Navy (MCPON). Each and everyone is with us today. Would you all stand and receive a well deserved and heartfelt round of applause.

One of those master chiefs who was just recognized is Master Chief Petty Officer Duane Bushey. Now Duane had his share of recognition while serving as MCPON and doing so in such an outstanding fashion. But Duane, as great as you were, as great as you are, today is Sue Bushey's day for she was chosen as this ship's sponsor. She, represents all the best that wives and husbands of Navy chief petty officer are and can be. In fact, she represents all sailors wives and husbands and, by sailors, I mean sailors of every rank and grade from admiral to the most junior first termer. Sue Bushey has done it all. From loving and encouraging - your petty

officer to be all that he could be and seeing him become our Master Chief of the Navy, to raising a wonderful Navy family, to looking after and taking care of countless thousands of Navy men and women and their families. No ship could have a more fitting sponsor, a more dedicated lady to put her spirit and soul into her ship. And, of course, a ship named USS *Chief* must have a superb chief petty officer's wife to start her on her voyage, to keep her safe in the storms she will encounter, to help set her course and keep her headed fair and to return her safely to harbor each and every time until the very last time when she completes her tour of duty in our Navy and, many years from now, is finally allowed to swallow the anchor and join her sister ships who have served well and earned their place in our history. That day, the end of her voyage and return to her final homeport, is a long way off for our new USS *Chief*. Throughout this fantastic voyage that she begins today, from her very first sea detail after this ceremony to the very last time her captain, a captain who is not even in our Navy as we meet her today, the very last time her final captain says, "Double up all lines and secure main engines" for the very last time, USS *Chief* will carry the prayers, the spirit and the soul of Sue Bushey. With Sue as her sponsor there can be no doubt that we commission a great ship today. Sue, please stand for a moment so we can all show just how very proud we are of you and all the dedication and service you have personally given to our Navy and our total Navy family.

Chief - what a great name for a ship. The title chief raises so many memories for all sailors. Everyone of us who has served for any length of time can tell you about his or her "chief." Most of us have more than one. New sailors and, even seasoned petty officers, given half a chance will tell you stories about their chief, a salty individual who took care of them, who taught them all the important things, who set the example, who cared about them and was not afraid to show it in so many ways.

Officers have their chief too. Usually somewhat older than they, always wiser in the ways of the Navy than they could possibly be as they started out on their careers. Ready with advice, with counsel and with the know how to make it happen, to get it done, no matter how difficult; and through it all, those great chiefs were and are also great teachers. They know, as only chief petty officers can know, that getting the job done today is important. It is the mission; but that task of helping that new officer become a real pro, a true naval officer in the finest sense of the term, is just part of the job—a key part, a critical role that chief petty officers have been playing for over 100 years now. Our Navy depends upon them, the chiefs know it and they thrive on the challenge.

I am a lucky man for I have served as a new seaman, a petty officer and an officer. In each phase of my career, a great chief petty officer appeared at just the right time, guiding me, pushing me when necessary, leading me and, when it was appropriate, letting me think I was leading him. There have been many chiefs in my life, all important, many personal friends, all professionals. Many of those chiefs are here today, but I want to recognize just one, a chief petty officer who led the way for me, who believed in me and who encouraged and even badgered me to be all that I could be. He is, as much as any other person here other than my wife, Bettie, the reason I am here today as Chief of Naval Operations. I am speaking of a special chief - my chief, George Everding. George, please stand so everyone here can applaud a chief petty officer who not only cared about his people but who went more than the extra mile to make a difference in so many peoples' lives. I am so proud of you - my chief.

This new USS *Chief* is the second ship to bear the proud name. I have a special affection for this ship and her predecessor for they are mine warfare ships and I had the good fortune to command a minesweeper many years ago as a young lieutenant. The first *Chief* was a proud ship, commissioned in 1943, she was a warrior. By the end of World War II she had earned five Battle Stars in the Pacific participating in the capture, occupation and mine clearing of Saipan, Tinian and Okinawa. She then performed minesweeping near Honshu in preparation for the occupation of Japan. After the war, our first *Chief* went back to San Pedro where she got a well deserved rest. But, she was not destined to lie silent alongside the pier for long; she was recalled for Korea and there, as always, she served proudly earning two more Battle Stars. Yes, that first USS *Chief* was a warrior and, because she was, I want to spend a moment or two talking about chief petty officers as warriors - the backbone of war fighting at sea for 100 years - no less today than in the last century. For, my friends, war fighting is what we do, fighting when we must, winning when we do fight - that is the stuff of chief petty officers in the past, today and in our future.

We know and every sailor knows, that the word leadership and the title chief petty officer go together. You cannot say one without thinking of the other. Oh yes, chiefs lead in peacetime too. They teach, they provide the technical expertise and experience, they know how to get the job done and they know how to make all the right things happen. They know that combat readiness is based on taking care of people, on keeping their ship and her equipment ready at all times, and on all those many attributes and abilities that make chiefs special as they lead their sailors and teach their young officers. But all of that is preamble to leadership in **war**. No chief, no sailor, seaman or admiral wants to fight in wars. We want to deter them. But when our nation calls, when the fight is no longer optional, no longer avoidable, but now is required - now is the order of the day - that is when all that chief petty officer leadership, commitment, honor and courage really pays off. For deep down, each and every chief petty officer know that we are warriors and that he or she is a leader of warriors as well. In the smoke and fury and, yes, the confusion of battle at sea, it is

then that the chief petty officer proves again and again that everything else was simply preparation for the moment of truth. The moment when all that work pays off. When those young sailors do what is required, do it almost instinctively and when they look to the chief for the example of all that is great in our Navy.

Just like all of you, I pray that this ship, our new USS *Chief*, will not have to enter a real minefield, endure the danger that goes with being a ship of this very special class, that she does not have to sail into danger to do what must be done. But, just like the USS *Chief* who went before her, just like the chief petty officers who went before all of today's chiefs, the day will probably come when she does just that. The day will probably come when it is not a drill, when the crew of this ship knows that the chips are down, that the mines lurking beneath the surface are real and they are deadly. The day when the landing craft, the larger ships, the sailors and Marines who will use the waters she has cleared, will follow her to fight and win on the land. They will depend on her to make their trip to the battle safe, to make it possible for them to get there so that they can engage the enemy and win. It is then that this ship and her crew will truly be tested. It is then that the chief petty officer leadership we are all so proud of will have it's biggest payoff. It is then we, as chiefs and sailors have always done, will prove, as we have proven so many times in so many distant waters before, that the words "Follow Me" and "Ask the chief" are not just words but part of a creed, a way of life, that sets the warrior apart from all others. It is then, in the difficult and dangerous times, in ways that only warriors can understand, that this ship, our USS *Chief*, her crew and chief petty officers everywhere will once again be tested in a special way. Even though I pray that day will not come, I know that if and when it does, they will make us proud - they will show the way.

Captain Garrold, you have a special charge. Your ship carries the special title, a very special title, as her name. You bear a special burden and you carry a special trust as the skipper of a proud shop with a proud name. Chiefs are leaders, chiefs really are the backbone, the heart and soul of our Navy. If you, or if a captain who follows you, must take this ship into the fight, you must prepare her to do her duty just as so many great chiefs have done their duty in so many places, in so many ways, in so many fights, before you. Chief is more than a name; it is a symbol, a tradition of all that is good and honorable in the greatest Navy the world has ever known. I envy you, every sailor here envies you. For you, as commanding officer of USS *Chief* will have the opportunity to carry on in the finest traditions of one of the proudest and most important groups of men and women in our Navy. I know that you, your ship and your crew will not let us down. You will, you must be the best; you carry a special trust for your ship bears the name, *Chief*, and she must embody all the great and inspiring things that her name symbolizes.

God bless this ship, her captain and her crew and the Navy and nation she serves, and God bless each and everyone of you here today. I am proud to be one of you. I am proud to be an American sailor. Thank you.

Sincerely,

J.M. Boorda
Admiral, USN

USS Chief *(MCM-14)*

State of Tennessee

House of Representatives

Proclamation

WHEREAS, the 1st of April in 1993 marks the First Centennial of the Chief Petty Officer in the United States Naval Service; and

WHEREAS, this historic occasion will be celebrated across the nation, but nowhere more spiritedly than at the CPO Club, NAS in Memphis; and

WHEREAS, there is one Memphis resident in particular who has contributed significantly to preserving the heritage and mystique of the Chief Petty Officer; and

WHEREAS, this man is retired U.S. Navy Senior Chief Hospital Corpsman Joe B. Havens; and

WHEREAS, this well-decorated officer served with distinction from 1949 until 1970 on board naval hospitals, naval air stations, a national naval medical center, basic and advanced hospital corps schools, USS FDR, CVA-42, Kermit Roosevelt (ARG-12), USS Estes (AGC-12), USS Cleveland (LPD-7) and the first marine division in Korea; and

WHEREAS, his respect for the men and women who serve as Chief Petty Officers led Joe Havens to begin compiling stories, pictures, anecdotes, humor, and heroism testimonials from Chiefs, as well as from those who have served with CPOs; and

WHEREAS, the end result of this labor will be CHIEF: The Evolution, Development and Role of the Chief Petty Officer in the U.S. Naval Service, a book dedicated to those exceptional individuals who attain this coveted status; and

WHEREAS, the profits from this timely publication will benefit the Naval Historical Center, The Navy Museum, Washington Navy Yard, Washington, DC, whose aim is the preservation, analysis and interpretation of naval experience in order to enhance its future effectiveness; and

WHEREAS, CHIEF's proceeds will also benefit The Navy-Marine Corps Relief Society, an organization which assists active duty and retired Navy and Marine Corps members and their dependents; and

WHEREAS, containing the reminiscences of Chief Petty Officers and the treasured memories of those who have been fortunate enough to serve alongside "the best of the best", CHIEF will serve as an invaluable resource now and in the years to come; and

WHEREAS, the varied observances of the first centennial of the Chief Petty Officer, as well as the wealth and diversity of information in the book-in-progress, CHIEF, all herald the paramount importance of this office and of those who have earned the privilege of serving in it; now therefore,

I, Jimmy Naifeh, Speaker of the House of Representatives of the 98th General Assembly of the state of Tennessee, at the request of and in conjunction with Representative Carol Chumney, do hereby proclaim that we join in the celebration of this First Centennial of the Chief Petty Officer in the United States Naval Service and we look forward to the publication of CHIEF, a book which is certain to preserve the rich traditions of this office, and inspire young and old alike to aspire to be among "the best of the best".

Proclaimed in Nashville, Tennessee on the 1 day of April 1993.

Jimmy Naifeh

SPEAKER OF THE HOUSE OF REPRESENTATIVES

Carol Chumney

REPRESENTATIVE CAROL CHUMNEY
89th DISTRICT

Seven "Old Salts" engaged in tall tales aboard the USS Enterprise *(1877-1909), circa 1887. (Courtesy of U.S. Navy)*

History of
Chief Petty Officers

Chapter 1
Flashback – Career Synopsis
Reflections from Chief

He tossed and tumbled all night, similar to the many nights spent prior to a long deployment. A feeling of foreboding engulfed him as he arrived at the base just in time to witness the "hoisting the colors." Seeing this ceremony he felt better…at home again. It was an unusually beautiful day. The sun with a red glow was slowly rising in the east. A day of reflections, to be sure. The commanding officer and family members joined him in the mess and the camaraderie of shipmates uplifted his spirits. What seemed only minutes later, the shrill sound of the boatswain's pipe sent chills up his spine. The reverie was broken when the presiding officer announced, "Chief _____, departing."

Thoughts were racing through his mind and he chuckled aloud when remembering being told in recruit training, "if it moves, salute it," if stationary, paint it," and the time when he naively responded to a request from those characters he had known to "fetch a bucket of steam from the engine room." He remembered the story of the shipmate sent to captain's mast for a misdemeanor. The commanding officer asked, "guilty or innocent?" He pleaded guilty. The captain remarked, "three days bread and water." The young shipmate said, "I can do that standing on my head." The commanding officer then said, "very well, three more days to get on your feet."

Reflecting on the time years ago when, at his hometown recruiting office, he began a career in the greatest Navy in the world. Now, too quickly, he realized he had crossed over to the other side to begin a second career in the civilian sector. Taking his spouse's arm, he picked up his briefcase to embark on a new challenge. He knew he would make it, but he couldn't shake the feeling of uncertainty and loneliness.

"Wake up, Joe, it's time to 'hit the field.'" That was the voice of my father summoning me to an unusual day of chores on a rainy Monday morning in early June 1949. Having decided to stay behind to help my dad out on the farm after graduation from high school that year, rather than "join the service" like several of my friends, I felt that alone gave me a little more independence. Wrong! My dad was set in his routine of tasks that required attention, and he set a time to all of those.

It was after a weekend of partying – as we then knew it – with friends that I didn't want to get up that Monday morning. How was he to know that the night before I had been stood up by a favorite girlfriend and was suffering from a mild case of rejection which today is known as depression? So and So had chosen one of my best friends over me because he got there first and had the privilege.

At breakfast, through bloodshot eyes caused by lack of sleep and one too many beers, I stated to Dad that it was raining. What were we going to do? He replied, "fix fences." Away we went. I sulked most of the morning. After trying his best to cheer me up, I announced at lunch I was going to Little Rock, some 40 miles away, to join the Navy. Dad was not totally shocked as I had been motivated to do so at the age of 14 when my brothers came back from World War II from both Army and Navy experiences.

I was the youngest of nine children and was only 17 at the time and enrolled in the county National Guard unit. I just knew that joining the Navy was what I wanted to do. I put on my best Sunday clothes and hitch-hiked to Little Rock. I was lucky, arriving about an hour before the recruiting office closed. That's when I met the first chief petty officer i had ever known. He had grown up a few miles from our community and knew my father and very quickly put me at ease from that frame of reference. His name was E.E. Matthews, a chief machinist mate. He explained that the county from whence I came only had one person left for the June 1949 quota and if I passed the physical and other tests, and the big "if," was if my father would also sign for me, I would be on a train heading to San Diego that week. I had no idea of the fact that my dad would have to sign for me so when Chief Matthews gave me a pass for room and board for the night I headed for the local boarding house.

I was away from home in a large city, alone, with a lot to think about. The old farm and the simple country life started to look pretty good. Well, at supper time, I met other young men who came from other parts of the state who were there for the same reason, and soon, a spirit of camaraderie prevailed that took my mind off the notion of being homesick. Then I thought, "What if Dad doesn't sign for me?" Years later, I would rue the day he did.

Introduction to the Navy

Welcome aboard! These words carry a world of significance. They mean that you have made one of the biggest decisions a young person can – you have volunteered to enlist in the United States Navy. By doing so, you have become a member of one of the most famous military services in the world and joined one of the biggest businesses in the United States. Not only have you proved your understanding of citizenship by offering your services to your country, but you have also taken the first step toward an exciting and rewarding career.

Today's Navy is a massive and complex organization, a far cry from the makeshift fleet that opposed the British in the Revolutionary War. For the fiscal period beginning October 1989, the Department of the Navy had budgeted 590,500 Navy and 347,700 civilian employees, and a total battle force of 551 ships and 5,035 average operating aircraft. It will cost a proposed $97.3 billion to operate the department this year; that 33.9 percent of the United States defense budget of 8.1 percent of total federal spending.

The Navy plays a vital role in maintaining our national security; it protects us against our enemies in time of war and supports our foreign policy in peacetime. Through its exercise of sea power, the Navy ensures freedom of the seas so that merchant ships can bring us the vital raw material we import from abroad, such as petroleum, rubber, sugar, and aluminum. Sea power makes it possible for us to use the oceans when and where our national interests require it, and denies our enemies that same freedom.

Boot camp – what? I suppose the correct terminology would be recruit training. Anyway, let me share with you a little about my introduc-

A rendition of the bookplate of the Navy Department Library. The finely-engraved plate was designed and executed in 1906 by Adolph C. Ruebsam of the Navy Hydrographic Office. The grigate Constitution, symbolic of the victorious exploits of the United States Navy, is depicted with a hippopus shell from the deep seas. The Navy Department Library, Washington Navy Yard, Washington, D.C. was established in 1800 by direction of President John Adams.

Uniforms of the United States Navy, 1898 showing Captain, Civil Engineer Corps, Service Dress; Boatswain's Mate 1st. Class Rear Admiral Special Full Dress; Bandmaster Full Dress; Commander, Undress; and Chief Master at Arms. (Courtesy of U.S. Navy)

tion to the Navy. Its customs, heritage, and traditions were gleaned from what was considered rigorous training for about 12 weeks. This training took place in San Diego, California. The first three weeks were really an orientation to the institution's way of life. Many didn't appreciate this sudden departure from the way of life to which we had previously been accustomed. For others, it was an improvement for they had not only food, shelter, health and dental care, but also new friends. Our first real look at a chief petty officer was a man by the name of Wyld, who became our company commander. I liked him right away and I saluted him every time he moved. He finally convinced me that such was not required. After getting off the buses, we were labeled such endearing terms as "knuckle head," "girls," and "clowns," to name a few. Our company commander, Chief Boatswain's Mate Wyld took his job seriously, trying to get us up on time for muster?????, calisthenics??????, breakfast, colors order drill, and any other task that would complete the "plan of the day" for his recruits. Actually, there were times in those early, formative days that the chief would actually "hiss a threat" when we didn't perform up to his expectations. Keep in mind, these first three weeks were comparable to a baby learning to walk.

Boot Camp – The Good Old Days – Not Quite!

One writer commented that most of what is written about boot training today (circa 1929-30) is concerned with war-time inductees. He further states that the system at Norfolk Recruit Training Center in 1926 was far more rugged. The Navy recruits then were all volunteers, having of their own free will put their necks out. The usual retort to any complaint of rough treatment was, "Listen, mate, no one pushed you into this man's Navy, so pipe down!" Because sounding off was very dangerous, recruits did a lot of piping down. The Recruit Training Center (boot camp) at Norfolk was really run by the chief petty officers who were stationed there, most of whom were grizzled old-timers (who had come this way before) and who harbored their own minds about what this training was intended to accomplish. We can all look back and remember the chief who most impressed us in our recruit training – the author writes that the chief he recalls was a chief boatswain mate, a real human type person when you discovered the chinks in his armor, but so blistering initially that he frightened most of us out of our wits. His job, of course, was to transform the "men" of our recruit training company into sailors. Looking back, most agreed that he hadn't much to work with. Most in the company were physically limp and mentally rigid. Some were barely age 15 while one person secretly admitted he was only 14 years old.

The chief boatswain mate did his best in a manner that seemed best to him. When drilling didn't suit him, he would march the whole unit across the training field to a desolate location and amidst ankle-deep sand, the chief ordered double-time under with arms in a tight rectangle. For seemingly hours on end, this unit marched in the hot sun of July and August while the chief stood, like a circus ringmaster, in the center. The

epithet he yelled were endured and well remembered by those in attendance.

The chief won the unit's profound respect with this type of treatment as we evolved into Navy men. Frequently thereafter, when members of the unit, scattered throughout the Navy, met, invariably the question was asked, "Remember old Chief So-and-So at Norfolk?" The answer was certain to be, with a grin, "That old S__ of a __! He sure knew his stuff!"

Navy trainees, he continued, bunked in hammocks then, and those at Norfolk were slung between jack-stays in the barracks were indeed something to remember. The hammocks had to be taut. None of the casual backyard slackness, with a comfortable canvas cradle slung for repose of the soul – no way! Should the hammocks in the unit not be as tight as fiddle strings, the chiefs, when inspecting them, would pass penalties with abundance.

Sailors were always falling out of their hammocks. It was indeed a sore night when some poor shipmate didn't hit the deck with a tremendous clatter. Two of the men actually suffered broken arms in nocturnal nose-dives.

Cleanliness was also the by-word. Sailors had to do their own washing and subsequently numerous bag inspections were held, requiring all items of uniform gear to be laid-out just so, or else. It was noted that if a boy became slack, others in the unit would go to work on him, sometimes on our own initiative, and sometimes in response to strong hints from the chief. The chief diplomatically turned his head while we led the culprit who allowed himself to become slack, out to the wash rock, stripped him down, and scrubbed him unmercifully with a Navy scrub brush (called Ki-Yi's) and sand. The cure was always effective.

From the book *I Took the Sky Road* by Commander Norman M. Miller, USN as told to Hugh B. Cave, Dodd, Mead & Company, New York, 1945.

First Enlistment

Your introduction to the Navy started at your hometown recruiting station, with interviews and processing conducted by a trained petty officer. Your enlistment (often called a hitch or cruise) may be for three, four, five, or six years. Four years is a normal hitch; five years is for those approved for training in one of 12 specific ratings; and six years is for those qualifying for advanced training. If you enlisted for four years and are a high school graduate, you may have selected one of the 60 or more Navy technical schools.

All recruits begin their naval careers at a Naval Training Center (NTC) in either Great Lakes, Illinois; San Diego, California; or Orlando, Florida. Women are trained in Orlando; men may be sent to any of the three centers, not necessarily the one nearest their home.

Inspection on the USS Franklin D. Roosevelt *(CVA-42). (Courtesy of U.S. Navy)*

Naval Training Center, Great Lakes

This center, located on Lake Michigan about 40 miles north of Chicago, was opened on July 1, 1911. During World War II, nearly a million men were trained at Great Lakes.

Naval Training Center, San Diego

Located on San Diego Bay, this center was opened on June 1, 1923. San Diego is also home port for many Pacific Fleet ships.

Naval Training Center, Orlando

The Orlando center, about 50 miles west of the Kennedy Space Center, was opened July 1, 1968. All enlisted women – about 9,000 a year – receive training here.

Each Naval Training Center consists of three commands:

Naval Training Station (NTS)–maintains buildings and grounds at Naval Training Center and provides housing, clothing, medical and dental care. Naval Training Station also handles recreational and Navy Exchange facilities, communications, postal and transportation service, and police and fire protection.

The Service School Command (SSC) operates the schools that provide technical training for various ratings. These schools train petty officers from the fleet and recruits who have finished boot camp.

The Recruit Training Command (RTC) is where you go first. The Recruit Training Command helps you make the transitions from civilian to military life with a busy schedule of lectures and drills on naval history, traditions, customs, and regulations. It also gives you instruction in basic military subjects.

The day of arrival at the center is called your receipt day, when your initial processing begins. The next three days will be referred as P-Days. The day after P-Days is one-one day, indicating the first week and first day of training. The rest of the first week consists of one-two day, one-three day, one-four day, and so on.

The First Weeks in the Navy

The procedure may vary from one Naval Training Center to another, but in general it goes like this: Report in, turn in orders, and draw your bedding and bunk assignment for your first night on board. Regulations now require that you be given a urinalysis test upon arrival. That same day, or the next, you will begin training. You will also fill out forms: A bedding custody card, a stencil chit, a receipt for a chit book (to be used instead of money for purchases at the Navy Exchange), a safe-arrival card for your parents, a clothing requisition, a packing slip to send your personal gear home (you must pay the shipping costs), and others.

You might have your first meal while still in civilian clothes. Here is a typical menu:

Breakfast: Hard or soft cooked eggs, grilled hotcakes, hot maple syrup, beef hash, broiled ham slices, pastry bar.

Breakfast Speed Line: Chilled prunes, chilled fruit juice, assorted dry cereal, hot oatmeal.

Dinner: Spaghetti, baked lasagna, lyonnaise green beans, toasted garlic bread, chilled peach halves.

Dinner Speed Line: Chicken noodle soup, salad bar, fruit gelatin.

Supper: Country-style chicken, chicken gravy, oven-browned potatoes, buttered green peas, salad bar, pineapple pudding.

Chief Petty Officer (CPO)

Chief Petty Officer (CPO), as defined by Webster's New Collegiate Dictionary, "an enlisted man (sic) in Navy or Coast Guard, ranking above a petty officer first class and below a senior chief petty officer."

When I first became interested in writing a book about the chief in the United States Navy, I thought it would be an easy task to assimilate outside research into my perceptions from my career experience in the Navy, dating from June 1949 to June 1970. Thinking simply I could go to the reference section of the library and commence my research, I was dumbfounded by the lack of available information in respect to the origin and evolution importance of this special position. I visualize the CPO as the epitome of an important leader in the chain of command structure. Lead-

ers are not always endowed with innate leadership qualifications, but I believe a point can be made for the CPO being made through leadership. Here is an enlisted man or woman who starts at the very bottom of the hierarchy and grows through a matriculation that begins at the recruiting station involving an interface with many individuals previously unknown to him/her. This journey takes years to yield a CPO. Involving an understanding of his/her own culture as well as the myriad of other cultures imposed on him/her during that career, often this development may be interrupted by a break in service and subsequent re-enlistment. The sources I will identify in the process of writing this book have been a source of inspiration and generous with their comments about the subject. While the intent of this publication originally related to the CPO (nowadays known as the enlisted grade 7), we certainly will incorporate the senior chief and master CPO (Enlisted grades 8 and 9), about which we are writing. The recognition by naval leaders in the advancement have rewarded the CPO by simplifying career advancement to deserving individuals. In Captain Edward L. Beach's book entitled *The United States Navy – 200 Years*, he alludes to an aura of idealism in the reformers and to the fact that by 1900, the Navy had become an established institution. My data reflects that the origin of the CPO came about in 1893. Captain Beach candidly presents the Navy's system of promotion, and this book proposes to discuss mainly those areas of promotion and structure relating to the CPO. I feel comfortable in my ability to outline briefly this evolution primarily for the casual reader who may be somewhat unfamiliar with the United States Navy. The number of such persons is sadly increasing due to the decrease in persons subjected to and/or patriotically inclined to become a member of the United States Naval or Reserve Naval Service.

The CPO is made from seamen (sic) apprentice to pay grade E7, somewhat analogous to creating from building blocks through progress, which evolution I will specifically address in a chapter devoted to evolution. Not nearly enough has been written about the CPO, and I intend to rectify that. While I do not expect this book to be utilized as a primer in naval schools and commands, I do feel that once published, there may be a trend toward its official adoption as a genuine resource guide.

Chapter 2

Definition of Chief Petty Officer – Origin

We Serve With Honor

Tradition, valor, and victory are the Navy's heritage from the past. To these may be added dedication, discipline, and vigilance as the watchwords for the present and the future.

At home or on distant stations we serve with pride, confident in the respect of our country, our shipmates, and our families.

Our responsibilities sober us; our adversaries strengthen us.

Service to God and Country is our special privilege. We serve with honor.

Historically Speaking

Origins of the Chief Petty Officer

According to naval records, the first mention of the chief petty officer (CPO) was on a shop's muster roll in 1775. This brief mention of the CPO title did not resurface in naval history for almost 100 years.

The history and design of the CPO's uniform date back to the 18th Century Continental Navy of 1776. With the colonization of the new world, a need for a navy became apparent. Many of the people that settled our nation learned their seafaring skills in England. These sailors brought not only their seafaring skills, but also their customs, traditions, and uniform similarities to this country. Many of our uniform styles can be traced to the British Royal navy.

In 1865 a navy regulation re-established the term CPO. The term was first used for the ship's master-at-arms, making him responsible for preservation of order and obedience to all regulations. An excerpt from an 1865 regulation tasked the senior enlisted person with the following responsibility:

The master-at-arms will be the CPO of the ship in which he shall serve. All orders from him in regard to the police of the vessel, the preservation of order and the obedience to regulations must be obeyed by all petty officers and others of the crew. But he shall have no right to succession in command and shall exercise no authority in matters not specified above.

This, however, did not establish the term chief as a rate. It was merely a function rather than a rate. Petty officers were divided into petty officers of the line and petty officers of the staff. Chief referred to the principal petty officers of the ship.

The next reference to the term chief was in United States Navy Regulation Circular Number 41 dated January 8, 1885. But this again refers to the term chief as a function or title rather than rate.

All evidence indicates that CPOs were first officially recognized by General Order 409 of February 25, 1893. This order, published for the naval service in an Executive Order of the same date, was issued by President Benjamin Harrison.

General Order Number 431, issued Sept. 24, 1894 changed the three rockers on the Master-At-Arms rating badge to one rocker. We know this as the rating badge of the CPO today. This general order also changed first, second and third class chevrons to their present-day form.

(Information based on 1988 edition of Military Requirements for CPOs.)
Source: Naval Affairs.

In considering the evolution of the CPO, I'm reminded of a passage in one of my favorite books, *Passages*, by Gail Sheehy (Bantam Books, February 1976). Ms. Sheehy states, "that men and women continue growing up from 18 to 50. There are predictable crises at each step. The steps are the same for both sexes but the developmental rhythms are not." This concept is analogous to the maturation of the seaman recruit – to obtaining the rating of CPO.

Chapter 3

USS *Tennessee*

Every Battleship A School
by Dr. Frank Crane

I recently had a very illuminating visit on board the battleship *Tennessee*.

It was illuminating because I found it not only a fighting machine but a schoolhouse.

All through it I saw classes of sailors studying lessons, the sailors were gathered in groups around tables.

The officers were enthusiastically acting as teachers. They were utilizing their superior advantages in helping the men get on.

I witnessed the early stages of growth of a magnificent transforming idea. That idea is that, instead of grinding up the lives of the common sailors as grist in the great mill of naval efficiency, the Navy was preparing a better efficiency by making a man's enlistment term a college course, a preparatory school for life.

A young man enlisting today in the Navy need not look upon that period as so much lost out of his best days, and expect to emerge a stunted thing, having learned only the one lesson of absolute obedience, but he may anticipate a few years of special training wherein he can learn that which will enable him to be an expert in some useful occupation, and if he wishes to remain in the service he can reasonably look forward to advancement.

Why not? A battleship is perhaps the most perfect laboratory in the world. All its machinery must be the best of its kind and up to date. The instructors are most capable. By using his spare time to improve his intelligence, a sailor can make his years of naval service a valuable life-asset.

And all this time he is fed, housed and clothed. He travels over the world. He learns team-play and the knack of getting along with his fellows as he can learn it nowhere else. He develops habits of efficiency and accuracy. And if he does not graduate a better and more capable man the fault is his own.

"Join the Navy" is an exhortation that ought to reach our best grade of young men. It stands to reason that our Navy will better represent the nation, and be a surer defense and a prouder asset, if it is recruited, not from the dregs of coast towns, but from the best product of our high schools.

As soon as this movement is generally understood there is going to be a waiting list of naval recruits, for young men will see that an enlistment spent in this best of schools will be a privilege.

Extract from an article by Dr. Frank Crane, reproduced through the courtesy of the New York Globe and Commercial Advertiser.

24 Outstanding Events in United States Naval History

Famous Sayings

"Give me liberty or give me death!" – Patrick Henry

"Surrender? I have not yet begun to fight!" - John Paul Jones

"Don't give up the Ship!" - James Lawrence

"We have met the enemy and they are ours!" - Oliver H. Perry

"Damn the torpedoes! Full speed ahead!" - David G. Farragut

"I will find a way or make one." - Robert E. Peary

"Millions for defense; not one cent for tribute!" - Expression of feeling which led to the building up of the Navy and establishment of the Navy Department in 1798.

"Remember the Maine!"

Evolution

One of the proudest days, if not the proudest, in an individual's naval service is the date on which a first class petty officer (PO1) dons the uniform and joins the CPO community. A special bond begins to develop

The USS Tennessee (BB-43)

The Great White Fleet about to enter the Golden Gate in San Francisco Bay, 1908.

with such a promotion for one's leadership and professional abilities displayed during previous service have been recognized by your superiors. Hopefully, these two qualities will continue to be honed with experience and maturity until the day of retirement.

The following information contained herein covers the history of the grade of Chief Petty Officer (CPO) and the evolution of that grade. Naval historians date the origin of this rating to 1893. For the readers edification, a synopsis of facts which relate to the Navy's purpose through history, traces the development of the CPO grade in the United States Naval Service.

The Navy's Purpose Through History

Throughout our history, the United States has depended upon the world's oceans for its security and economic well-being. For more than 200 years, the roles played by the United States Navy have remained remarkably consistent: to guard our shores from foreign attack, preserve freedom of the seas for the passage of trade and commerce, protect our overseas interest, support our allies and serve as an instrument of America's foreign policy. The bravery, dedication and hard work of generations of American sailors have ensured the Navy's success in these diverse and challenging tasks.

The Revolution (1775-1785): The Navy's Vital Role in the Birth of Our Nation

America's origins are intimately linked to the sea. North America was discovered and colonized by Europeans who took passage across the Atlantic Ocean to the New World. Overseas trade was a mainstay of the economies of the 13 English colonies for more than a century before the War of Independence. During that war, the Continental navy, privateers and commerce raiding squadrons chartered by individual American states, and the navy of our French ally all played vital roles in our fight against the British.

The Continental navy's squadrons and individual ships attacked British sea lines of communications and seized transports laden with munitions, provision and troops. Continental and state Navy ships and privateers also struck at enemy commerce, taking nearly 200 British ships as prizes, forcing them to divert warships to protect convoys and trade routes. In one of those shipping raids, off Flamborough Head on the coast of England in 1779, Captain John Paul Jones, commanding an old, half-rotten former merchant man, *Bonhomme Richard*, gave the new Navy one of its first battle cries, "I have not yet begun to fight!" as he defeated a much superior British ship, the frigate *Serapis*.

The Battle of Yorktown in 1781 was a near-perfect example of how naval forces can support an army. At the Battle of the Virginia Capes off Chesapeake Bay, the French navy under Admiral De Grasse prevented the British from evacuating their troops under siege at Yorktown. This led to the surrender of British forces under General Cornwallis to General George Washington and, shortly thereafter, final victory for the newborn United States.

The War of 1812 and the Rebirth of the United States Navy (1785-1815): Protecting Free Trade and Preserving Freedom of the Seas as an Instrument of Foreign Policy

In the 1780s and 1790s, pirates from the Barbary states on Africa's north coast attacked our defenseless merchant ships, stealing their cargoes and enslaving their crews. Determined to protect the freedom of the seas, the new American Congress authorized the building of a naval force to be sent to the Mediterranean. After a series of sea fights and operations ashore between 1801 and 1807, and another expedition in 1815, the Barbary rulers agreed to stop their attacks on American shipping.

Conflict with revolutionary France in the so-called Quasi-War prompted the establishment of the permanent Navy Department in 1798. French attacks on United States merchant men led to intermittent hostilities between American and French warships through 1800. American warships captured more than 80 French vessels and defeated two French men-of-war in combat on the high seas, giving the world a convincing demonstration of both the new navy's force and capability and United States determination to protect its commerce.

In the early 1800s, as the Napoleonic Wars in Europe wore on, Britain also interfered with United States merchant shipping, boarding our ships and forcibly "pressing" United States sailors into the Royal navy. Congress declared war on Britain in June 1812, in part over the issue of freedom of the seas and free trade. Outnumbered by the powerful British navy, American sailors nevertheless distinguished themselves in a series of ship-to-ship engagements on the high seas, in squadron combat on Lakes Erie and Ontario and in coastal waters defending New Orleans.

Following the War of 1812, Navy ships participated in the suppression of piracy in the Caribbean, anti-slavery patrols off Africa and Brazil, diplomatic initiatives such as Commodore Matthew Perry's expedition to open relations with Japan in 1852 to 1854, naval exploration in the Pacific and the Arctic and amphibious and blockade operations during the Seminole and Mexican Wars.

Civil War (1861-1865): Blockade and Joint Riverine Operations

The Navy's principal role during the Civil War was to blockade the South's coastline and support Union Army operations on inland rivers. Over the course of the war, these joint operations with the Army cut the South off from outside support and gradually constricted its trade and commercial livelihood.

Rapid improvements in engineering and weaponry led to the beginning of a revolution in naval technology, illustrated by the construction of ironclad warships by both sides. The Union Navy's *Monitor* contained more than 40 patentable inventions. The *Monitor* was built to counter the Confederate *Virginia*, an armored ship built on the hull of the former USS *Merrimac*. Although the 1862 battle between the *Monitor* and *Virginia* ended in a draw, this first battle of ironclads signaled a profound change in the nature of naval warfare. The war also saw innovations in mines, mine countermeasures and submarines.

In the years following the Civil War, the Navy was reduced in size utilized the 1880s when, with the settlement of the American West essentially complete, the United States became increasingly interested in overseas trade and foreign affairs. The Navy had undergone considerable decline since the Civil War. Many of the technological innovations introduced from 1861 within 1865 had been adopted and improved upon by foreign navies, but the United States fleet was essentially a force of antiquated wooden-hulled gunboats. Construction began in the late 1880s and early 1890s on a new Navy of all-steel ships. The ideas of Captain Alfred Thayer Mahan about the role maritime power played in building great nations and how battle fleets were critical components of those nations' defense provided a useful framework for the resurrection of American sea power.

The Spanish-American War: The Navy's First Two-Ocean Conflict

American support for Cuban independence from Sapin escalated into conflict when the battleship USS *Maine* exploded and sank in Havana harbor in February 1898. The United States blamed Spain for the explosion and declared war on April 25. Five days later, the Spanish navy was completely defeated in the Pacific at the Battle of Manila Bay, and on July 3, the United States Atlantic Squadron devastated the bulk of Spain's remaining naval power in a fleet engagement off Santiago, Cuba. A subsequent naval blockade of Santiago enabled the United States Army to capture the city.

As a result of the Spanish-American War, Spain ceded the Philippines and Puerto Rico to the United States. Congress subsequently annexed Hawaii, Wake Island and part of the Samoa Islands giving the United States far-flung possessions to protect and overseas bases from which to defend its interests.

Evolution

In order to logically trace the development of the CPO grade, one must understand the original Federal Navy structure to establish a foundation of the related grades and classifications which lead to its ultimate establishment. Interestingly, one must look back to the Revolutionary War to discover any similar promotion. Jacob Wasbie, a cook's mate, serving aboard the very first Continental navy warship, the *Alfred*, was promoted to "chief" cook on June 1, 1776. "Chief cook" was construed to mean cook or "chip's" cook which was the official rating title at that time.

The Federal navy was originally authorized by the Act of March 27, 1794. The fledgling navy consisted of four 44-gun vessels, two ships of 36 guns, and the necessary operational personnel. The action taken by congress on that date was based upon the need to counter the piratical Algerian situation. However, prior to completing any of the vessels, a treaty was reached between the United States and Algeria, and the act was allowed to expire. The Act of July 1, 1797, a date now often considered to be the true date of establishment of the federal or United States Navy directed the construction of three frigates. These ships were the *Constitution* and *United States* each with 44 guns and the *Constellation* with 36 guns. The first and last named ships are on exhibit in Boston and Baltimore respectively. Personnel, by ratings and numbers, assigned to the two sizes of warships were the same in both acts and consisted of the following: petty officers, who were appointed by the captain, captain's clerk, two boatswain's mates, one coxswain, one sailmaker's mate, two gunner's mates, one yeoman of the gun room, nine quarter gunners (11 for the two larger vessels), two carpenter's mates, one armorer, one steward, one cooper, one master at arms, and one cook. Non-petty officers, as listed in an act of 1797, named 150 seamen and 103 midshipmen and ordinary seamen for the larger frigates. Conversely, the numbers for the *Constellation* were set at 130 midshipmen and able seamen and 90 ordinary seamen. None of those figures included the United States Marines who were numbered at three sergeants, three corporals, one drummer, one fifer, and 50 marines (privates) for the larger ships. The 36-gun frigate was allowed one less sergeant and corporal and 40 versus 50 marines.

Generally speaking, precedence of petty officers was not really introduced until the publication in 1853 of the United States Navy regulations that were approved on February 15. These regulations were declared invalid on May 3, 1853, by the attorney general and were rescinded only because the President, as opposed to congress, favored them. However, it

is important to note the accuracy of the information and the guidelines. Further, the Secretary of the Navy submitted a set of naval regulations for congressional acceptance on December 8, 1858, which was never approved. According to pay tables of the period, those regulations of 1853 appear to have contained the then current rating structure.

Prior to 1853, one possibly could establish a semblance of precedence of rating based upon the sequence in which ratings were listed within complement charts. Further, these charts were substantiated by the differences in the amount of pay given to various petty officers. Another relevant issue is the fact that the petty officer's name appearing first on the master rolls generally could be considered senior to another individual of equal rating. Precedence of ratings was explicitly outline in navy regulations approved on March 12, 1863. For clarification, a review of the early Civil War petty officer rating structure follows. This outline was chosen because it was immediately prior to the official usage of "chief" with rating titles. Petty officers in the Civil War were listed under two categories: petty officer of the line and petty officers of the staff.

March 12, 1863

Petty Officers of the Line	Petty Officers of the Staff
1. Master's Mates (not warranted)	Masters at Arms
	Yeomen
	Surgeon's Stewards
	Paymaster's Stewards
	Masters of the Band
	Schoolmasters
2. Boatswain's Mates	Ship's Corporals
	Armorers
	Painters
3. Gunner's Mates	Carpenter's Mates
	Sailmaker's Mates
	Firemen, First Class
4. Coxswains to Commanders in Chief	
5. Captain's of the Forecastle	
6. Quartermasters	
7. Quarter Gunners	
8. Captains of the Maintop	
9. Captains of the Foretop	
10. Captains of the Hold	Coopers
	Ship's Cooks
	Armorer's Mates
11. Captains of the Mizzentop	
12. Coxswains	Stewards to Commander in Chief
	Cabin Stewards
	Wardroom Stewards
13. Captains of the Afterguard	Cooks to the Commander in Chief
	Cabin Cooks
	Wardroom Cooks

To provide a sound understanding of precedence, the paragraph on Page 7 of the 1863 Regulations follows:

"Precedence among petty officers of the same rate, if not established particularly by the commander of the vessel, will be determined by priority of rating. When two or more have received the same rate on the same day, and the commander of the vessel shall not have designated one of that rate to act as a chief, such as chief boatswain's mate, chief gunner's mate, or chief or signal quartermaster, their precedence shall be determined by the order in which their names appear on the ship's books. And precedence among petty officers of the same relative rank is to be determined by priority of rating; or in case of ratings being of the same date, by the order in which their names appear on the ship's books."

That lengthy paragraph was shortened in the 1865 regulations to simply read, "Precedence among petty officers of the same rate shall be established by the commanding officer of the vessel in which they serve."

Precedence by rating was a fact of Navy life for the next 105 years and was basically substantiated by rating priority and the date of an individual's promotion. Precedence for ratings remained in effect until the issue of Change No. 17 of August 15, 1968, to the 1959 *Bureau of Naval Personnel (BuPers) Manual*.

Through the years many have been audience to an appreciable number of boiling point arguments on the ship's fantail and in the chiefs' messes concerning seniority of ratings. As one can determine from the foregoing information boatswain's mates have not always been the senior rating in the Navy. For aviation machinist's mates it is noted that, likewise, they have not always been the senior rating within the aviation branch. During the periods of 1927-1933 and from 1942-1948 the rating of aviation pilot was the superior of all aviation ratings.

It is not the intention of this synopsis to present an extended dissertation on individual ratings. However, clarification of a stubborn controversy and its resultant misconceptions regarding the chief boatswain's mates, chief gunner's mates, and chief or signal quartermasters of the 1863-1864 era is a necessity. These three ratings have at one time or another been erroneously identified and argued as being CPOs. General Order No. 36 on May 16, 1864, effective July 1, 1864, listed navy ratings along with monthly pay for each rating. Among the ratings shown therein were the rates of chief boatswain's mate, boatswain's mate in charge, boatswain's mate, chief gunner's mate, gunner's mate in charge, gunner's mate, chief quartermaster and quartermaster. Boatswain's mates and gunner's mates received $27 monthly and quartermaster's, $25. Chief boatswain's mates and chief gunner's mates were paid $30 per month and were listed for service only aboard vessels of the first and second rates. Chief quartermasters were paid the same except for a $2 reduction while serving in ships of the third and fourth rates. Boatswain's mates and gunner's mates in charge were also paid $30 per month.

The primary difference between the chief boatswain's mate and boatswain's mate in charge, the chief gunner's mate, and the gunner's mate in charge lay in their assignment. Chief boatswain's and chief gunner's mates were permitted aboard ships of the first two classes of vessels (first and second rates with 100 or more crewmen). The boatswain's mate in charge and the gunner's mate in charge could be assigned to any of the four classes of vessels (first, second, third, and fourth rates) and specifically only when a warrant boatswain or warrant gunner was not assigned to the ship. Boatswain's mates and gunner's mates in charge remained with the rating structure for only five years. They are last listed in the pay table included in the Navy register for July 1, 1869. They were eliminated with the issue of January 1, 1870. Thereafter, chief boatswain's mates and chief gunner's mates were assigned to vessels of all four classes. Approximately five years later, according to the allowance list of 1877, they were assigned only to ships without a warranted boatswain or gunner.

The previously quoted 1863 regulations covering precedence mentioned of chief or signal quartermaster, and these rates required a considerable amount of explanation.

The title of signal quartermaster was utilized from at least the early 1800s. Identified were those quartermasters who were principally involved with signaling and the care of flags, halyards, markers, lanterns, and other paraphernalia as opposed to quartermasters mainly concerned with navigational and steering duties. During the years of 1863 to 1865 the rating title of chief quartermaster and signal quartermaster were basically synonymous. Further, the 1863 navy regulations and the 1864 pay order did not present a distinction between those two titles. In 1865, by United States Navy Regulations approved April 18, 1865, under petty officers of the line, delineations between quartermaster, no chief quartermaster (unlisted) and signal quartermaster is presented. Signal quartermaster is listed as Number 3 in precedence (after gunner's mate) whereas, quartermaster is indicated as Number 6 (after coxswain to commander in chief "of a squadron or fleet"). Those two ratings continued to be carried in successive issues of *Navy Regulations* until 1885. Signal quartermaster is never listed as a separate rate from chief quartermaster in the pay tables covering those 20 years. Therefore, the title of signal quartermaster, instead of chief quartermaster, can be considered as the official title from April 18, 1865, to January 8, 1885. The title of chief quartermaster, primarily found in Navy pay tables for that same period, can be judged to be an alternate title for signal quartermaster. In other

directives and correspondence, these two titles were often used interchangeable, depending on the author.

It is necessary to reflect back to chief boatswain's mates and chief gunner's mates to define their exact status. Navy Regulations of 1865, 1870, and 1876 fail to show chief boatswain's mate and chief gunner's mate as a different rate or level from boatswain's mate and gunner's mate. It therefore follows that justification for calling the chief boatswain's mate and the chief gunner's mate alternative rates depends upon General Order 36 of May 16, 1864, effective July 1, 1864, and Tables of Allowances for the 1870s which lists them as rates or ratings along with boatswain's mate and gunner's mate. To argue whether the chief boatswain's mate, chief gunner's mate, and chief quartermaster, or signal quartermaster of the 1863-1865 era were or were not actually CPOs is elementary. They *were not* CPOs simply because the grade had not yet been established.

On January 1, 1884, at which time new pay rates became effective, there existed the three aforementioned rates carrying the word chief. The boatswain's mate, gunner's mate, and quartermaster were all paid at $35 per month. It is noted that several other rates were paid higher amounts – ranging from $40 to $70 per month.

Fifty-three weeks later on January 8, 1885, the Navy classed all enlisted personnel as, or equal to, petty officers first, second, or third class, and non-petty officers at seaman first, second, or third class. Chief boatswain's mates, chief quartermasters and chief gunner's mates were positioned at the petty officer first class level within the seaman class; master-at-arms, apothecaries, yeomen (equipment, paymasters, and engineers), ships' writers, schoolmasters and band masters were also first class petty officers but came under the special branch. Lastly, machinists were carried at the top grade within the artificer branch. Included under the special branch at the second class petty officer level was the rate of chief musician who was junior to the band master. That rate was changed to first musician under the 1893 realignment of ratings and carried, until 1943, a petty officer first class rating.

On April 1, 1893, two important steps were taken. First, most enlisted men received a pay raise, and, second, the grade of CPO was established. The question is often asked, "Who was the first CPO?" The answer is flatly: "There was no first CPO due to the fact that nearly all ratings carried as petty officers first class from 1885 were automatically shifted to the CPO level." Exceptions were the schoolmasters who stayed at first class, ship's writers the same but expanded to include second and third class, and, finally, carpenters' mates who were carried as second class petty officers but subsequently extended to include chief, first, second, and third class. Therefore, the CPO grade on April 1, 1893, encompassed the following eight rates:

Seaman Branch – chief master at arms, chief boatswain's mate, and chief quartermaster.

Artificer Branch – chief machinist, and chief carpenter's mate.

Special Branch – chief yeoman, apothecary, and band master.

Precedence among ratings was eliminated and changed to a single system for military and non-military matters based on pay grade and time in grade on July 1, 1968. For over 45 years the acronyms (PA) and/or (AA) were written beside the rate titles and abbreviations. The letters meant (permanent appointment) and (acting appointment) and were used to signify the CPO's status. Incidentally, pay differential status was in effect until 1949.

Prior to the establishment of the CPO grade, and for many years thereafter, petty officers were promoted and given acting appointments by the commanding officer of the vessel to fill vacancies in complement. Before July 1, 1903, men served various lengths of time with an acting appointment, generally lasting six months. If his service was satisfactory, the captain recommended to the Bureau of Navigation (Bureau of Personnel (Bupers) from October 1, 1942) that the individual be given a permanent appointment for the rate. Otherwise the commanding officer could reduce the individual to the grade or rate held prior to promotion the acting appointment.

The following is quoted from a general order dated June 1903 which became effective on July 1, 1903: "CPOs whose pay is not fixed by law and who shall receive permanent appointments after qualifying therefore by passing such examination as the Secretary of the Navy may prescribe shall be paid at the rate of $70 per month."

Pay for CPOs, in 1902, ranged from $50 to $70 depending upon the specific specialty. Pay levels for enlisted men at that time was and had been regularly established by executive order. From July 1, 1908, initially by the Act of May 13, 1908, the United States Congress has set pay for the enlisted men. However, during the Great Depression, President Roosevelt, temporarily decreased the pay of all Armed Forces personnel by 15 percent from April 1, 1933—June 30, 1934, and five percent from July 1, 1934—June 30, 1935, by executive order.

Those CPO's holding permanent appointment dated prior to July 1, 1903, were required to requalify by standing an examination before a board of three officers. Upon successful completion, they were issued a permanent appointment by the Bureau of Navigation. If they did not requalify, their pay remained as listed in the then current pay tables instead of increasing to the $70 level.

To identify the status of permanent CPO's the words (Permanent Appointment) or the letters (PA) were normally included with the individual rate title or rate abbreviation. Such identification was discontinued on June 16, 1944. CPOs normally served for one year with an acting appointment prior to the issue of a permanent appointment. It may be noted that the time in service under an acting appointment fluctuated from time to time and certain requirements for sea service and marks were applied by the Chief of the Bureau of Naval Personnel. CPOs serving with acting appointments were identified with those words or the letters "AA" next to their rate title or abbreviation until April 2, 1948. From that date the letter "A" was used together with the rate abbreviation. For example, chief boatswain's mate acting appointment was abbreviated as CBMA. Paygrade 1-A no longer signified acting appointment for CPOs after October 1, 1949, which was affected by the Career Compensation Act of October 12, 1949. Since that time CPOs have received the same amount of pay regardless of whether they held a permanent appointment or an acting appointment.

The change in status from acting to permanent appointment was always a "breathe-easier" occurrence. This meant that the commanding officer could not reduce you to first class without due process. It took a court-martial and the Bureau's authorization to be demoted from permanent appointment chief. (On November 1, 1965, acting appointments were dropped from use.)

Before addressing the issues of senior and master chiefs, the issue of pay grades should be clarified. The Act of May 18, 1920, effective January 1, 1920, standardized pay at all levels. Base pay established for Permanent Appointment chiefs was $126 per month and for those with an acting appointment at $99. (Such pay rates remained effective until June 1, 1942. Under the Act of June 16, 1942, pay was increased to $138 and $126 for CPOs with permanent and acting appointments, respectively). Two years later, by the Act of June 10, 1922, which became effective July 1, 1922, the pay grades of 1 and 1-A to 7 were established. CPOs (PA) and mates were carried in pay grade 1; whereas, chiefs with acting appointments were listed in pay grade 1-A. On October 1, 1949, by the Career Compensation Act of October 12, 1940, pay grades were reversed and the letter E, for enlisted, was added setting all CPOs at E-7 vice pay grades 1 and 1-A.

In regards to retirement or transfer to the Fleet Naval Reserve, many active duty and retired personnel are probably unaware that several years ago such transfer could be made by voluntary application upon completion of 16 years service. Public Law 241 (Chapter 417) approved August 29, 1916, becoming effective on July 1, 1916, authorized personnel to leave active duty with one-third of their pay and permanent allowances. Title 34, Chapter 374, approved on February 28, 1925, generally eliminated that privilege. However, personnel who had been in the service on July 1, 1925, or had been discharged prior to that date and re-enlisted within three months from that date were still eligible to transfer to the Fleet Naval Reserve with 16 years active service. Normally, under those circumstances all such men would probably have been transferred to the Fleet Naval Reserve by October 31, 1929. Still, if an individual had been given special permission and executed an extension of enlistment for two years on his four-year enlistment he would have been able to remain on active duty, at the latest, until October 31, 1931, to complete 16 years total service. Regardless, the Act of June 30, 1932, (Chapter 318) repealed the applicable section of the February 28, 1925 Act that permitted men to transfer to the Fleet Reserve with 16 years service under any circumstances.

Congress directed that the Navy be reduced to 86,000 men during Fiscal Year 1923 by Public Law 264 (Chapter 259) approved July 1, 1922. That law also permitted men with 18 years of active naval service to transfer to the Fleet Naval Reserve with the pay and allowances of 20 years service. Such provision was allowed only until the foregoing limitation of 86,000 men was reached and in no case after January 1, 1923. Men with

20 to 25 years of service could automatically be transferred to the Fleet Naval Reserve at the discretion of the Secretary of the navy. As disseminated in Bureau of Navigation Circular Letter 16-22 of April 25, 1922, those men with 25 or more years were permitted to continue on active duty until they had completed 30 years service before full retirement from the United States Navy. Currently (1922), information concerning normal transfer to the Fleet Naval Reserve and retirement is contained in Article 3855180 of the Bureau of Naval Personnel Manual (NMPC Manual).

It is of some interest that the only two ratings established on July 1, 1797, to remain in continuous use since that date are the boatswain's mate and the gunner's mate.

The readers attention is invited to the fact that April 1, 1993, marks the first centennial for the grade of CPO, a real cause for all CPOs whether on active duty, retired, or simply having served with that grade, to plan for a gala celebration.

Officers and CPO's Gray Working Uniforms

General – A working uniform is authorized for use on all ships and shore stations when prescribed by the senior officer present.

Coat, working uniform (for all officers and CPOs) - This garment shall be single-breasted with roll collar and notched lapels of gray cotton, Palm Beach, tropical worsted, lightweight woolen, or wool gabardine material, with three front buttons (bottom button to be in the middle of the belt), four outside pockets to be patch type, to have a two-inch belt of same material as coat, seared down all around, and coat to be 30 inches long for a man 5 feet 6 inches, and grades accordingly.

Trousers, working uniform (for all officers and CPOs) - These items shall be designed to conform to combination cap (blue and white) and garrison cap as prescribed for commissioned, warrant and the cloth of the latter shall be of the same color as the uniform.

Shirt, working uniform (for all officers and CPOs) - These garments shall be gray cotton with attached collar.

Tie, working uniform (for all officer and CPOs) - The tie shall be plain black, four-in-hand, made of silk, rayon, or wool.

Belt - A plain black or gray belt, approximately 1 1/2 inches wide, fitted with a nontarnishable buckle, shall be worn with all working uniform trousers.

Shoes - Black shoes shall be worn with all working uniforms.

Socks - Plain black or gray socks shall be worn with all working uniforms.

Insignia of rank - CPOs shall wear black rating badges on a gray background.

During the necessary transition period officers and CPOs will be permitted to wear khaki uniforms now in their possession of manufactured until the supply of these uniforms is exhausted or those in possession are worn out. Cap cover or garrison cap of the same color as the uniform may be worn.

The wearing of gilt buttons or blue-black plastic buttons is optional. Until such time as gray cloth rating badges are available, CPOs may wear the blue rating badge.

Credit: PT Boats, Inc.

13 Oct 1775	Continental Navy established; considered birth of United States Navy. (1)
10 Nov 1775	United States Marine Corps established.
1775-1783	Revolutionary War.
25 Mar 1776	Washington grants first Congressional Medal award.
1 Jun 1776	Jacob Wasbie, a cook's mate, is promoted to "chief cook" aboard continental warship *Alfred*, in the earliest known use of the word "chief" in association with sailors.
18 Jun 1777	John Paul Jones flies the "Stars and Stripes" on his vessel.
24 Apr 1778	Captain John Paul Jones defeats the *HMS Drake*, the first major British warship to be taken by the United States Navy.

1778	France openly enters the war on our side.
Nov 1778	John Paul Jones receives the ship, *Bonhomme Richard*. (2)
22 Sep 1779	Battle between the *Bonhomme Richard* and the *Serapis*. (3)
20 Jan 1783	American Revolution ended.
3 Sep 1783	British sign the Treaty of Paris. (4)
1 Aug 1785	The Alliance is sold. (5)
25 May 1787	A Constitutional Convention is called in Philadelphia. (6)
1788	The Federal Constitution is ratified.
1789	Legal power is established to create a navy. (7)
4 Aug 1790	The United States Coast Guard is created.
1793	England, Spain and France molest American shipping. (8)
27 Mar 1794	Congress authorizes the building of six frigates. (9)
1796	The United States pays $56,000 in tribute to Tripoli. (10)
May 1797	The *United States* is launched, the first one of the six frigates. (11)
1 Jul 1797	The United States Navy is born. Congress directs construction of frigates *Constitution*, *United States* and *Constellation*. Crews include "petty officers," sailors with special skills and responsibilities, appointed by the captain of each vessel.
21 Oct 1797	The *Constitution* is launched; christened by Captain James Sever.
30 Apr 1798	The United States Navy Department is established by Congress. (12)
18 May 1798	President Adams appoints Benjamin Stoddert the first Secretary of the Navy.
23 Jul 1798	The *Constitution* puts to sea, commanded by Captain Samuel Nicholson.
Nov 1798	The British impress the crew of the *Baltimore*. (13)
2 Oct 1799	The Washington Navy Yard is established. (14)
1799	The United States Navy captures, or sinks, almost 30 French privateers.
1800	Sailors and Marines had a choice for their grog ration. (15)
1798-1801	Quasi, or undeclared war with France. (16)
10 May 1801	Tripoli declares war on the United States (17)
1801-1805	Barbary Coast Wars. (18)
6 Feb 1802	Declaration of war against Tripoli.
16 Mar 1802	West Point Military Academy established.
4 Jun 1804	Treaty of Peace with Tripoli.

1806	Whiskey was allowed to be substituted for rum in the grog ration.
1801-1809	The United States Navy is reduced to six ships. (19)
Jun 1807	The United States Frigate *Chesapeake* is boarded by the British from Frigate *Leopard*. (20)
1812-1815	War of 1812; our second war with Great Britain. (21)
1812	The first women to serve aboard United States Navy ships are contract nurses brought aboard during the War of 1812.
19 Jun 1812	War with England proclaimed.
19 Aug 1812	The *Constitution* battles *Guerriere*, and earns the nickname "Old Ironsides"; America's first major naval battle. (22)
10 Sep 1813	Perry's victory on Lake Erie; "We have met the enemy and they are ours." (23)
1815-1816	Our troubles with the Barbary pirates is ended.
1813-1816	The United States Congress authorizes 12 74-gun ships of the line.
24 Aug 1814	The United States Capitol building and White House are burned by the British.
1814	The introduction of canned food aboard Navy ships.
24 Dec 1814	England signs a Peace Treaty with the United States
1815-1842	The beginning of a steam-powered Navy.
8 Jan 1815	The Battle of New Orleans.
1815	Drinking water is now stored in metal tanks aboard Navy ships. (24)
1815	The British abandon impressment and go to voluntary enlistments.
4 Apr 1818	The United States flag is adopted by Congress.
19 Aug 1818	Captain James Biddle, United States N., takes possession of the Oregon Territory. (25)
1825	The introduction of fresh vegetables, beef, and bread aboard Navy ships.
1828	The Navy Museum building is built as part of the Navy Gun Factory.
1830	The Frigate *Constitution* is determined to be unseaworthy and is ordered to be condemned.
1833	The poem "Old Ironsides" is published. (26)
21 Apr 1836	The Battle of San Jacinto.
19 Jan 1840	Lieutenant Charles Wilkes conducts sub-polar explorations, proving conclusively that Antarctica is a continent.
1841	The first "shell guns" come into use.
1842	The enlisted men's grog ration was reduced. (27)
10 Oct 1845	The Naval Academy is opened at Annapolis. (28)
1846-1848	The Mexican War.

1850	Flogging is ended in the United States Navy by an act of Congress. (29)
1850	The introduction of chocolate, sugar, tea and preserved potatoes aboard ship.
8 Jul 1853	Commodore Matthew Perry calls on Japan to open trade negotiations. (30)
Feb 1854	Perry returns to Japan to complete the trade treaty.
31 Mar 1854	The treaty opening Japan to United States trade is approved and signed by Commodore Perry.
16 Apr 1856	The Declaration of Paris abolishes privateering. (31)
1858	The French order four frigates of 5,000 tons, to be literally clad in iron.
24 Nov 1858	The French burn Fort Duquesne.
1850s	Improved smooth-bore muzzle loaders are developed. (32)
1860	The United States Navy had 90 ships, 1,300 officers and 7,600 enlisted men. (33)
1861-1865	The Civil War. (33)
12 Apr 1861	Fort Sumter fired on by Confederates – the Civil War begins.
19 Apr 1861	President Lincoln proclaims a blockade of Southern ports.
20 Apr 1861	The Norfolk Navy Yard is partially destroyed and is abandoned by Union forces to prevent the facilities from falling into Confederate hands.
21 Jul 1861	The Confederate army in the East won the first of a string of victories over the Union army at Bull Run.
29 Aug 1861	Union forces received the surrender of Confederate-held Forts Hatteras and Clark; closing Pamlico Sound, North Carolina.
7 Nov 1861	Union naval forces under Flag Officer DuPont captured Port Royal Sound
1862	The United States Navy abolishes grog aboard ship. (34)
30 Jan 1862	Ericsson's *Monitor* is launched.
3 Mar 1862	DuPont's forces took the deepwater port of Fernandia, Florida, and the surrounding area.
8 Mar 1862	Ironclad ram *C.S Virginia* destroyed wooden blockading ships USS *Cumberland* and *Congress* in Hampton Roads.
9 Mar 1862	Duel between the *Monitor* and the *CSS Virginia* at Hampton Roads, Virginia.
6-7 Apr 1862	The Union armies in the western theater won the Battle of Shiloh.
8 Apr 1862	Union forces capture Island 10, which was vital to the Confederate defense of the upper Mississippi.

29 Apr 1862	New Orleans surrenders to Admiral Farragut.
11 May 1862	The CSS *Virginia* is blown up by the Confederates. (35)
5 Jun 1862 - 1 Jul 1862	General Robert E. Lee's Army of Northern Virginia fought the Union Army of the Potomac to a standstill outside Richmond, the Confederate capitol.
19 Jun 1862	Slavery is abolished by all the United States Territories.
1 Sep 1862	The grog ration for the United States Navy is outlawed by Congress. (36)
17 Sep 1862	The Army of the Potomac fought the Army of Northern Virginia to a draw at Antietam.
31 Dec 1862	The *Monitor* founders and sinks off of Cape Hatteras.
12 Mar 1863	Congress approves new Navy regulations, including precedence of petty officer ratings, which can be set by ship's captain or the "priority" of a rating, which goes to deck skills.
7 Apr 1863	A Union ironclad squadron engaged Confederate forts in Charleston harbor in an attempt to penetrate the defenses and capture the city. The ironclads suffered heavy damage; USS *Keokuk* sank the next day.
16-17 Apr 1863	Admiral Porter's gunboats, escorting army transports, successfully passed the Vicksburg batteries.
1-3 Jul 1863	The Army of the Potomac decisively defeated the Army of Northern Virginia at Gettysburg.
4 July 1863	Vicksburg, Mississippi, surrendered after a lengthy siege by Union naval and military forces.
9 July 1863	Port Hudson, Louisiana, surrendered after a prolonged attack by Northern warships and troops.
10 Jul 1863-	Union sea and land forces cooperate in attacking Charleston's defenses.
16 May 1864	General Order #36 appends the word "chief" to three ratings with precedence: boatswain's mate, gunner's mate and quartermaster.
15 Jun 1864	The siege of the vital Confederate transportation hub at Petersburg, Virginia, began.
19 Jun 1864	The last sailing-ship gunnery duel in history. (37)
5 Aug 1864	Admiral David G. Farragut's victory at Mobile Bay, when he said, "Damn the torpedoes, full speed ahead."
2 Sep 1864	General William T. Sherman's Union army occupied Atlanta, a major transportation and industrial center.
13-15 Jan 1865	The joint amphibious assault took Fort Fisher, the key to defense of Wilmington, the last port by which supplies from Europe could reach Lee's army.
3 Apr 1865	Lee's Army of Northern Virginia abandoned Petersburg and Richmond after massive attacks by Grant's Army of the Potomac.
9 Apr 1865	Lee surrenders to grant at Appomattox.

14 Apr 1865	President Lincoln was shot at Ford's Theater, Washington, D.C. He died the next morning.
27 Apr 1865	The body of John Wilkes Booth, President Lincoln's assassin, was delivered on board the USS *Montauk*, anchored in the Anacostia River off the Washington Navy Yard.
3 Mar 1867	The purchase of Alaska.
16 Nov 1869	The Suez Canal is opened.
1870s	Invention of the automotive torpedo. (38)
1870s-1880s	Warships are designed with underwater rams. (39)
1800s	Explosive shells and smokeless gun powder are invented. (40)
1866-1900	Malaria and yellow fever are conquered by modern medicine.
6 Jan 1885	The Navy classes all enlisted personnel as petty officers first, second and third class, or as seamen first, second and third class. Chief quartermaster's boatswain's mates and gunner's mates are positioned with petty officers first class.
1886	The USS *Maine* is built at a cost of $2,000,000.
1885-1900	The breech-loading gun came into general use. (41)
1890s	Rise of the all-big-gun battleship.
1875-1906	The tonnage of capitol ships raises only moderately.
16 Oct 1891	Bosun Charles W. Riggin is killed in Valparaiso, Chile. (42)
25 Feb 1893	President Benjamin Harrison signs an executive order that provides new pay scales for sailors and divides them into rates. The executive order and subsequent supporting documents are the first to list CPO as a separate grade.
1 Apr 1893	The new enlisted structure takes effect. Nearly all sailors rated as petty officers first class since 1885 are automatically named chiefs, so the Navy won't have a "first" CPO.
17 Feb 1898	The USS *Maine* is blown up while at anchor at Havana, Cuba. (43)
19 Apr 1898	Congress recognizes Cuba as an independent nation.
20 Apr 1898	The United States severs diplomatic relations with Spain.
21 Apr 1898	President McKinley asks Congress to declare war on Spain.
1 May 1898	Dewey's victory at Manila. (44)
3 Jul 1898	Battle of Santiago, Cuba. (45)

USS Baltimore (C-3)–Two Chief Petty Officers enjoy a game of "Acey-Deucy" on deck, circa 1904-06. Note coiled fire hose and sewing machine in background. Man at left wears an ex-apprentice badge on his sleeve. (Courtesy of U.S. Naval Historical Center)

1898-1914	The effective range of naval guns is increased from 6,000 yds. to nearly 20,000 yds.
1899	The Hague Peace Conference is conducted in that city in the Netherlands.
1900	Introduction of the completely steel armored and steel-hulled battleship.
12 Oct 1900	Commissioning of the submarine *Holland*, SS-1. (46)
1902	The first edition of *The Blue Jackets Manual* is published by the Naval Institute.
1902	The first destroyer is commissioned; Bainbridge, DD-1.
6 Nov 1903	Panama is recognized by the United States as an independent country.
17 Dec 1903	First powered aircraft flight by the Wright Brothers at Kitty Hawk, North Carolina; four successful powered flights.
8 Feb 1904	Japan attacks the Russian Fleet at Port Arthur in a surprise attack without warning.
27 May 1905	Japan destroys the Russian Fleet at Tsushima under Admiral Togo.
1905	Introduction of the battle cruiser and the dreadnought. (47)
1906	Salt beef phased out of food service on board Navy ships.
1906	The Atlantic Fleet is created, (The "Great White Fleet").
1907	The Pacific Fleet is created.
16 Dec 1907	The "Great White Fleet" departs Hampton Roads, Virginia. (48)
1908	The gyrocompass is perfected, making it easier for submarines to navigate.

Chief Yeomen Eloise Fort and Lassie Kelly. When America entered the war against Germany in 1917, Navy Secretary Josephus Daniels authorized the enlistment of women into the Navy. These women were called Yeomanettes, and more than 11,000 served honorably as chief yeomen. It was not until their service during World War II that women were to become a permanent and integral part of the U.S. Navy. (Courtesy of U.S. Naval Historical Center)

John Henry "Dick" Turpin, CGM, USN (Retired)–He was the first negro chief petty officer in the U.S. Navy and one of the Navy's best known characters for many years. Turpin enlisted in the U.S. Navy 4 Nov. 1896. He survived the explosions of USS Maine *(1898) and USS* Bennington *(1905). Qualified as a master diver, he was appointed Chief Gunners Mate of USS* Marblehead *on 1 Jun. 1917. Transferred to the Fleet Reserve on 8 Mar. 1919 and retired as a Chief Gunners Mate on 5 Oct. 1925. From 1938 throughout World War II he served voluntarily making inspirational visits to naval training centers and defense plants. When not on active duty, he worked as a master rigger at the Puget Sound Naval Shipyard. He was a truly dedicated and great Navy man. (Courtesy of U.S. Navy Historical Center)*

1908	The USS *Cheyenne* (BM-10), is converted from use of coal to that of fuel oil. (49)
1908	The first 20 nurses are recruited for the newly established Navy Nurse Corps. They are the first Navy women.
1909	USS *Smith*, DD-17, is the first United States destroyer to be commissioned with turbines.
1909	Costs of USS *Michigan*, BB-27 and USS *South Carolina*, BB-26, are $27,000,000 each.
22 Feb 1909	The "Great White Fleet" returns to Hampton Roads, Virginia. (48)
6 Apr 1909	Commander Robert E. Peary and Matthew Henson reach the North Pole.
1910	USS *Paulding*, DD-22, is the first destroyer to burn fuel oil rather than coal.
31 Aug 1910	Japan annexes Korea.
1906-1925	The tonnage of capitol ships almost doubles. (50)
14 Nov 1910	*The Beginning*. Eugene Ely flies an airplane off from a Navy ship. (51)
23 Dec 1910	Lieutenant T.G. Ellyson is the first naval officer to undergo flight training.
18 Jan 1911	Eugene Ely lands the first airplane on a Navy ship. (52)
12 Apr 1911	Lieutenant T.G. Ellyson and Lieutenant Jack Towers become the Navy's first aviators.
8 May 1911	The birthdate of naval aviation; coincides with the ordering date of the first naval aircraft.
1 Jul 1911	First flight of the A-1, the first airplane built for the Navy.
1 Jun 1914	The United States Navy abolishes all use of alcohol aboard Navy ships. (53)
15 Aug 1914	The Panama Canal is opened to traffic after 10 years and three months of work effort.
1 Jan 1916	Ten enlisted men start flight training as the first enlisted class at Pensacola.
1914-1918	World War I. (54)
1917	After Secretary of the Navy Josephus Daniels asks if there's a law requiring a yeoman to be a man, the Navy authorizes enlistment of women. Though designated yeomen, they quickly became known as "yeomanettes." When World War I ends a year later, 11,275 have enlisted.
21 Mar 1917	The first Navy woman to be designated a chief yeoman is one of the "yeomanettes" brought into the Navy to support the war effort, permitting more male sailors to be sent on World War I combat tours.
6 Apr 1917	The United States declares war on Germany.
14 Jun 1917	The United States Navy institutes convoying of merchant ships.

12 Nov 1917	Rear Admiral Earle proposes use of 14," 50-caliber naval railway mounts, in France to the Chief of Naval Operations.
17 Nov 1917	Destroyers *Nicholson* and *Fanning* become the first United States ships to sink an enemy submarine.
26 Nov 1917	The Navy Department approves the construction and use of five, 14" railway mounts.
10 Dec 1917	The Naval Gun Factory provides description of gun mounting to Rear Admiral Earle.
25 Jan 1918	One-hundred-thirty-six standard drawings and 11 sketches are available for submission to bidders.
13 Feb 1918	The bids are opened and contracts are awarded to the Baldwin Locomotive Works and the Standard Street Car Co.
25 Apr 1918	The first naval railway gunmount is completed.
6 Sep 1918	Five, 14", 50-caliber, naval railway batteries start bombarding German bases from positions behind the lines in France.
24 Sep 1918	Lieutenant jg David S. Ingalls shoots down his fifth enemy plane, becoming the United States Navy's first and only fighter ace in World War I.
1918	Fuel oil is adopted as the main source of energy in navy ships. (55)
1918	Opha M. Johnson becomes the first woman Marine. By the end of World War I, 276 other women have also enlisted in the Corps.
20 Oct 1918	Germany accepts peace terms.
11 Nov 1918	The end of World War I; armistice is signed by Germany. (Length of United States portion, one year, seven months, and five days).
8 May 1919	The NC-4 departs Rockaway, Long Island. (56)
31 May 1919	The NC-4 arrives at Plymouth, England. (56)
28 Jun 1919	The Peace Treaty is signed by Germany.
11 Jul 1919	Reconfiguration of the USS *Jupiter* to USS *Langley*, CV-1. (57)
10 Dec 1919	Graduation of the first class of enlisted pilots. They are designated Naval Aviation Pilots (NAPs). See 1 Jan 1916 entry.
1920-1921	Six Battle Cruisers are laid down by the United States (58)
11 Jul 1921	President Harding proposes a naval disarmament conference to be held between the United States, Great Britain, France, and Japan.
1921	A battleship is sunk by Brigadier General Billy Mitchell. (59)
25 Aug 1921	Germany signs peace treaty with United States officially ending state of war.
12 Nov 1921- 6 Feb 1922	Beginning and ending dates of the Limitation of Naval Armaments conference held in Washington, D.C. (60)

Chief Yeoman Daisy May Pratt pictured in the white regulation uniform of 1918. (Courtesy of U.S. Naval Historical Center)

An Artist's conception of the CPO uniform from 1920 to 1945. (Courtesy of U.S. Naval Historical Center)

20 Mar 1922	The USS *Langley* is commissioned as CV-1. (57)
17 Oct 1922	Lieutenant Commander V.C. Griffin takes a Vought biplane, VE-7-5F, off the *Langley*. The first United States Navy aircraft to fly from an aircraft carrier.
18 Nov 1922	Lieutenant Kenneth Whiting makes the first carrier catapult launch from the USS *Langley*, CV-1.
27 Dec 1922	The Japanese complete (commission) the CVL Hosho which was designed from the keel up as an aircraft carrier. This was 11 1/2 years ahead of our first, the USS *Ranger*, CV-4.
1926	Hirohito becomes Emperor of Japan.
9 May 1926	Floyd Bennett and Lieutenant Commander Richard E. Byrd circle the North Pole in a Fokker-trimotor aircraft.
2 Jul 1926	The Army Air Corps is created by Congress.
1926	Salt pork is phased out of food service on board United States Navy ships.
24 Mar 1927	The USS *Noah* helps rescue 52 besieged Westerners from Nanking during the "Nanking Incident."
20 Jun 1927 – 4 Aug 1927	The United States sponsors a three-power treaty between the United States, Great Britain, and Japan. (61)
16 Nov 1927	The USS *Saratoga*, CV-3, is commissioned at Camden, New Jersey (62)
14 Dec 1927	The USS *Lexington*, CV-2, is commissioned at Quincy, Massachusetts. (62)
1 Jul 1928	Congress passes a law that requires that 30 percent of all naval aviators to be enlisted pilots.
29 Nov 1929	Commander Richard E. Byrd with Bernt Balchen as pilot fly a Ford Trimotor over the South Pole.
17 Dec 1929 – 16 Jan 1930	The USS *Lexington*, CV-2, supplies 4 1/4 million kilowatt hours of electricity to Tacoma, Washington, as their only power source.
21 Jan 1930 – 22 Apr 1930	London International Arms parley. (63)
8 Aug 1931	The USS *Akron* is christened.
19 Sep 1931	Mukden in S. Manchuria is occupied by Japan.
Oct 1931	The USS *Akron* is commissioned.
2 Jan 1932	Japanese occupying forces in Manchuria set up a puppet government of Manchukuo, (a violation of the Nine Power Treaty of China).
24 Feb 1933	Japan withdraws from The League of nations.
11 Mar 1933	The USS *Macon* is christened.
Mar 1933	Not a single warship was authorized under President Hoover. The United States Fleet stands at 65 percent of treaty strength, while that of Japan stands at 95 percent.

4 Apr 1933	The USS *Akron* is lost off the coast of New Jersey. Seventy-three are killed; one of which was Chief of the Bureau of Aeronautics, Rear Admiral William A. Moffett.
16 Jun 1933	President Roosevelt includes shipbuilding as a means of fighting the depression. $238,000,000 is apportioned to build two aircraft carriers, five cruisers, and many smaller craft.
Jun 1933	The USS *Macon* is commissioned.
29 Dec 1933	Japan advises the world that the Washington Naval Treaty of 1922 is no longer acceptable.
Mar 1934	Admiral Byrd flies to Bolling Advance Base, 123 miles south of Little America.
27 Mar 1934	The Vinson-Trammel Act empowers the President to bring the United States Fleet to full strength by 1942, (but it didn't happen until 1944). (64)
4 Jun 1934	The USS *Ranger*, CV-4, is commissioned at Norfolk, VA (65)
Aug 1934	A four-man rescue party sets out from Little America to try to effect the rescue of Admiral Byrd.
Oct 1934	Admiral Byrd is airlifted back to Little America.
29 Dec 1934	Japan notifies the world that it will no longer be a party to the terms of the 1922 and 1930 London Naval Limitation Agreements after 31 Dec 1936.
1934-1935	Cost of an average destroyer is $3,500,000.
12 Feb 1935	The USS *Macon* crashes off Point Sur, California.
May 1935	Japan cancels its membership in The League of Nations.
25 Mar 1936	A second London Treaty is signed by Great Britain, United States, and Japan. (66)
15 Sep 1936	The USS *Langley* has the forward part of her flight deck removed and is reconfigured for duty as a seaplane tender. Her designation is changed to AV-3.
31 Dec 1936	The first London Treaty expires; Japan no longer recognizes the 1930 treaty also.
20 Aug 1937	The USS *Augusta*, CA-31, is hit by shrapnel while moored off Pootung Point, Shanghai, during the Sino Japanese War. One sailor is killed, 18 are wounded.
30 Sep 1937	The USS *Yorktown*, CV-5, is commissioned at Norfolk, VA.
Oct 1937	The first TBD-1 is delivered to Torpedo Squadron Three.
12 Dec 1937	The Japanese attack and sink the USS *Panay*, a gunboat, on the Yangtze River, in China. (67)
1938	Congress passes a naval reserve act that allows qualified women to enlist.
Mar 1938	The USS *Saratoga*, CV-3, performs a surprise attack on Pearl Harbor, T.H. (68)
12 May 1938	The USS *Enterprise*, CV-6, is commissioned at Newport News, Virginia.

17 May 1938	The United States Congress passes a Naval Expansion Act, that will finance the plan to build a two-ocean-navy.
6 Sep 1938	CNO Admiral Leahy forms the Atlantic Fleet, that consists of seven cruisers and seven destroyers.
1 Sep 1939	The German Nazis invade Poland; bomb Warsaw. The United States forms Special Squadron 40-T, for patrol duty in European waters.
3 Sep 1939	At 06:00, England and France declare war on Germany.
5 Sep 1939	The United States initiates the *Neutrality Patrol.*
6 Sep 1939	President Roosevelt orders the 110, "four-stacker" destroyers, that date back to the World War I time frame, in mothballs at San Diego, to be rehabilitated.
25 Apr 1940	The USS *Wasp*, CV-7, is commissioned at Boston, Massachusetts. This is the last carrier to be laid down under *Treaty* restrictions.
10 May 1940	Fleet Problem XXI terminates in Pearl Harbor, T.H., to be held there indefinitely by orders of President Roosevelt.
29 May 1940	The prototype XF4U-1 flies for the first time.
18 Jun 1940	France capitulates to Germany.
Jul 1940	The United States Navy consists of 160,997 officers and men.
2 Jul 1940	Japan conscripts 1,000,000 men for military service.
19 Jul 1940	Congress passes, and the President signs a $4,000,000,000, "Two-Ocean Act," funding the expansion of the United States Navy by 70 percent, (11 BBs, 11 CVs, 40 CAs and CLs, 115 DDs, and 67 SSs).
Jul 1940	President Roosevelt cuts off export of top-grade scrap iron and aviation fuel to Japan.
Aug 1940	The United States delivers 50 World War I, four-stacker destroyers to Great Britain. (69)
25 Sep 1940	United States Intelligence breaks Japan's supersecret *Purple* diplomatic code.
27 Sep 1940	Japan signs a Tripartite Pact with Germany and Italy. (70)
Nov 1940	The Grumman F4F Wildcat, goes into service.
12 Nov 1940	Congress approves dropping the minimum draft age from 21 years to 18. (71)
5 Jan 1941	The regulation for display of the National Insignia on aircraft placed the star on the upper right and lower left wing surfaces and revised rudder striping to 13 red and white horizontal stripes.
1 Feb 1941	The Atlantic and Pacific Fleets are reorganized and Admiral J.O. Richardson is replaced at Pearl Harbor by Admiral Husband E. Kimmel, in charge of the Pacific Fleet.
Mar 1941	Japanese naval spy, Ens. Takeo Yoshikawa, is secreted on Oahu, to spy for Japan.
11 Mar 1941	The Lend Lease Law is signed by President Roosevelt.
Apr 1941	USS *Niblack*, DD-424, engaged in active combat with a German U-boat in the North Atlantic. *Niablack* attacked the U-boat with depth charges and in so doing, fired the first shot for the United States in World War II.
Apr 1941	United States Naval and air bases are opened at Bermuda and air bases are opened on the east coast of Greenland.
Spring 1941	Admiral Ernest J. King heads the Atlantic Fleet; the Neutrality Patrol is extended to 26 degrees west, to include the Azores Islands.
May 1941	The United States takes over the naval base at Argentia, Newfoundland.
Jun 1941	The Grumman Aircraft Company is tasked to develop an improved version of the F4F Wildcat. (72)
2 Jun 1941	The USS *Long Island*, AVG-1, is our first escort aircraft carrier to be commissioned. She was converted from the cargo ship *Mormacmail* in 67 working days.
30 Jun 1941	The strength of the United States Navy is now 358,021 officers and men.
26 Jul 1941	The United States and Britain freeze Japanese assets.
28 Jul 1941	Japan freezes United States, British and Dutch assets.
30 Jul 1941	The United States gunboat *Tutulia* is hit by a bomb dropped by Japanese naval aircraft near Chung King, China.
1 Sep 1941	Admiral King orders United States Navy ships to convoy merchant ships in the Atlantic.
4 Sep 1941	The USS *Greer*, DD-145, and German submarine, U-652 attack each other; however, neither ship appears to suffer any damage.
15 Oct 1941	Admiral Yamamoto's plan to attack the United States Fleet at Pearl Harbor is adopted and approved by the Emperor.
17 Oct 1941	The USS *Kearny*, DD-432, is torpedoed by a German submarine. Eleven men are dead or missing.
20 Oct 1941	The USS *Hornet*, CV-8, is commissioned as the first CV to have a Radar Plot installed in the island structure to protect the fleet from air attack.
30 Oct 1941	The USS *Salinas*, AO-19, is attacked by a German submarine.
31 Oct 1941	The USS *Reuben James*, DD-245, is sunk by a German submarine while escorting convoy HX 156; 115 men are killed; 44 survive. (73)
1 Nov 1941	The United States Coast Guard is placed under orders of the Secretary of the Navy by Executive Order of President Roosevelt.
6 Nov 1941	A committee of the United States National Academy of Sciences recommends the immediate construction of an atomic bomb.
10-18 Nov 1941	The Japanese, Pearl Harbor attack, task force, leave their various bases in Japan.

5 Dec 1941	The Yangtze Patrol is formally discontinued.
6 Dec 1941	President Roosevelt approves research funds for the development of an atomic bomb and a promise of "enormous" resources if the project shows promise.
7 Dec 1941	Japanese carrier forces strike Pearl Harbor, Territory of Hawaii. (74)
8 Dec 1941	The United States declares war on Japan.
8 Dec 1941	Thailand falls to Japanese forces.
10 Dec 1941	Captain Colin Kelley, United States Army Air Force, reports that he has bombed a Japanese *Kongo* Class battleship. (75)
11 Dec 1941	United States forces on Guam surrender to the Japanese.
11 Dec 1941'	Germany and Italy declare war on the United States Congress declares war on Germany and Italy.
16 Dec 1941	The Secretary of the Navy approved an expansion of the pilot training program going from 800 students per month to 2,500 per month; leading to 20,000 pilots annually by mid-1943.
17 Dec 1941	Vice Admiral William S. Pye relieves Rear Admiral Husband E. Kimmel as Commander-in-Chief of the Pacific Fleet at Pearl Harbor.
23 Dec 1941	United States forces on Wake Island surrender to the Japanese.
25 Dec 1941	Hong Kong surrenders to the Japanese.
31 Dec 1941	Admiral Chester W. Nimitz takes command of the Pacific Fleet at Pearl Harbor, relieving Vice Admiral Pye.
1942	With World War II boiling, the 1938 act is amended to include a woman's auxiliary reserve. More than 27,000 women already are on active duty, soon they will be known as "WAVES," for Women Accepted for Volunteer Emergency Service. By the end of the war, 86,000 women will serve.
3 Jan 1942	Manila and Cavite, Philippine Islands, surrender to the Japanese.
10 Jan 1942	President Roosevelt orders General MacArthur to leave the Philippines.
11 Jan 1942	The USS *Saratoga*, CV-3, is hit by enemy torpedoes and put out of action for six months. She returns to Bremerton, WA, for repairs and modifications.
12 Jan 1942	The Japanese invade the East Indies.
22-24 Jan 1942	Battle of Makassar Straits. Navy "four-stacker" destroyers, *Ford, Pope, Parrott,* and *Paul Jones* sink four Japanese transports. This is the United States Navy's first surface engagement since the Spanish American War.
23 Jan 1942	Rabaul surrenders to the Japanese. (76)
24 Jan 1942	Balikpapan, Borneo is occupied by the Japanese.
10 Feb 1942	Gasmata, New Britain, is occupied by the Japanese.

CPO initiation during World War II.

15 Feb 1942	British forces in Singapore surrender to the Japanese.
17 Feb 1942	The athwartshp catapult is removed from aircraft carriers CV-5, CV-6, CV-7, and CV-8.
19 Feb 1942	Japanese carrier aircraft devastate Darwin, Australia.
19-20 Feb 1942	Battle of Badoeng Strait. One United States destroyer damaged.
20 Feb 1942	Lieutenant "Butch" O'Hare, flying a Grumman Wildcat of VF-3 from the USS *Lexington*, shoots down five Japanese Betty bombers in five minutes.
20 Feb 1942	Timor is invaded by the Japanese.
23 Feb 1942	The Bureau of Aeronautics expanded the pilot training course from 7—11 1/2 months for pilots of single and twin-engine aircraft and 12 1/2 months for four-engine pilots as follows: 3 months at Induction Centers, 3 months primary training, 3 1/2 months Intermediate training, and 2 to 3 months operational training, depending on type of aircraft to be operated.
23 Feb 1942	Japanese submarine I-17, shells the Elwood oil field, west of Santa Barbara, CA. There were no casualties.
Feb 1942	Aviation Machinist's Mate Bruno Gaido shoots down an attacking Japanese aircraft with the rear gun of an SBD aircraft while it was sitting on the deck of the aircraft carrier *Enterprise*.
27 Feb 1942 – 1 Mar 1942	Battle of the Java Sea. USS *Langley*, AV-3 (formerly CV-1), is sunk. USS *Houston*, CA-30, is damaged.

1 Mar 1942	The Battle of Sunda Strait. (77)	16 Jun 1942	Congress authorizes 200 LTA (Lighter than Air) craft for the Navy.
1 Mar 1942	Java is invaded by the Japanese. (78)	25 Jun 1942	The first production F4U flies.
8 Mar 1942	Lae-Salamoua surrenders to the Japanese.	25 Jun 1942	A decision is reached for a combined United States and British research and development of the atomic bomb.
12 Mar 1942	General Douglas MacArthur and Rear Admiral F.W. Rockwell depart Corregidor for Australia. General Jonathan Mayhew Wainwright is left in charge.	4 Jul 1942	The United States discovers that the Japanese are constructing an airfield on Guadalcanal.
17 Mar 1942	General MacArthur and his entourage arrive at Darwin, Australia.	21-22 Jul 1942	Japanese troops are landed at Gona, Papua, New Guinea.
2-5 Apr 1942	Admiral Yamamoto's plan to attack and invade Midway is formally approved in Tokyo.	23 Jul 1942 – 23 Jan 1943	The Papuan Campaign
7 Apr 1942	Buin (Southern Bougainville) is occupied by the Japanese.	30 Jul 1942	The United States Navy establishes the Women Accepted for Volunteer Emergency Service (WAVES) program.
9 Apr 1942	United States Forces on the Bataan Peninsula surrender to the Japanese, (after four months and two days). The "Death March" begins. (79)	30 Jul 1942	A protoype F6F Hellcat flies.
16 Apr 1942	Japanese forces are landed at Capiz and Iloilo, Panay Island, in the Philippines.	7 Aug 1942	Capture of Tulagi, Guvutu. The initial United States Marines landing on Guadalcanal in the Solomon Islands. The Japanese abandon the island on 7 Feb 1943. (86)
18 Apr 1942	Doolittle's group hits Tokyo from Shangri-La, (USS *Hornet*, CV-8). (80)	8-9 Aug 1942	The Naval Battle of Savo Island; Japanese title of *First Battle of the Solomon Sea*. (87)
30 Apr 1942	Nine thousand, three hundred American and 48,000 Filipino POWs arrive at Camp O'Donnell on the island of Luzon in the Philippines.	12 Aug 1942	The USS *Wolverine*, IX-64, was commissioned at Buffalo, NY. It was a Great Lakes excursion ship converted to aviation training to provide a flight deck for operations on Lake Michigan.
2 May 1942	Tulagi (in the Solomon Islands) is occupied by the Japanese.	20 Aug 1942	The designation of escort carriers was changed from AVG to ACV.
7-8 May 1942	Battle of the Coral Sea. USS *Lexington*, CV-2, USS *Neoshio*, AO-23, and USS *Sims*, DD-409 are sunk. (81)	23-25 Aug 1942	The Battle of the Eastern Solomons (Stewart Islands); Japanese title of *Second Battle of Solomon Sea*. (88)
6 May 1942	Hollandia is occupied by the Japanese.	Sep 1942	Completion of the building of the Pentagon. Colonel Leslie R. Groves was the engineer responsible for its construction. Following this assignment he was promoted to brigadier general and assigned to overseeing the construction of the atomic bomb.
May 1942	The Japanese occupation of Port Moresby is scheduled for early May this year; however, they do not make it in May, or ever.		
6 May 1942	General Wainwright surrenders Corregidor, and *all* United States Forces in the Philippines.	15 Sep 1942	USS *Wasp*, CV-7, is sunk. The USS *Hornet*, CV-8, is the only operational carrier remaining in the South Pacific. (89)
15 May 1942	The design of the National Star Insignia was revised by eliminating the red disc in the center of the star, and use of horizontal red and white rudder striping was discontinued.	17 Sep 1942	The Japanese drive over the Owen Stanley Mountains is halted at Imita Range, 20 airline miles from Port Moresby.
3-6 Jun 1942	Battle of Midway. The USS *Yorktown*, CV-5, and USS *Hammann*, DD-412 are sunk. (82)	1 Oct 1942	The XP59A, a twin-jet powered aircraft, built by Bell Aircraft Co. and the General Electric Corp., under a(n) United States Army Air Forces contract, makes its first successful low altitude flight.
3-6 Jun 1942	First combat action of the TBF aircraft.		
Jun 1942	RADAR is installed on most of our United States warships.	3 Oct 1942	First flight of a production model of the F6F Hellcat aircraft.
7 Jun 1942	Japanese troops are landed on Adak, Attu, and Kiska, Alaska.	11-12 Oct 1942	The Naval Battle of Cape Esperance; Japanese title of *Second Savo*. (90)
9 Jun 1942	Lt. Commander Lyndon B. Johnson's flight merits the Silver Star. (84)	20 Oct 1942	Admiral Bill Halsey relieves Admiral Ghormley as Command of the South Pacific area and South Pacific Forces.
12 Jun 1942	George Bush enlists in the United States Navy as a SN2/c, on his 18th birthday. (85)		
Jun 1942	A damaged Japanese Zero is recovered intact on an Alaskan island.		

25 Oct 1942	*Hughes* (DD-40) is accidently damaged by United States Naval gunfire in the Solomon Islands area.
25-26 Oct 1942	The Naval Battle of the Santa Cruz Islands. (91)
27 Oct 1942	USS *Hornet*, CV-8 is sunk. The USS *Enterprise*, CV-6, is now the sole remaining aircraft carrier in the Pacific battle zone and it has been damaged. (92)
30 Oct 1942	The Japanese land a second invasion force at Attu, Aleutilan Islands.
2 Nov 1942	Naval Air Station Patuxent River is established.
8 Nov 1942	North Africa invasion. United States Naval vessels damaged: *Massachusetts*, BB-59, *Wichita*, CA-45, *Brooklyn*, CL-40, *Ludlow*, DD-438, *Murphy*, DD-603, *Palmer*, DMS-5, by gunfire from coastal defense guns, North Africa landings; *Stansbury*, DMS-8, by mine; *Leedstown*, AP-73, by aircraft torpedo.
12-15 Nov 1942	The Naval Battle for Guadalcanal; Japanese title of *Third Savo*. This is the fourth attempt by the Japanese to retake Henderson Field. (93)
29-30 Nov 1942	The Naval Battle of Tassafarongo; Japanese title of *Fourth Savo*. (94)
2 Dec 1942	The first nuclear chain reaction is achieved. (95)
31 Dec 1942	The USS *Essex*, CV-9, is commissioned. The first of the *Essex* Class carriers.
31 Dec 1942	Grumman Aircraft Co., Bethpage, L.I., has now built 12 production model F6F Hellcat aircraft.
1943	The Marine Corps activates the Women's Reserve. Within 16 months, it has reached its full strength of 1,000 officers and 18,000 enlisted women. They are led by Colonel Ruth C. Streeter.
1 Jan 1943	VF-17 is commissioned at Norfolk, VA, equipped with Grumman F4F, Wildcat aircraft.
5 Jan 1943	First combat use of a proximity fused projectile by the USS *Helena*, CL-50, off the southern coast of Guadalcanal. (96)
15 Jan 1943	Grandpa Pettibone is introduced in the Bu Aer News Letter.
1 Feb 1943	Regulations for the National Insignia on aircraft was revised to remove those on the upper right and lower left wing surfaces.
8 Feb 1943	Over 11,000 Japanese troops have completed their evacuation from Guadalcanal.
9 Feb 1943	Organized Japanese resistance is ended on Guadalcanal.
13 Feb 1943	VF-17 is now equipped with F4U aircraft; they eventually had 36 aircraft and 39 pilots.
13 Feb 1943	First combat action of the F4U, Corsair aircraft, with VMF-124, at Kahili Field, Bougainville.
22 Feb 1943	USS *Iowa*, BB-61 is commissioned; wt., 50,000 tons; cost, $70,000,000; personnel, 2,500.
1 Mar 1943	The aircraft complement of airgroups assigned on board *Essex* Class aircraft carriers is, 21 VF, 36VSB, and 18VTB, totaling 75 aircraft.
4 Mar 1943	Installation of a Combat Information Center and Fighter Director Station, additional anti-aircraft batteries, a second flight deck catapult with removal of one athwartships were incorporated into the characteristics of *Essex* Class carriers.
2-5 Mar 1943	Battle of the Bismarck Sea. (7)
26 Mar 1943	Battle of the Komandorski Islands; Japanese title of Battle of Attu Island. (98)
1 Apr 1943	Naval Air Station, Patuxent River is established.
1 Apr 1943	The first Navy night fighter squadron VF (N)-75, was commissioned at Quonset Point, RI.
18 Apr 1943	Admiral Yamamoto is shot down at Ballale, Bougainville. (99)
10 May 1943	*MacDonough*, DD-351, and *Sicard* DM-21, are damaged by collision, Aleutian Islands area.
11-30 May 1943	Occupation of Attu, Alaska by the United States 7th Infantry Division.
19 May 1943	Tokyo decides to evacuate their troops and abandon the Aleutians; 2,600 Army troops and 3,400 Navy personnel.
30 May 1943	Japanese organized resistance ends on Attu, Aleutian Islands.
9 Jun 1943	George Bush is awarded his wings of gold at the age of 18 years, 11 months, 26 days.
28 Jun 1943	A change in the design of the National Star Insignia added white rectangles on the left and right sides of the blue circular field to form a horizontal bar, and a red border stripe around the entire design. The following September, Insignia Blue was substituted for the red.
29-30 Jun 1943	Nassau Bay and New Georgia Group landings.
5-6 Jul 1943	The Naval Battle of Kula Gulf. (100)
13 Jul 1943	Naval engagement off Vella Lavella.
13 Jul 1943	The Battle of Kolombangara, Solomon Islands. (101)
15 Jul 1943	New designation symbols are assigned to aircraft carriers: CV-*Saratoga*, *Enterprise*, and carriers of the *Essex* Class; CVB-aircraft carrier large (45,000 tons now under construction); CVL-aircraft carrier small (10,000 tons, built on light cruiser hulls); CVE-escort carrier (formerly ACVs).
15 Jul 1943	VF-17 is hoisted aboard the USS *Bunker Hill*, CV-17, at Norfolk.
18 Jul 1943	Naval airship K-74 is shot down by a German submarine in the Florida Straits; this is the only naval airship lost to enemy action.
22 Jul 1943	Arresting gear and related equipment for landing over the bow of aircraft carriers is approved for removal by the Vice Chief of Naval Operations.

28 Jul 1943	The Japanese garrison on Kiska Island is evacuated. (102)	25 Oct 1943	VF-17 is catapulted off the *Prince William* for Espiritu Santo, COMFAIR SOUTH HQ and told to report to Guadalcanal tomorrow.
2 Aug 1943	Lieutenant jg John F. Kennedy and crew survive ramming of PT-109 by Commander Kohai Hanami, in Japanese destroyer *Amagiri*.	27 Oct 1943	VF-17 moves to Ondongo, New Georgia Island, at Diamond Narrows.
6-7 Aug 1943	Battle of Vella Gulf. (103)	27 Oct 1943	Troops are landed on Mono and Stirling Islands in the Treasury Island Group, Solomon Islands. United States Navy vessels damaged: *Cony*, DD-508, by horizontal bomber; LST-399 and LST-485, by coastal mortar.
16 Aug 1943	United States forces land on Kiska Island, only to discover that there are no enemy there.		
31 Aug 1943	Second strike on Marcus Island.		
31 Aug 1943	First combat use of Grumman F6F Hellcat aircraft, by VF-5, aboard the USS *Yorktown*, CV-10, during the attack on Marcus Island.	28 Oct 1943	VF-17s first combat mission. A 90 minute mission covering the 3rd New Zealand Division's landing in the nearby Treasury Islands. Tom Blackburn emphasized that he only wanted news of the enemy planes positively shot down. No "probable," or "damaged" claims.
8 Sep 1943	The Allies sign an armistice with Italy.		
10 Sep 1943	The USS *Bunker Hill*, with VF-17 on board, departs Norfolk and heads for the Pacific.	31 Oct 1943	The first kill by a radar-equipped night fighter of the Pacific Fleet is the downing of a Betty bomber off Vella Lavella.
Sep 1943	George Bush is assigned to VT-51, aboard the USS *San Jacinto*, CVL-30.	2 Nov 1943	Battle Empress Augusta Bay; another night action battle. (106)
1943	Radar is installed on Japanese warships well into this year.	Nov 1943	The authorized complement of aircraft aboard CVL Class carriers is changed to 24 VF, 9 VB, and 9 VT, totaling 42 aircraft.
11 Sep 1943	MacArthur recaptures Salamaua from the Japanese.		
15 Sep 1943	French Patrol Squadron 1, VFP-1, manned by "fighting French" naval personnel, trained under United States Navy control, was commissioned at NAS Norfolk, VA.	8 Nov 1943	Air Group 51, composed of VF-51 and VT-51 assemble at Norfolk, VA. Flyaway cost of a TBF is $96,159.00.
		5 Nov 1943	First Rabaul strike. (107)
16 Sep 1943	MacArthur recaptures Lae from the Japanese.	11 Nov 1943	Second Rabaul strike; SB2C Curtiss Helldivers are used in combat for the first time. (108)
28 Sep 1943	The *Bunker Hill*, with VF-17 aboard, departs San Diego and heads for Pearl Harbor. (104)	11 Nov 1943	VF-17 participates in the Battle of the Solomon Sea. (109)
1 Oct 1943	The authorized complement of airgroups aboard *Essex* Class carriers is changed to 36-VF, 36-VB, and 18-VT, (an increase of 15 fighters), totaling 90 aircraft.	18-26 Nov 1943	Occupation of the Gilbert Islands. (110)
2 Oct 1943	VF-17 is off-loaded from the *Bunker Hill* at Pearl Harbor.	23 Nov 1943	Betio, Tarawa Atoll, and Makin in the Gilbert Islands are declared secured.
2 Oct 1943	MacArthur recaptures Finschafen from the Japanese.	25 Nov 1943	Battle of Cape St. George. (111)
5-6 Oct 1943	The second Wake Island raid. A task force of six aircraft carriers, seven cruisers, and 24 destroyers bomb and bombard the island.	26 Nov 1943	Commander Butch O'Hare, flying a Hellcat at night, Air Group 6, from the *Enterprise* is lost under mysterious circumstances.
6-7 Oct 1943	A night action naval battle of Vella Lavella. (105)	6 Dec 1943	General Dwight D. Eisenhower, USA, is named commander of Allied Expeditionary Forces for the invasion of Europe.
12 Oct 1943	VF-17, with 36 F4U Corsairs, departs Pearl Harbor aboard "jeep" carrier *Prince William*, AVG-19.	15 Dec 1943– 29 Nov 1944	The Bismarcks Campaign.
13 Oct 1943	Italy declares war on Germany.	29 Jan 1944– 22 Feb 1944	Occupation of the Marshall Islands.
16 Oct 1943	The United States Navy accepts its first helicopter, a Sikorsky YR-4B (HNS-1).	7 Feb 1944	Kwajalein Atoll, Marshall Islands is declared secured.
Oct 1943	The Japanese install Jose Laurel as "their President" of the Philippines.	16-17 Feb 1944	A naval carrier task force strikes Truk; 200,000 gross tons of shipping is sunk or disabled.
Oct 1943	Lieutenant jg Roger W. Duncan, VF-5, is the first Hellcat pilot to shoot down a Japanese Zero.	23 Feb 1944	A naval task force strikes the Marianas.

24 Feb 1944	The first detection of an enemy submarine by use of MAD gear was that installed in PBYs of VP-63 at the Strait of Gibraltar in attacking U-761 with retrorockets.	15 Jul 1944	George Bush participates in carrier air strikes against the Palaus.
3 Mar 1944	President Roosevelt announces that the Italian Fleet will be distributed among the United States, Great Britain, and Russia.	18 Jul 1944	Premier Tojo and the Japanese cabinet resign; General Koiso forms a new cabinet.
		21 Jul 1944	United States forces are landed on Guam.
Mar 1944	Japanese Admiral Mineichi Koga (who replaced Yamamoto), is killed. Admiral Soemu Toyoda succeeds him.	24 Jul 1944	United States forces are landed on Tinian.
		25-27 Jul 1944	Operation "Snapshot"; George Bush participates in photographing Peleliu for the Marines.
26 Mar 1944	*Tullibee*, SS-284, is sunk by a circular run of its own torpedo, north of Palau Islands, Caroline Islands.	25-28 Jul 1944	Strikes on the Western Carolines (Palau, Yap, Ulithi).
27 Mar 1944	PT-121 and PT-353 are sunk by a "friendly" bomber, Bismarck Archipelago area.	1 Aug 1944	Organized Japanese resistance ends on Tinian, Marianas Islands.
30 Mar 1944	*Tunny*, SS-282, is damaged by "friendly" aircraft off the Palau Islands.	6-7 Aug 1944	First air attacks on the Philippines since 1942 (at Sassa Naval Airdrome near Davao).
30 Mar 1944– 1 Apr 1944	Strikes on Western Caroline Islands.	10 Aug 1944	Organized Japanese resistance ends on Guam, Marianas Islands.
21-24 Apr 1944	Landings at Hollandia.	10 Aug 1944	The operating complement of carrier Air Groups is revised to: 54VF, 24VB, and 18VT, tot. 96. Four night fighters and two photo aircraft are included in the VF complement.
28 Apr 1944	Secretary of the Navy, Frank Knox, dies at Washington, D.C.		
5 May 1944	Admiral S. Toyoda is appointed CIC of the Japanese Combined Fleet, succeeding Admiral Koga, who was killed in an aircraft crash on 31 Mar 1944.	2 Sept 1944	George Bush and crew are shot down over Chi Chi Jima. They are rescued by the *Finback*, SS-230.
19 May 1944	James Forrestal of New York, Under Secretary of the Navy since 1940, becomes Secretary of the Navy.	3 Sep 1944	Fourth Wake Island raid. A naval task group of one carrier, three cruisers and three destroyers strike enemy positions on Wake Island.
19-20 May 1944	Third raid on the Marcus Islands.	8 Sep 1944	The first German V-2 rocket bomb hits in England.
23 May 1944	Third raid on Wake Island.	13 Sep 1944	*Warrington*, DD-383, sunk by hurricane off the Bahama Islands.
29 May 1944	USS *Block Island*, CVE-21, is torpedoed and sunk in the Atlantic, near the Azores. This is the only aircraft carrier to be lost in the Atlantic.	31 Aug 1944 – 30 Sep 1944	Occupation of Palau and Morotai.
4 Jun 1944	German submarine, U-505, is captured off Cape Blanco, Africa, after being attacked by VC-8 aircraft from CVE Guadalcanal and five DEs.	15 Sep 1944	United States troops are landed on Morotai; Marines are landed on Peleliu.
6 Jun 1944	Allied Expeditionary Force landings are made on the beaches of Normandy France. United States Naval vessels sunk: *Corry*, DD-463 and PC-1261, both by mines. LST-375 was damaged by collision.	10 Oct 1944 – 30 Nov 1944	Occupation of Leyte.
		12 Oct 1944	United States Marines capture Peleliu, aided by an Army regiment, after 27 days of battle.
11 Jun 1944 – 10 Aug 1944	Occupation of the Marinanas. (112)	12-14 Oct 1944	A fast carrier task force raids the Ryukyu's, Taiwan, and Luzon.
15 Jun 1944	First B-29 raid on Japan (Kyushu), from Chengtu, China.	20 Oct 1944	General MacArthur returns to the Philippines, near Palo, on Leyte.
19 Jun 1944	The *Barbara* suffers a terrible engine oil leak. Bush and crew crash land in the water. They are picked up by *Clarence K. Bronson*, DD-668, in eight minutes.	23-26 Oct 1944	The Naval Battle of Leyte Gulf. (114)
		24 Oct 1944	*Tang*, SS-306, is sunk by a circular run from her own torpedo, north of Formosa.
19-20 Jun 1944	Battle of the Philippine Sea; Japanese title of *Battle of Marianas*. (113)	25 Oct 1944	Battle of Surigao Strait.
1 Jul 1944	Radar is installed on all Japanese battleships and cruisers at Kure, by this date.	25 Oct 1944	Battle of Samar.
		25 Oct 1944	Battle off Cape Engano. The first use of planned Kamikaze attacks. (115)
9 Jul 1944	Organized resistance is ended on Saipan.	29 Oct 1944	USS *Intrepid*, CV-11, is damaged by kamikaze attacks.

30 Oct 1944	USS *Franklin*, CV-13, and USS *Bellau Wood*, CVL-24, are damaged by kamikaze attacks.	3 Feb 1945	United States Forces recapture Manila.
30 Oct 1944	George Bush is returned to the San Jacinto, after being shot down eight weeks earlier.	10 Feb 1945	*Batfish*, SS-310, sinks Japanese submarine RO-55; this, is the first of three submarines to be sunk (RO-55, RO-112, and RO-113), by *Batfish* in four days.
5 Nov 1944	USS *Lexington*, CV-16, is damaged by kamikaze attacks.	16 Feb 1945 – 16 Mar 1945	Capture of Iwo Jima, Volcano Islands.
12 Nov 1944	German battleship, *Tirpitz,* is sunk by British aircraft in Tromso Fjord, Norway.	21 Feb 1945	Kamikaze attacks sink USS *Bismarck Sea*, CVE-95.
20 Nov 1944	Kamikaze torpedoes (Kaiten), are used by the Japanese for the first time at Ulithi. (116)	21 Feb 1945	Kamikaze attacks damage USS *Saratoga*, CV-3.
21 Nov 1944	United States Submarine *Sealion*, SS-195, sinks the Japanese battleship *Kongo*.	21 Feb 1945	Kamikaze attacks damage USS *Lunga Point* CVE-94.
24 Nov 1944	B-29s start bombing Japan from airfields in the Marianas.	9 Mar 1945	B-29s firebomb Tokyo, destroying 250,000 homes and 16 1/2 sq. miles of the city, killing 80,000 to 300,000 people. In missions that follow they destroy 170,000 sq. miles in 66 target cities. (117)
25 Nov 1944	Aircraft from two carrier groups bomb enemy shipping and aircraft in central Luzon, P.I. Japanese suicide aircraft attack and damage United States carriers, *Essex*, CV-9; *Intrepid*, CV-11; *Hancock*, CV-19; *Independence*, CVL-22; *Cabot*, CVL-28.	18 Mar 1945 – 21 Jun 1945	The Okinawa Campaign.
		18 Mar 1945	Kamikaze attacks damage USS *Enterprise*, CV-6, USS *Intrepid*, CV-11, and USS *Yorktown*, CV-10.
28 Nov 1944	United States Submarine *Archerfish*, SS-311, sinks the new Japanese aircraft carrier *Shinano*, and is awarded a Presidential Unit Citation.	19 Mar 1945	Kamikaze attacks damage USS *Franklin*, CV-13 and USS *Wasp*, CV-18.
29 Nov 1944	The aircraft complement of *Essex* Class Carrier Air Groups is revised to: 73VF, 15VB, and 15VT, totaling 103 aircraft.	21 Mar 1945	The Japanese employ a new "Baka Bomb," a rocket powered 1,800-kilogram missile carried aloft by a "mother" bomber.
29 Nov 1944	VT-51 departs the *San Jacinto*; George Bush has accumulated 924.8 pilot hours and 116 carrier landings.	25 Mar 1945	United States forces land on the Kerama Islands in preparation for Okinawa.
14-16 Dec 1944	Naval support of landings on Mindoro, P.I.	30 Mar 1945	Kamikaze attacks damage USS *Indianapolis*, CA-35, in the Okinawa area.
18 Dec 1944	United States Navy vessels sunk by typhoon, east of the Philippine Islands; *Hull*, DD-350; *Monaghan*, DD-354; *Spence*, DD-512.	1 Apr 1945	United States forces begin landings on Okinawa; the beginning of 80 days of fierce battle.
20 Dec 1944	Organized enemy resistance ends on Leyte, Philippine Islands.	3 Apr 1945	The JCS appoint General MacArthur, Commander of all United States Army forces and Admiral Nimitz Commander of all United States Navy forces in the Pacific.
31 Dec 1944	F4U Corsairs did not regularly operate from United States aircraft carriers until the end of this year.	3 Apr 1945	Kamikaze attacks damage USS *Wake Island*, CVE-65.
3-22 Jan 1945	Invasion of Luzon, Philippine Islands. Seventeen escort carriers cover the approach of the Luzon Attack Force.	6 Apr 1945	Japanese BB *Yamato*, CL *Yahagi*, and two DD are sunk by carrier aircraft, while enroute to Okinawa.
4 Jan 1945	USS *Ommaney Bay*, CVE-79, is sunk by kamikaze attack off the Philippines.	6 Apr 1945	Kamikaze attacks damage USS *San Jacinto*, CVL-30.
5 Jan 1945	Kamikaze aircraft damage USS *Manila Bay*, CVE-61 and USS *Savo Island*, CVE-78.	7 Apr 1945	Battle of the East China Sea.
8 Jan 1945	Kamikaze aircraft damage USS *Kadashan Bay*, CVE-76 and USS *Kitkun Bay*, CVE-71.	7 Apr 1945	Kamikaze aircraft damage the USS *Hancock*, CV-19, and the USS *Maryland*, BB-46.
9 Jan 1945	United States troops are landed at Lingayen Gulf on Luzon.	11 Apr 1945	Kamikaze aircraft damage the USS *Enterprise*, CV-6, and the USS *Essex*, CV-9, the first ships to be hit with the new "Baka Bomb" at Okinawa. (118)
12 Jan 1945	The second kamikaze attack using manned torpedoes (*Kaiten*), from six submarines, and 24 *Kaiten*. (116)	12 Apr 1945	President F.D. Roosevelt dies of a cerebral hemorrhage at Warm Springs, GA. Harry S. Truman is sworn in as President by Chief Justice Harlan Stone.
13 Jan 1945	Kamikaze aircraft damage the USS *Salamaua*, CVE-96.	16 Apr 1945	Kamikaze aircraft damage the USS *Intrepid*, CV-11 and the USS *Missouri*, BB-63.

28 Apr 1945	Mussolini and his mistress are executed.	10 Aug 1945	Japan accepts the provisions of the Potsdam Declaration and agrees to surrender. They notify world news in the late evening.
30 Apr 1945	Hitler and his mistress commit suicide.	15 Aug 1945	General MacArthur receives official notification that he is to act as Supreme Commander for the Allied Powers during the surrender ceremony.
4 May 1945	Kamikaze aircraft damage escort carrier USS *Sangamon*, CVE-26.		
8 May 1945	V-E Day. Germany surrenders unconditionally to the Western Allies and Russia at Reims, France.	15 Aug 1945	The F6F *Hellcat* shoots down its 5,000th Japanese aircraft.
11 May 1945	Kamikaze aircraft damage USS *Bunker Hill*, CV-17.	28 Aug 1945	Air Force technicians land at Atsugi Airdrome, near Tokyo; these are the first United States troops to land in Japan.
14 May 1945	Kamikaze aircraft damage USS *Enterprise*, CV-6.		

28 Apr 1945 — Mussolini and his mistress are executed.

30 Apr 1945 — Hitler and his mistress commit suicide.

4 May 1945 — Kamikaze aircraft damage escort carrier USS *Sangamon*, CVE-26.

8 May 1945 — V-E Day. Germany surrenders unconditionally to the Western Allies and Russia at Reims, France.

11 May 1945 — Kamikaze aircraft damage USS *Bunker Hill*, CV-17.

14 May 1945 — Kamikaze aircraft damage USS *Enterprise*, CV-6.

17 May 1945 — The last Japanese sea battle. IJN heavy cruiser *Haguro* is sunk by five Royal Navy destroyers off Penang Malaya.

25 May 1945 — Three hundred B-29 bombers firebomb Tokyo, creating more destruction than the atomic bombing of Hiroshima and Nagasaki. Over 100,000 were killed and 2,725,000 were left homeless.

6 Jun 1945 — Kamikaze aircraft damage escort carrier USS *Natoma Bay*, CVE-62.

16 Jun 1945 — The Naval Air Test Center is established at Naval Air Station, Patuxent River, Maryland.

20 Jun 1945 — Fifth Wake Island raid.

21 Jun 1945 — Okinawa, Ryukyu Islands is declared secured, 82 days after the landing on 1 April.

30 Jun 1945 – 3 Jul 1945 — Landings at Balikpapan, Borneo.

5 Jul 1945 — General Douglas MacArthur announces the liberation of the Philippine Islands.

10 Jul 1945 – 15 Aug 1945 — Carrier operations against Japan.

13 Jul 1945 — Italy declares war on Japan.

16 Jul 1945 — The first atomic bomb is exploded at Alamogordo, New Mexico. (119)

18 Jul 1945 — Sixth Wake Island raid.

24 Jul 1945 — President Truman tells Stalin that the Atomic Bomb will be used against Japan.

27 Jul 1945 — The Potsdam Declaration calling for unconditional surrender is delivered to Japan.

30 Jul 1945 — The USS *Indianapolis*, CA-35 is sunk by Japanese submarine I-58 in the Philippine Sea. (120)

1 Aug 1945 — Seventh Wake Island raid.

6 Aug 1945 — The first Atomic Bomb ever used in war is dropped on Hiroshima, Honshu, Japan; 78,000 people are killed. (121)

6 Aug 1945 — Eighth Wake Island raid.

9 Aug 1945 — Russia declares war on Japan.

9 Aug 1945 — A second Atomic Bomb is dropped on Japan. This one is dropped on Nagasaki by the B-29, *Baux Car*. Casualties are estimated at 35,000.

10 Aug 1945 — Japan accepts the provisions of the Potsdam Declaration and agrees to surrender. They notify world news in the late evening.

15 Aug 1945 — General MacArthur receives official notification that he is to act as Supreme Commander for the Allied Powers during the surrender ceremony.

15 Aug 1945 — The F6F *Hellcat* shoots down its 5,000th Japanese aircraft.

28 Aug 1945 — Air Force technicians land at Atsugi Airdrome, near Tokyo; these are the first United States troops to land in Japan.

2 Sep 1945 — V-J Day. Japanese surrender documents are signed on board the battleship *Missouri*, BB-63, at anchor in Tokyo Bay, Japan. General Douglas MacArthur signs for the Allied Powers, and Fleet Admiral Chester W. Nimitz signs for the United States.

2 Sept 1945 — The enlisted Naval Aviation Pilot Training Program is discontinued.

3 Sept 1945 – 31 Dec 1945 — The period between the signing of the capitulation documents on board the battleship *Missouri*, BB-63, and the end of the year 1945, witnessed the surrender of Japanese garrisons on the Asiatic mainland, and on by-passed islands scattered throughout the western Pacific. Occupation of Japan progressed and the administrative organization of the United States Naval forces in the area was adjusted where necessary to enable the Navy to carry out its assigned occupation and demilitarization duties.

10 Sep 1945 — The USS *Midway*, CVB-41, is commissioned at Newport News, VA; the fist of the 45,000 ton class of aircraft carriers.

Sep 1945 — George Herbert Walker Bush leaves the Naval Service. (85)

10 Oct 1945 — Headquarters of the Commander in Chief, United States Fleet (Fleet Admiral E.J. King) was disestablished.

Nov 1945 — The 12,275th *Hellcat* is delivered to the Navy.

28 Dec 1945 — The President directed that the Coast Guard be transferred from the Navy and returned to the jurisdiction of the Treasury Department.

1946 — Beginning of the Navy's Blue Angels, who started with F6F Hellcats.

26 Jun 1946 — The knot and nautical mile are adopted by the Army Air Forces and Navy by order of the Aeronautical Board.

21 Jul 1946 — The *F.D. Roosevelt*, CV-42, is the first carrier to adapt a jet aircraft to shipboard operation. A FD-1 *Phantom*, piloted by Lieutenant Commander James Davidson made successful landings and takeoffs on board the ship.

29 Sep 1946 — *The Truculent Turtle*, a Lockheed P2V Neptune sets a new world's record for distance by flying from Perth, Australia to Columbus, Ohio; a distance of 11,235,600 miles in 55 hours and 17 minutes, without refueling.

14 Jan 1947 — A horizontal red stripe, centered on the white horizontal bar, was added to the National Star Insignia.

A chief boatswain's mate and four midshipmen first class swap sea stories aboard USS Dyess (DD 880) *during a midshipman cruise in 1948. (Courtesy of U.S. Navy National Archives)*

3 Feb 1947	Admiral Marc Mitscher dies.
4 Jun 1947	Project 27A incorporated new aircraft carrier characteristics, giving the capability to operate aircraft up to 40,000 lbs.: Installation of two H-8 catapults, strengthening the flight deck and clearing it of guns, increasing elevator capacity, adding jet aircraft blast deflectors, increased fuel capacity and jet fuel mixers.
17 Sep 1947	James Forrestal, Secretary of the Navy, took the oath of office as the first Secretary of Defense. The following day, the National Security Act of 1947 became effective and the Departments of the Army, Navy, and Air Force were constituted as integral parts of the National Military Establishment.
Dec 1947	The last enlisted man to be designated a NAP is made this date.
1948	Women are permitted to enlist in the active Navy as well as the reserves. But their advancement opportunities and career options remain limited. Women Marines are integrated into the Regular Corps.
24 Apr 1948	The Berlin Airlift begins; the United States starts flying in supplies.

12 May 1949	After one year and 18 days, the Russian blockade of Berlin is ended.
13 Jun 1949	Congress repeals the 30 percent enlisted pilot requirement that began on 1 July 1928.
25 Jun 1950	North Korea attacks South Korea.
1 Jul 1950	United States troops arrive in Pusan and move north to engage the enemy.
10 Sep 1950	North Korea pushes south to the "Pusan perimeter."
15 Sep 1950	MacArthur begins the "Inchon Landings."
26 Sept 1950 – 29 Sept 1950	MacArthur recaptures Seoul. Syngman Rhee's government returns.
8 Oct 1950	United States forces push across the 38th Parallel.
15 Oct 1950	President Truman and General MacArthur meet at Wake Island.
19 Oct 1950	Chinese troops cross the Yalu River into North Korea.

24 Nov 1950 – 27 Nov 1950	North Koreans and Chinese attack United States forces at Chosin; United States forces start retreating toward South Korea.
24 Jan 1951	The southern most Chinese advance, south of the 28th Parallel.
11 Apr 1951	Truman fires MacArthur.
May 1951	South Korean and United States forces push north of the 38th Parallel, near Pyongyang. Large numbers of Chinese surrender.
10 Jul 1951	Truce talks start at Kaesong and then move to Panmunjom.
May 1952 – Oct 1952	Stalemate, but heavy fighting continues along the 38th Parallel.
1953	Enlisted female hospital corpsmen are assigned aboard hospital ships and transports carrying dependents.
Jan 1953	The USS *Antietam*, CVS-36, our first angled-deck aircraft carrier, begins test operations.
27 Jul 1953	A truce is signed by United Nations and Communists negotiators; all fighting stops.
5 Aug 1953	The Korean War, Prisoner of War, exchanges being.
Sep 1954	All United States Navy fighter aircraft in production are outfitted with in-flight refueling gear.
17 Jan 1955	The USS *Nautilus*, SSN-571, our first nuclear powered submarine departs on her maiden voyage: "Underway on nuclear power."
Aug 1955	Admiral Arleigh A. Burke becomes Chief of Naval Operations. He serves an unprecedented three terms, from this date until August 1961.
1 Oct 1955	The USS *Forrestal*, CVA-59, our first super carrier, is commissioned. This is our first aircraft carrier designed to handle jet aircraft.
1 Jun 1958	Congress approves creation of senior chiefs and master chiefs, with the establishment of the E-8 and E-9 pay grades. But responsibilities aren't clear, and the move is met with resentment from both senior officers and noncoms, particularly in the Navy.
16 Nov 1958	Based on service-wide exams, the first E-7 chiefs are advanced into the new super-grades. Those with at least six years as chief and 13 years of service make E-9, and those with at least four years in grade make E-8. A second group of chiefs is elevated in the same way May 16, 1959. After these selections, only senior chiefs will qualify for E-9.
16 Aug 1959	Admiral William F. Halsey dies.
30 Dec 1959	The USS *George Washington*, SSBN-598, our first ballistic missile submarine is commissioned.
25 Apr 1961	The Bay of Pigs fiasco.
9 Sep 1961	The USS *Long Beach*, CGN-9, our first nuclear powered surface ship is commissioned.

14 Nov 1961	The United States now has 700 "United States Advisers" in Vietnam.
25 Nov 1961	The United States Navy's first nuclear powered aircraft carrier, USS *Enterprise*, CVN-65, is commissioned. (122)
27 May 1962	The Pearl Harbor Memorial is dedicated on Memorial Day.
1962	The Navy Museum is established by the Chief of Naval Operations, Admiral Arleigh Burke.
18 Sep 1962	A joint Army-Navy-Air Force regulation was issued establishing a uniform system of designating military aircraft similar to that previously in use by the Air Force.
22 Oct 1962	President John F. Kennedy imposes a naval quarantine on Cuba.
Jul 1963	There are still 61 Naval Aviation Pilots on active duty. (See 1 Jul 1928)
5 Aug 1964	The Vietnam War starts.
20 Feb 1966	Admiral Chester Nimitz dies.
13 Jan 1967	Master Chief Gunner's Mate Delbert Black, a decorated World War II veteran and Pearl Harbor survivor, becomes the Navy's first senior enlisted adviser. Among his first achievements: getting his title changed in April 1967 to master CPO of the Navy.
1967	There are still 34 Naval Aviation Pilots on active duty. (See 1 Jul 1928)
8 Jun 1967	The USS *Liberty* is attacked by Israel off the Sinai; 34 men are killed, 117 wounded.
6 Nov 1967	The last operational flight of a seaplane is made by a Marlin SP-5B of VP-40, at NAS North Island, California.
23 Jan 1968	The USS *Pueblo* is captured by North Korea. (123)
1 Jul 1968	Precedence of sailors by rating is eliminated, so paygrade and time in grade become the single means for determining level of authority. The term "leading chief" now sets the most senior chiefs above their peers.
24 Dec 1968	The crew of the *Pueblo* is released after 11 months of captivity.
27 Dec 1968	Special cap insignias are authorized for senior and master chiefs, with E-8s getting a single star atop the fouled-anchor device and E-9s getting two stars. The MCPON wears three stars.
25 Jan 1969	The United States Navy launches the NR-1, our first nuclear powered deep submergence research and ocean engineering vessel.
29 Jan 1971	The Navy Uniform Board says all chiefs will wear working khakis.
Jul 1971	The second MCPON, Jack Whittet, convinces Admiral Elmo R. Zumwalt Jr., Chief of Naval Operations, to establish the command master chief program to deal with continuing morale problems. Zumwalt also authorizes fleet and force master chief to advise senior admirals in the fleet.

| 1972 | Alene Duerk, director of the Navy Nurse Corps, is the first woman promoted to rear admiral. The term Wave is dropped as an official title. Women are cleared to command shore activities. |

| Late 1972 | The Grumman F-14, *Tomcat* joins the fleet. |

| 27 Jan 1973 | After 8 years, 5 months, and 22 days, the Vietnam War ends. |

| 1973 | Women start fight training. Pregnancy no longer means automatic discharge. |

| 20 Feb 1974 | The Lockheed S-3A *Viking*, joins the fleet. |

| 22 Feb 1974 | Barbara Allen Rainey is designated a Naval Aviator in ceremonies held at Corpus Christi, Texas. She was the first female Naval Aviator in the United States Navy. She was assigned to VR-30 at NAS Alameda, California. Lieutenant Commander Barbara Allen Rainey was killed in a crash while practicing touch-and-go landings at Middleton Field, Alabama on 13 Jul 1982. |

| 1975 | Women are assigned to service craft, and Congress authorizes admission of women to the service academies. |

| 20 Sep 1975 | The USS *Spruance*, DD-963, the United States Navy's first gas turbine-powered destroyer is commissioned. |

| 1976 | The foremast and fighting top of the USS *Constitution* is replaced and installed in the Navy Museum. |

| 1976 | Fran McKee becomes the Navy's first female unrestricted line flag officer. |

| 1978 | Coast Guard Commandant John B. Hayes lifts assignment restrictions in his service, allowing women into all billets. Congress amends the law to permit women to serve on some Navy ships. |

| 1978 | The last United States Marine Corps Naval Aviation Pilot retires. (See 1 July 1928) |

| 1979 | The last United States Coast Guard Naval Aviation Pilot retires. (See 1 July 1928) |

| 1979 | The first female naval aviator qualifies for carrier landings. The first woman qualifies as a surface warfare officer. Lieutenant (jg) Beverly Kelly assumes command of the 95-foot cutter *Cape Newagen*, becoming the first woman to command a United States warship. |

| 30 Oct 1979 | The F/A-18 *Hornet* made its first landing at sea aboard the USS *America*, CV-66, for five days of sea trials. A total of 32 catapult and arrested landings were completed. |

| 5 Mar 1980 | Congress authorizes the building of the United States Navy Memorial. |

| 1981 | The last United States Navy Naval Aviation Pilot retires. (See 1 July 1928) |

| 19 Aug 1981 | Two F-14 aircraft from the USS *Nimitz* shoot down two SU-22 Libyan fighter bombers in a dogfight across Colonel Muammar Quadhafi's "Line of Death." |

| 14 Sep 1981 | The Senior Enlisted Academy opens in Newport, Rhode Island. The first class of 16 senior and master chiefs works on improving communication skills, knowledge of Navy programs and understanding of national security matters. |

The Navy's diving community is small, to say the least. With only 1,753 divers, there are few professional "specialty" groups that can claim smaller numbers. Though the work is hard, Senior Chief Electrician's Mate (SW/MD) Mary Bonnin managed to work her way up higher than any of her peers could ever have imagined. She is the Navy's first and only certified woman master diver.

| 1 Sep 1983 | Korean Air Lines Flight 007, with 269 on board, is shot down by Soviet aircraft off Soviet island of Sakhalin. |

| 1985 | Marine training for women is expanded to include marksmanship qualification and drill under arms. Gail M. Reals is the first woman to become a Marine brigadier general on a competitive basis. |

| 14-15 Apr 1986 | A three-ship Service Action Group, composed of USS *America*, USS *Saratoga* and USS *Yorktown* and their aircraft destroy several Libyan ships, radar sites, terrorists headquarters, and other targets in Tripoli. |

| 1987 | Combat logistics force ships are opened to women. |

| 17 May 1987 | An Iraq F-1 *Mirage* fires two Exocet missiles that hit the port side of the USS *Stark*, FFG-31, 37 are killed. |

| 1988 | The first women are selected for command at sea and command of an aviation squadron. A year later, the Navy selects its first female at-sea command master chief. |

| Jan 1989 | Japan's Emperor Hirohito dies. |

| 19 Apr 1989 | At 09:55, 650 pounds of gunpowder explode in No. 2 turret, USS *Iowa*, BB-61, killing 47 men. (124) |

| 16 January 1992 | The Persian Gulf War begins by Allied aircraft launching bombing raids against Iraq in Operation *Desert Storm*. |

41

Feb 1992	The Navy redesignated its first super carrier, the *Forrestal*, CV-59, as a training carrier, AVT-59. Commissioned in 1955, the carrier was the first built with an angle deck to permit simultaneous take-offs and landings. It replaced the *Lexington*, AVT-16, which will become a museum in Corpus Christi, Texas.
12 Feb 1992	The Navy successfully completed the deepest salvage recovery on record, when the *Salvor*, ARS-52, retrieved the forward section of a helicopter from a depth of 17,250 feet off the coast of Wake Island. The previous record was 14,800 feet off the coast of Madagascar in 1988.
24 Feb 1992	The land forces portion of operation *Desert Storm* go into action.
26 Feb 1992	Kuwait is liberated; 120 United States personnel are killed; 213 are wounded; 87 other personnel are killed in non-combat incidents and accidents.
1992	More than 2,600 women served in Operation *Desert Shield* and *Desert Storm*. Congress authorizes the services to put women in combat aircraft, leaving a final decision to the Pentagon.
30 Sep 1992	The naval station at Subic Bay in the Philippines was disestablished after 47 years of service to the United States Pacific Fleet. Until the end of the year, United States NAS Cubi Point was the last United States facility to remain in the Philippines.
1 Oct 1992	During Display Determination 92, a breakdown in communications (failure to communicate), the aircraft carrier USS *Saratoga* fired two Sea Sparrow missiles, hitting Turkish destroyer *Mauvenet*, killing her captain and four of the crew. Thirteen others were injured.

1. Congress, in Philadelphia, decrees that two war vessels be outfitted to capture British shipping. The two are followed by others; merchantmen commissioned: *Alfred, Columbus, Andrew Doria,* and *Cabot*, followed by *Wasp, Fly, Providence* and *Hornet*. Congress purchased the ships and ordered others built. Approximately 60 ships eventually are involved, two-thirds are converted merchantmen. A seven-man naval committee ran the Navy, independently from Washington's control. Navies created by 11 of the 13 colonies had no appreciable effect on the outcome of the Revolution. American Privateers (approximately 2,000 sloops, schooners, converted fishing vessels) captured over 2,200 British vessels.

2. The King of France gives John Paul Jones an old East India ship named *Le Duc de Duras*. Jones renames the ship *Bonhomme Richard* in honor of Benjamin Franklin's famous almanac, *Les Maximes du Bonhomme Richard*.

3. This is the battle where John Paul Jones makes the immortal statement, "I have not yet begun to fight."

4. A naval force is considered prohibitively expensive and superfluous. All ships are sold except the *Alliance*.

5. All ships of the Continental Navy are now sold. America has no men-of-war. An American; merchantman, the *Maria*, is captured by Algerian pirates and the crew is thrown into prison.

6. The government established under the Confederation had no real taxing power to fund a National Defense. The Articles of the Confederation need to be altered. A Constitutional Convention is called in Philadelphia.

7. The Articles of Confederation are altered out of existence. A new central government is created under a new constitution and put into effect. A Navy is still not considered necessary.

8. Their privateers cruise the United States coast, capturing our merchant vessels.

9. Six new 204 foot frigates are authorized and money allocated to build. Four of 44 guns each, the other two, 36 guns. These ships are designed for one purpose, fighting. They are designed by Joshua Humphreys, an experienced Philadelphia shipbuilder, with extra strength, durability, swiftness of sailing, provisions for heavier batteries, thicker timber, finer lines, longer stronger spars. Strong in construction and swift of sail to enable escape when outgunned or outclassed. They were: 1. *United States*; 2. *Constitution*; 3. *Constellation*; 4. *President*; 5. *Chesapeake*; 6. *Congress*. This is the beginning of a permanent Navy to protect our interests.

10. The tribute is paid to protect our shipping. A larger sum is paid to Algeria.

11. *Constellation* and *Constitution* follow shortly thereafter.

12. President John Adams asks Congress to create the Department of the Navy.

13. The *Baltimore*, returning from convoy duty near Havana is set upon by two British frigates. The British remove 55 members of the *Baltimore's* crew and impress them into service. This is the first major incident leading to the War of 1812.

14. The Navy yard is authorized by Benjamin Stoddert, and is the Navy's oldest shore establishment.

15. Sailors and marines had a choice for their grog ration: 1/2 pint of rum mixed with water, or a quart of beer.

16. In April 1798, the ship *Nonpareil* is taken off the Florida coast by a Frenchman from Guadeloupe. On May 1, 1798, the ship *Favorite*, two days out of Charleston, South Carolina, is taken by a French privateer. On May 8, 1798, the schooner *Liberty* is stopped by the Frenchman Cape Henlopen. The French are attacking American vessels because of our trading with Britain. The United States Navy has one year enlistments. Yellow Fever is our biggest killer.

17. When the United States Government refuses to consent to new demands for tribute money, the Pasha of Tripoli declares an open war on the United States.

18. On October 31, 1803, the United States Frigate, *Philadelphia*, runs aground in blockading the harbor of Tripoli. The enemy captures the ship's complement of 22 officers and 315 enlisted men. They free the ship at flood tide and tow it into a dead part of the harbor. On February 16, 1804, Lieutenant Stephen Decatur and 84 volunteers slip into the harbor aboard the ketch *Intrepid*, board the *Philadelphia*, load it with powder and set it afire. The ship is destroyed; the Pasha of Tripoli demands that the United States pay him $500,000 for the loss of the ship. On July 14, 1804, Commodore Preble sets sail with the *Constitution, Nautilus, Enterprise,* six gunboats, and two mortar boats, to attack Tripoli shipping and fortifications. They complete attacks on five separate areas. In August 1804, Captain Richard Somers loads the *Intrepid* with gunpowder and sails into the same harbor, to destroy the Tripolitan fleet. He is sighted by the Tripolines, who open fire and blow up the ship, killing Somers and his entire crew. The peace treaty was signed on June 10, 1805.

19. Jefferson is President, and Robert Smith, Secretary of the Navy. The Navy is reduced to six ships and a hand-full of officers and men.

20. The *Chesapeake* was approached by the *Leopard*. The captain of the British ship said that he was sending a boarding party over. Captain James Barron of the *Chesapeake* asked for what purpose. The reply was that the United States ship had British seamen aboard that had illegally deserted. Captain Barron said that was a problem to be settled by diplomatic channels. The British reply was a "broadside" that killed or wounded 21 men. The *Chesapeake* was boarded and four seamen removed. One was a British deserter; three were United States citizens. Two of the men were black.

21. The war began on June 18, 1812. The United States Navy had 16 warships which included 44-gun frigate *Constitution, United States,* and *President*; the 36-gun frigates *Chesapeake, Congress* and *Constellation*; and the 32-gun *Essex*. The British had over 600 warships, which included 124 ships-of-the-line; *Africa* had 64 guns. On June 1, 1813, off Boston, the American frigate *Chesapeake* was captured by the British ship *Shannon*. As wounded Captain James Lawrence, of the *Chesapeake*, lay dying, he said, "Don't give up the ship!"

22. *Constitution*, mounting 44 guns, in a fight with *Guerriere*, which mounted 49 guns, is given the name "Old Ironsides." The shots glanced ineffectively from the hull of the *Constitution* or fell harmlessly into the sea. The *Guerriere's* foremast and mainmast went by the board, and she was left a helpless hulk.

23. Oliver Hazard Perry assumed command of the American contingent on Lake Erie in March 1813. By July he had two brigs completed, the *Lawrence* and the *Niagra*. The *Lawrence* was named for Captain James Lawrence. Perry used it as his flagship. His battle pennant carried the

The battle ensign onboard USS Constitution. Photo by PH3 David J. Hallimore.

inscription, "Don't Give Up The Ship." The British had the *Detroit*, 19 guns; the *Queen Charlotte*, 17 guns; *Lady Prevost*, 13 guns; *Hunter*, 10 guns; *Little Balt*, three guns; and *Chippewa*, one gun. The United States had *Scorpion* and *Ariel*, each with four guns; *Lawrence*, 20 guns; the brig *Caledonia*, three guns; the brig *Niagra*, 20 guns; the schooners, *Somers*, two guns; *Porcupine*, one gun; *Tigress*, one gun; and the sloop *Trippe*, one gun. The battle started about noon. By 15:00, Perry wrote, "We have met the enemy and they are ours." In August 1814 the city of Washington was captured. The Capitol, the White House, and other public buildings were burned due to our failure to organize an adequate defense. The Battle of Lake Champlain was fought on Sept. 11, 1814. The British commander was Commodore George Downie. The United States force was commanded by Captain Thomas Macdonough. Downie was killed during the action, with the British ships all surrendering. The United States had 52 men killed and 59 wounded; the British had 84 men killed and 110 wounded. A treaty of peace was signed at Ghent on December 24, 1814.

24. Until this time, water had been stored in foul wooden casks.

25. Captain Biddle, in the American sloop *Ontario*, lands at Cape Disappointment, at the mouth of the Columbia River in the Pacific Northwest and takes possession for the United States.

26. "Old Ironsides," by Oliver Wendell Holmes arouses public sentiment. The vessel is rebuilt and restored to service.

27. The enlisted men's ration of grog was reduced to one gill (4 oz.) of spirits or 1/2 pint of wine per day. Men under 21 were allowed no liquor, but received a few cents extra per day as compensation. The Confederate Navy stayed "wet" throughout the Civil War.

28. Congress was suspicious of naval officers because they were conservative and aristocratic by nature. Supreme naval authority was kept in the civilian Secretary of the Navy. They refused to create a rank as high as rear admiral. Commodore was temporary and honorary. They allowed the creation of a naval academy only because of the demands for education in steam engineering. American naval strategy in the early 19th century was subordinate to the needs of the Army.

29. Flogging was not ended by most other navies until the 1880s. It did not end by British law until 1939.

30. Commodore Matthew Calbraith Perry, younger brother of the famed hero of Lake Erie, arrived at Yedo Bay, Japan (now called Tokyo), aboard the side-wheeler *Susquehanna* to open trade negotiations. He had three other vessels with him, the *Mississippi, Plymouth* and *Saratoga*.

31. The United States had not commissioned a single privateer since 1815, but had long depended on privateers as a naval weapon in lieu of a standing fleet, and did not sign the declaration. It was signed by England, France, Russia, Prussia, Austria, Turkey, and Sardina. It did not totally pass out of existence in the world's navies until 1863, after the Civil War. (The South used it, but the North didn't during the Civil War.)

32. They were made of stronger metals, such as the "soda-bottle" gun of the American, John A. Dahlgren.

33. About 1/4 resigned and joined the Confederate forces. When the Civil War ended with Lee's surrender at Appomattox, in April 1865, the Union Navy had more than 600 ships, of which about 60 had some armor, including a number of well-armored monitors and a considerable number of gunboats with some armor protection in vital places. In addition, there were 57,900 officers and men.

34. The British custom of the grog ration introduced in the mid-18th century, despite the alcoholism it bred, survived the temperance crusades of many decades. It was cut in half in 1850 and finally abolished.

35. The *Merrimac* (CSS *Virginia*), was blown up by the Confederates in May 1862, since it was a deep draft boat and could not be moved up the James River to help in the defense of Richmond, VA.

36. In the days of "wooden ships and iron men," the authorized consumption of various liquors by officers on board ship was mostly at their own expense, while the daily rationing of spirits, wine, or beer to the enlisted men was a government expense. Brandy was issued until 1687, when West Indies rum was substituted on ships visiting the Caribbean. In 1731, each member of the ships company was entitled to a half pint of straight rum per day. In 1740, Admiral Edward Vernon substituted "Old Grog." In 1781, the daily issuance of grog was required by regulation. British tars received a daily ration of one-half pint (8 oz.), rum mixed with four times that volume of water. A temperance-minded Congress, acting partly under the guise of coping with Civil War exigencies ruled that "the spirit ration in the Navy of the United States shall forever cease." Wine messes were continued by the officers until 1914.

37. The raider *Alabama*, under Raphael Semmes, after 69 captures, is sunk by the steam frigate *Kearsarge* off the coast of Cherbourg. The Civil War signaled the end of wooden sailing ships and much more.

38. The propelled automotive torpedo was invented by the Scotsman Robert Whitehead.

39. Well-armored warships were designed with underwater rams for line abreast formation tactics, dormant since the last Mediterranean galley battles. The mobility of steamships allowed this technique to be used for 25 years or so.

40. During the late 1800s, explosive shells were invited by the Frenchman Henry J. Porxans and smokeless powder by the Swedish inventor Alfred Nobel.

41. Muzzle-loading guns had to be short, so that the muzzle of the weapon could be brought inside the gun-port for loading. Breech-loading guns could be made any length, consonant with rigidity.

42. Bosun Charles W. Riggin, of the USS *Baltimore*, was killed on a street in Valparaiso, Chile, while on liberty. This was due to anti-American feelings arising from the United States decision to grant political asylum to former Chilean leaders who had been overthrown in a recent revolution. In response to a suggestion in the *New York Recorder*, the public contributed over 26,000 silver dimes to build three memorial statues. The statues were cast from a model made by Alexander J. Doyle of New York City. One of the statues casted was presented to the President of the United States, Benjamin Harrison, a second to the Secretary of State, James G. Blaine, and the third one was presented to the Secretary of the Navy, Benjamin F. Tracy. The statue in the Navy Museum is the one that was presented to the Secretary of the Navy.

43. The *Maine* sailed into the harbor on January 25, 1898 to protect American citizens living in Cuba. Cuba was dissatisfied with Spanish rule and there were a number of revolutionary uprisings in the colony.

44. Commodore Dewey aboard his flagship *Olympia*, plus light cruisers *Baltimore, Boston, Raleigh* and *Concord*, 892-ton gunboat *Petrel*, supply steamers *Nanshan* and *Zafiro*, escorted by USCG cutter *McCulloch* attack Admiral Montojo's squadron. Dewey waits until they are less than 5,000 yards from the nearest Spanish ship, then turns to Captain Gridley, standing next to him, and says, "You may fire when you are ready Gridley." Admiral Montojo lost 381 men killed, or badly wounded out of 1,200 men. The United States had no lives lost and eight men slightly wounded.

45. Rear Admiral William T. Sampson's fleet attacked Admiral Pascual Cervera's fleet outside the harbor at Santiago, Cuba, Cervera had 323 killed in action and 151 wounded. The United States had one death and one wounded. With the winning of the Spanish American War, we obtained Puerto Rico, Guam, and the Philippines.

46. The *Holland*, SS-1, is the United States Navy's first submarine to be commissioned. The *Holland* was named after her designer and builder, John Holland.

47. The United States Navy is the third largest in the world. The Battle Cruiser is used as part of an advanced scouting squadron. From 1906-1916 was the decade of American Dreadnoughts. Armour plate up to 18" in turrets; 14" side armor belts.

48. This was President Theodore Roosevelt's idea. The ships were all painted white. They visited 20 nations, stopping at Japan. This demonstration was to prove that the United States fleet was second, only to Britain. The 16 battleships departed Hampton Roads, Virginia, on December 16, 1907, and returned on February 22, 1909, after 14 months at sea and steaming 46,000 miles. Following is their itinerary: December 16, 1907, departed Hampton Roads VA; December 23-29, 1907, Port of Spain, Trinidad; January 12-22, 1908, Rio de Janeiro, Argentina; February 1-7, 1908, Punta Arenas, Chile; February 14, 1908, Valparaiso, Chile; February 20-29, 1908, Callo Bay, Peru; March 12—April 11, 1908, Magdalena Bay, Mexico (fleet and gunnery practice); April 14—July 7, 1908, San Diego, San Pedro, Santa Barbara, and San Francisco; July 16-23, 1908, Honolulu, Hawaii; August 8, 1908, Auckland, New Zealand; August 20, 1908, Sydney, Australia; August 29, 1908, Melbourne, Australia; September 11, 1908, Albany, Australia, (275 men deserted in the two countries); October 1-10, 1908, Manila, Philippines and Yokohama, Japan; October 30—December 5, 1908, Amoy, China; January 1, 1909, Manila, Philippines; December 14-20, Columbo-Suez Canal and Mediterranean; February 6, 1909, 1st Division: Italy; 2nd Division: France; 3rd Division: Eastern Mediterranean; 4th Division: African Coast; February 22, 1909, Hampton Roads, VA. The First Division was: *Connecticut*, BB-18; *Kansas*, BB-21; *Louisiana*, BB-19; *Vermont*, BB-20. The Second Division

was: *Georgia*, BB-15; *Virginia*, BB-13; *New Jersey*, BB-16; *Rhode Island*, BB-17. The Third Division was: *Minnesota*, BB-22; *Ohio*, BB-12; *Missouri*, BB-11; *Maine*, BB-10; The Fourth Division was: *Alabama*, BB-8; *Illinois*, BB-7; *Kentucky*, BB-6; *Kearsarge*, BB-5. Not included on the cruise: *Indiana*, BB-1; *Massachusetts*, BB-2; *Oregon*, BB-3; *Iowa*, BB-4; *Wisconsin*, BB-9; *Nebraska*, BB-14. At San Francisco, *Alabama* and *Maine* dropped from the fleet due to excessive coal consumption and defective engines. They were replaced by *Wisconsin* and *Nebraska*. The fleet was re-organized: 1st Division, *Connecticut*, *Kansas*, *Vermont*, and *Minnesota*. 2nd Division, *Georgia*, *New Jersey*, *Rhode Island*, and *Nebraska*. 3rd Division, *Louisiana*, *Ohio*, *Missouri*, and *Virginia*. 4th Division, *Wisconsin*, *Illinois*, *Kentucky*, and *Kearsarge*. (This information was extracted from *One and Only Kentucky*, page 15, *Sea Classics Magazine*, December 1989.)

49. The monitor *Cheyenne* (ex *Wyoming*), was converted from coal to fuel oil as the Navy's first experiment with this fuel.

50. The cost has increased almost five times. The caliber of the principal gun has gone from 12" to 16".

51. First operation of a land-plane from a Navy ship. Eugene Ely, a Curtiss Aircraft Company demonstration pilot, took off (flew) an aircraft from a temporary deck erected on the forward part of the cruiser *Birmingham*, CL-2. The ship was at anchor at Hampton Roads, Virginia. Ely landed on Willloughby Spit.

52. Eugene Ely, a Curtiss Aircraft Company demonstration pilot, landed on a deck erected on the after part of the USS *Pennsylvania*, which was at anchor in the San Francisco Bay. The deck was 120 feet long by 32 feet wide. The only thing that separated the forward end of the deck from the ship's mast was a piece of canvas stretched vertically at the end of the platform. Lines stretched athwartship the deck, weighted at each end by sandbags, were spaced throughout the length of the deck and lifted away from the deck in order that they might engage hooks installed on the aircraft. In his landing, the platform was stationary (the ship was at anchor), and the wind was blowing across the deck. Because of the skill of the pilot, the landing was successful. Fifty-seven minutes later, he took off and returned to Selfridge Field, near San Francisco.

53. General Order No. 99, by Josephus Daniels, Secretary of the Navy under President Woodrow Wilson. "The use or introduction, for drinking purposes, of alcoholic liquors on board any naval vessel, or within any yard or station, is strictly prohibited, and commanding officers will be held directly responsible for the enforcement of this order."

54. The United States enters the war against Germany. We only have 67 destroyers to combat their fleet of U-boats. One-fourth of British shipping has already been sunk. Losses for April 1917 amount to almost 900,000 tons. The armistice was signed on November 11, 1918.

55. There was considerable controversy over the adoption of fuel oil. A considerable amount of work was entailed in coaling a ship. In addition, a considerable amount of time was expended in washing down and clean up work after the operation was performed. Strokers were also required to trim the coal and shovel it into the furnaces. It was also extremely difficult to perform the coaling operation at sea. Oil can be kept in ships' double bottoms, lower down than coal can be stored. The greatest danger to use of fuel oil is that of having its supply being cut off.

56. The letters "NC" stood for "Navy-Curtiss," i.e., a joint Navy-Curtiss venture. Three NC flying boats departed the United States East Coast on May 8, 1919, in an attempt to reach Europe via Newfoundland and the Azores. The NC-2 was sacrificed for spare parts. The NC-1, NC-3, and NC-4 all reached Trepassey, Newfoundland. The NC-1 and NC-3 both had to land on the water before reaching the Azores. The NC-4 reached Plymouth, England on May 31, 1919, and 4,105 miles later.

57. Conversion of the USS *Langley* from a collier to a flush deck carrier is initiated. The *Langley* was initially named the *Jupiter*.

58. The USS *Lexington*, *Ranger*, *United States*, *Constitution*, *Constellation*, and *Saratoga* are laid down as battle cruisers.

59. The German battleship *Ostfriesland* is sunk off the Atlantic Coast by bombs dropped from an aircraft flown by Brigadier General Billy Mitchell. The well-publicized tests were to prove to the Navy that an aircraft could sink a battleship.

60. The Five-Power Treaty was signed by the United States, Great Britain, Japan, France, and Italy. It pertained to battle ships and aircraft carriers only.

The British insisted that *Standard Displacement* tonnage be used. *Standard Displacement* was defined in terms of condition of the ship ready for sea in wartime. It was considered possible to deduct items used or carried only in peacetime such as: drill ammunition, the nominal "standard" supply of ammunition, the nominal potable water allowance per man, etc.

Navies were to be limited by a ratio calculated on existing strength. Existing capital ship strength as of November 21, 1921 was:

United States battleships (BBs) 18, battle cruisers (BCs) 0, total 18.
Great Britain BBs, 33, BCs, 10, total 43.
Japan BBs, 6, BCs, 4, total 10.
Under construction at this time:
United States BBs, 9, BCs, 6, total 15.
Great Britain BBs, 0, BCs, 0, total 0.
Japan BBs, 3, BCs, 4, total 7.
Authorized but not begun:
United States BBs, 0, BCs, 0, total 0.
Great Britain BBs, 0, BCs, 4, total 4.
Japan BBs, 4, BCs, 4, total 8.

Specific tonnage quotas for the different categories would be met by: scrapping vessels afloat; halting construction on the ways; regulating building in the future.

The following capital ships were scrapped (BBs and BCs):

United States, 15 built plus 15 being built, to. 30 @ 845,740 tons, valued at $277,695,000.
Great Britain, total 19 @ 583,375 tons.
Japan, total, 17 @ 448,929 tons.
The following tonnage quotas were set:
United States 525,850 gross tons to be distributed among 18 ships.
Great Britain, 558,950 gross tons to be distributed among 20 ships.
Japan, 301,320 gross tons to be distributed among 10 ships.
France, 221,170 gross tons to be distributed among 10 ships.
Italy, 182,800 gross tons to be distributed among 10 ships.

The text of the treaty contained the name of every vessel to be retained or given up through 1942. Replacements after that date would place the quotas to the United States and Great Britain at 525,000; Japan 315,000; France and Italy, 175,000 each, i.e., 5:5:3:1.67:1.67.

Aircraft carriers were considered to be in the experimental stage. No carrier could exceed 27,000 tons, or mount guns of more than 8 inch caliber. Tonnage quotas established were: United States and Great Britain, 135,000; Japan, 81,000; France and Italy, 60,000 each. Each signatory was permitted to convert two capital ships as large as 33,000 tons to aircraft carriers. Congress would not authorize any new carrier construction, so we converted two BCs (*Lexington* and *Saratoga*) to CV-2 and CLV-3, that joined the fleet in 1927.

The treaty also incorporated two other agreements. The Nine Power Treaty, to preserve the territorial integrity of China and the Four-Power Pact, where Japan, United States, Great Britain, and France agreed to respect one another's rights in the Pacific.

Capital ships were restricted to 35,000 tons in weight and gun caliber could be no greater than 16 inches.

The United States agreed to not increase fortification of any bases west of Hawaii (Guam, Midway, Philippines), and Japan agreed to maintain the status quo in the Pacific Territories under the cognizance.

Each signatory could request a new conference whenever it believed its security was affected. One had to be scheduled after eight years.

This treaty and the subsequent ones that followed contained no provisions to determine verification of accomplishment, nor procedures or provisions for handling treaty violations.

The Senate approved the treaty by a vote of 74 to 1.

After 20 years of indecision and neglect, the United States had passed control of the Western Pacific to Japan.

61. The United States sponsored a Three-Power Treaty with the United States, Great Britain, and Japan that met at Geneva. It was a fiasco. Mostly because of cruisers. They varied in size, speed, armament, and duties. Different nations used them for different missions. The treaty was poorly conceived, badly executed, and was a dismal failure.

62. The Five-Power Treaty described in No. 60 above, eliminated the six battle cruisers described in No. 58. However, the *Lexington* and *Saratoga* were two that were allowed to be converted to aircraft carriers.

63. A summary of the London Treaty:

The 10 year holiday in construction of battleships (BBs) and battle cruisers (BCs) was extended to December 31, 1936.

The tonnage allowance of BBs and BCs was reduced; still holding a 5:5:3 ratio:

The United States went from 525,000 to 464,000.

Great Britain went from 558,000 to 474,500.

Japan went from 315,000 to 270,000.

Ships that were scrapped or demilitarized: United States, 3: Great Britain, 5; Japan, 1. The number of BBs and BCs allowed were: United States, 15; Great Britain, 15; Japan, 9.

Cruisers were settled by classifying them as either *heavy*, or *light*, with both restricted to 10,000 tons. Vessels mounting guns with calibers of 6.1 inches or less were known as light cruisers (CLs). Those mounting guns with calibers of 8 inches were known as heavy cruisers (CAs). They were limited in both quantity and total tonnage as follows:

	United States	Great Britain	Japan
Heavy Cruisers	(15) 180,000	(15) 146,800	(12) 108,400
Light Cruisers	143,000	192,000	100,450
Total tonnage	323,500	338,800	208,850

Destroyers (DDs), and submarines (SSs), also received tonnage quotas. For DDs, the United States and Great Britain were each allotted 150,000 tons, while Japan was allotted 105,000 tons. Each country received an allotment of 52,000 tons for their submarines. Aircraft carriers were not changed.

Article XXII forbade submarines of attacking without prior warning of an attack, or without caring for the care and safety of those aboard merchant vessels. This article did not contain a time limit, although the rest of the treaty expired on December 31, 1936.

64. The Vinson-Trammel Act is approved and signed by President Roosevelt. It established the composition of the Navy at the limits prescribed by the Washington and London naval treaties. The Vinson-Trammel Act did not appropriate any funds for a single vessel; however, it empowered the *Executive* to bring the fleet to full treaty strength by 1942. The USS *Wasp*, CV-7, is approved for construction as the last carrier meeting the restrictions of the Washington and London treaties. She was commissioned at Boston on April 25, 1940.

65. The USS *Ranger*, CV-4, was the first Navy ship to be designed from the keel up as an aircraft carrier. Like the *Langley*, the *Ranger's* smoke stacks were designed to fold down out of the way during landing operations, which was supposed to eliminate turbulent heat currents that retarded earlier carrier experiments.

66. This Second London Treaty contained three restrictions and one requirement that are binding until December 31, 1936, i.e., no 8 in. gun cruiser can be built for six years, and no 6 in. gun cruiser can be built that weighs more than 8,000 tons. No BB is allowed to weigh more than 35,000 tons or mount guns exceeding 14 in. caliber. CVs are limited to 23,000 tons and 6.1 in. caliber guns. Information on construction programs must be exchanged annually. Escape clauses were included so that signatories could renounce specific curbs when they believed their security required it.

67. When Japanese invaders began to invest the city of Nanking in early December 1937, the gunboat took on the staff of the American Embassy, as well as American and foreign journalists. When shelling in and about the city became heavy, the *Panay* and several ships owned by the Standard Oil Company moved a few miles up the Yangtze River. On the morning of December 12, a heavily armed party of Japanese soldiers boarded the Panay and demanded that her commanding officer, Lieutenant Commander James Joseph Hughes, tell them whether he had seen any Chinese troop movements on his way up the river. Hughes politely explained that he could not answer because the United States considered itself neutral. The Japanese disembarked in what appeared to be a surly frame of mind. Shortly after noon, the gunboat was attacked by a flight of dive bombers. The ship was bombed and strafed, taking two direct bomb hits, shrapnel and machine-gun fire. As they were abandoning ship they suffered continual strafing. There were 30 wounded and two deaths. The Japanese paid $2,214,007.36 in reparations. Nanking fell in an orgy of killing known as the "Rape of Nanking."

68. The USS *Saratoga* approached to within 100 miles of Pearl Harbor, and launched her aircraft, that attacked the installation in a simulated fleet problem exercise. The defenders were taken by complete surprise. Less than four years later, the base was again attacked in a complete surprise by the Japanese; (this was not a drill) although the installation was equipped with radar at this time.

69. The 50 four-stacker destroyers were of the World War I vintage that had been in "mothball" at San Diego. They were provided to Britain in exchange for the rights to the use of six sites in the Caribbean suitable for air and naval stations. That is, England would lease the six sites to the United States without charge for 99 years. In addition, Great Britain granted, without compensation, the right to develop similar facilities in southern Newfoundland and eastern Bermuda. Simultaneously, but independently, Britain pledged that its fleet would never be allowed to fall to Germany. The eight sites were located in the Bahamas, Antigua, St. Lucia, Trinidad, Jamaica, British Guiana, Newfoundland, and Bermuda.

70. This allowed Japanese troops to be based in French Indo-China, (Vietnam today), since France is under German control. Reference *The American Magic, Codes, Ciphers and the Defeat of Japan*, by Ronald Lewin, Copyright 1982.

71. On November 12, 1940, the United States Congress approved dropping the minimum draft age from 21 years to 18 years of age. Able-bodied men were eligible for the draft until they reached 37 years of age. A man within that age group could be classified by his draft board anywhere between I-A and IV-F, with I-A being the most eligible for draft. Some of the things that could lower a man's rating to "Four-F" were being too short (less than 5-feet tall), underweight (less than 105 lbs.), venereal disease, flat feet, not enough teeth, poor eyesight or mental deficiency. In addition to physical and mental deferments, occupational deferments were available to people who had jobs judged to be essential to the war effort, and some men were deferred because they had too many dependent children. As the war progressed, many of these standards fell by the wayside.

72. The Navy asks the Grumman Aircraft Company to develop an improved version of the F4F Wildcat, due to problems with the F4U Corsair. Four problems were affecting F4U operations: 1. Inadequate cockpit visibility. Answer: Redesign the canopy and cockpit location. 2. Adverse stall characteristics; particularly dangerous when turning at low levels. Answer: Install a wing spoiler; 3. Excessive bounce on landing. Answer: Modify the landing gear shock system. 4. Frequent failure of the tailhook to engage the carrier's arresting wires and the carrier's decking being destroyed by the extended tailhook. Answer: Perform modifications to the tailhook.

73. The *Reuben James*, DD-245, was torpedoed by the German submarine U-562. This was the first United States Navy ship to be sunk by hostile action in World War II; 44 survived and 115 died.

74. The Japanese attacked with 360 aircraft: 81 fighters, 104 horizontal bombers, 40 torpedo bombers and 135 dive bombers. They took flight from six aircraft carriers (*Akagi, Hiryu, Kaga, Shokaku, Soryu, Zuikaku*), supported by two battleships (*Hiei, Kirishima*), two heavy cruisers (*Chikuma, Tone*), one light cruiser (*Abakuma*), nine destroyers (*Akigumo, Arare, Hamakaze, Isokaze, Kagero, Kasumi, Sazanami, Shiranuhi, Tanikaze, Urakaze, Ushio*), plus escorts. The aircraft were launched 200 miles north of the islands. They executed two separate attacks. The first at 07:55-08:25 with dive bomber and torpedo bombers, which were followed by horizontal bombers at 08:40. The second attack occurred from 09:15-09:45. They lost 29 aircraft. United States losses were: sunk or severely damaged, eight battleships, three cruisers, three destroyers, and four other vessels. We lost 188 aircraft. Casualties were: Navy and Marine Corps, over 2,000 men killed and over 700 wounded; Army, almost 200 killed and over 400 wounded. Japan delivered a declaration of war on the United States at 2:20 p.m., Washington time. This was about one hour after the actual attack. United States ships in the harbor: eight battleships (*Nevada, Oklahoma, Pennsylvania, Arizona, Tennessee, California, Maryland, West Virginia*); two heavy cruisers (*New Orleans, San Francisco*); six light cruisers (*Detroit, Helena, Honolulu, Phoenix, Raleigh, St. Louis*); 29 destroyers (*Allen, Aylwin, Bagley, Blue, Case, Cassin, Chew, Conyngham, Dale, Dewey, Downes, Farragut, Helm, Henley, Hull, Jarvis, McConough, Monahan, Mugford, Patterson, Phelps, Ralph Talbot, Reid, Schley, Selfridge, Shaw, Tucker, Worden*); four submarines (*Cachalot, Dolphin, Plunger, Tautog*); one gunboat; nine mine-layers; 10 mine sweepers; 10 seaplane tenders; three repair ships; two oilers; two ocean tugs; one hospital ship; one surveying ship; one supply vessel; one ammunition vessel; one submarine rescue ship; one antique cruiser; one target ship; one submarine tender; total: 94 vessels. Sunk: six battleships; one heavy cruiser;

two oilers. Seriously damaged: two battleships; one heavy cruiser. Heavy damages: one heavy cruiser; six light cruisers; three destroyers; three auxiliary vessels. Aircraft destroyed: 99 Navy; 65 Army. Killed: 2,403 Army, Navy, and civilian. Wounded: 1,178. Japanese losses: eight fighters, 15 dive bombers, five level bombers. Killed: one pilot; wounded: several; 74 aircraft "holed"; failed to return: one submarine and five midget submarines. The water at Pearl Harbor was 12 meters deep and 500 meters wide. The Japanese code for a successful attack were the words, TORA, TORA, TORA. The Japanese time table called for the occupation of the Philippines in 50 days, and securing the whole southern area in 90 days from the start of hostilities. This short period was intended as a precaution against Soviet interference and the United States moving in reinforcements.

75. A B-17, piloted by Captain Colin Kelly, United States Army Air Force, attacked the Japanese heavy cruiser *Ashigara* at Aparri, Luzon, in the Philippines. He did not survive the mission, but reported by radio that he had hit and set afire a *Kongo*-Class battleship.

Chapter 4

Above and Beyond – Heroism

The Congressional Medal of Honor
Statutory and Administrative History

I. Establishment by Congress

The Medal of Honor, our highest decoration for valor, was established in the early years of the Civil War. A bill passed by both Houses of Congress and signed by President Lincoln on December 21, 1861, authorized the Secretary of the Navy to prepare 200 medals which were to be awarded to enlisted men of the United States Navy "as shall most distinguish themselves by their gallantry in action and other seamanlike qualities during the present war.[1] The following year the President of the United States, was authorized to make similar awards to enlisted men of the United States Army "as shall most distinguish themselves by their gallantry in action, and other soldier-like qualities during the present insurrection."[2] In 1863 Congress made officers of the Army eligible for the medal but added the qualification that the award would only be made to those who "most distinguish themselves in action."[3]

II. Standards and Procedures of Award

A. 1861-1918: Between 1861 and 1918 the Army awarded 2,612 Medals of Honor[4] and the Navy and Marines, 735. A number of factors must be borne in mind in considering the award for the medal during these periods. The statutes establishing the Medal of Honor were quite broad in scope. "Gallantry in action," "seamanlike" and "soldier-like" qualities being the sole criteria laid down by Congress to govern the award of the medal.[5] Such language gave authority for awards for non-combatant exploits and many such awards were made before the legislation was modified. The fact that the Medal of Honor was the only decoration[6] given for valor by the armed services gave it an "all or nothing at all" aspect. The basic fact that there are degrees of valor could not be expressed by the use of this lone decoration. The diverse standards used by individual commanders tended to increase the disparity in the types of heroism that were recognized by the same medal. A comparison of award statistics of this era with those of the World Wars may indicate, to some extent, the situation that resulted. The Army awarded over 2,000 Medals of Honor in the Civil War (67,058 men killed in action). In World War II (175,407 men killed in action) it awarded only 292. During the Indian Campaigns (919 men killed in action) the Army awarded about 425 Medals of Honor but in World War I (37,568 men killed in action) it gave only 95. The Navy, similarly, awarded 66 Medals of Honor in the Spanish-American War (10 killed in action, while in World War II (6,950 men killed in action) it awarded only 57.

The quality of the award was of considerable concern to some officers after the Civi[l] War. Former soldiers were making applications for Medals of Honor on the basis of alleged exploits which had taken place years previously and about which they had little or no documentation. Likewise, there was the tendency of some officers to recommend medals in great numbers of their commands. This was noted by Brigadier General

The Congressional Medal of Honor

Alfred A. Terry who, in disapproving a large number of recommendations for the medal in 1876, wrote that it appeared to him that company commanders were attempting to award the medal to everyone who had "behaved ordinarily well" during an engagement with some Indians. "Medals of Honor," the General explained, "are not intended for ordinarily good conduct, but for conspicuous acts of gallantry."[7] A board set up to examine the rejected recommendations was "of the opinion that only such persons should be recommended for Medals of Honor as displayed in the discharge of duty a zeal, energy, and personal during which far exceeded any just demand of duty."[8] These statements ultimately became the crux of administrative regulations on the subject. Toward the end of the 19th and the beginning of the 20th century, regulations were promulgated with particular emphasis on the problems of delayed applications and evidential standards for the award.

The final legislative development before World War I was the creation of a "Medal of Honor Roll" and the payment of $10 a month to those medal winners who had attained the age of 65. The law permitted on the roll only those men who had been awarded a Medal of Honor "for having in action involving actual conflict with an enemy distinguished (themselves) conspicuously by gallantry or intrepidity, at the risk of ...life, above and beyond the call of duty."[9] The law required that the Secretaries of War and Navy should examine prior awards of the Medal of Honor and determine which holders were entitled to the benefits of the law. A subsequent act[10] authorized the Secretary of War to appoint a board of five retired officers who were to conduct the examination of the records of past awards. If the board found that a holder was not entitled to the Medal of Honor, the War Department was to strike his name from the roll and it would be a misdemeanor for him to continue to wear his medal. The board found itself in an awkward position, since the legislation that had established the medal was not nearly as strict as the requirements of the 1916 act. Awards which could be justified under the Civil War acts often fell below the requirement of the later legislation. After an unsuccessful attempt had been made to get Congress to modify its terminology, the board examined all the cases that did not appear per se valid. It found that 911 persons could not hold the Medal of Honor under the 1916 act.[11] Of these, 864 had been members of the 27th Main volunteer Infantry. This regiment's enlistment was to have expired in June 1963. President Lincoln, as an inducement to keep its personnel on active duty in a critical time, authorized Medals of Honor for those men who volunteered to stay on. All but 309 of the regiment went home but through some inadvertence all of its men were given Medals of Honor. In addition to these, 47 other persons lost their Medals of Honor. Among the latter were William F. Cody (Buffalo Bill) and Mary Walker, a Civil War surgeon and the only woman ever to receive the award. The Navy took no comparable action and maintains it has never taken away a Medal of Honor. However, as it will appear later, the status of some of its awards for non-combatant heroism has been questioned by other government agencies.

B. World War I and II: The difficulties encountered in the administration of the 1916 legislation highlighted the need for clarification of the rules governing the award of the Medal of Honor. On July 9, 1918, Congress passed a law that set a new legislative standard for the Army Medal of Honor. The act authorized the President, in the name of the Congress, to present "a medal of honor only to each person who, while an officer or an enlisted man of the Army, shall hereafter, in action involving actual conflict with an enemy, distinguish himself conspicuously by gallantry and intrepidity at the risk of his life above and beyond the call of duty."[12] This law remains in force today as the legislative definition and authority for the Army Medal of Honor. As a result of congressional realization that the quality of the medal was threatened by the lack of secondary awards, the act further provided for the establishment of the Distinguished Service Corps. Distinguished Service Medal, and the Silver Star and provided additional monthly pay for active-duty holders of the decorations. The Senate Military Affairs Committee commented in its report on a similar bill that if this had been done in the past the Medal of Honor "would have been much more jealously guarded than it was for many years."[13] This act, along with the Navy act in the following year, with the genesis of what has been called the "Pyramid of Honor" – a hierarchy of military decorations with the Medal of Honor at the top.

On Feb. 4, 1919, Congress passed an act for the Navy Medal of Honor and established the Distinguished Service Medal and the Navy Cross with an extra pay proviso.[14] In contrast to the Army legislation, the act, in some respects caused confusion rather than clarification in the award of the Navy medal. Though phrased in almost identical terms, the Navy Medal of Honor act did not expressly repeal the prior acts governing the medal. The Judge Advocate General of the Navy interpreted this to mean that the acts of 1901[15] and 1915[16] were still in effect and that the Navy could still award the Medal of Honor for extraordinary heroism during non-combat duty. The Comptroller of the Treasury believed that this was not the intent of Congress and that the legislation of 1919 "is to be regarded as a complete substitute for the earlier law."[17] The upshot of this controversy was the strange situation in which the Navy continued to award Medals of Honor under earlier legislation, but the Treasury refused to pay the gratuities provided by these laws. This confusion persisted until 1942 when Congress amended the 1919 act and stated that awards for Naval personnel could be made for non-combat heroism.[18]

The first World War provided the basic administrative standards and award procedures which were used in World War II and in the Korean conflict. The most important development, in this respect, was the establishment of a board in France whose function it was to review recommendations and evidence for the award of the Medal of Honor, General Pershing checked the findings of the board to satisfy himself that standards were being maintained. Seventy-eight Medals of Honor were awarded in this manner. Altogether in World War I, 95 Medals of Honor were awarded to Army personnel and 28 to members of the Navy and Marine Corps.

In 1921 the adjutant general took over the operating functions in connection with Army awards and decorations. A War Department Decorations Board was established in Washington in the same year. The Navy Department did away with their Decorations Board after World War I. But by 1927 the department was convinced that such a board was necessary for the effective supervision and standardization of award actions and it was re-established. The time limits for recommendations and award of the Medal of Honor were extended by Congress in 1922[19] and 1928.[20]

During the second World War, there were no significant changes in the mechanism for awarding the Medal of Honor. In 1943, regulations were promulgated by the Army which called for very detailed information as to the circumstances of the exploit and affidavits of eyewitnesses were required.[21] The Decorations Board ruled in 1944 that every recommenda-

tion for the Medal of Honor must be endorsed by the theater commander. Naval regulations called for prompt transmittal of Medal of Honor recommendations through channels, required at least one eyewitness to the action in question, and that the "individual must clearly render himself conspicuous above his comrades by an act, the omission of which would not subject him to censure or criticism for shortcomings or failure in the performance of his duty. Incontestable justification will be exacted."[22] In World War II, awards of the Medal of Honor were made to the armed services as follows: 282 to the Army, one to the Coast Guard, 79 to the Marines and 57 to Navy personnel.

C. Korean War and the Present: The Act of Congress, July 9, 1918,[23] remains as the statutory authorization for Army awards of the Medal of Honor. Navy and marine awards of the Medal of Honor. Navy and Marine awards are authorized by the act of Congress of Aug. 7, 1942.[24] This act is basically the same as that for the Army except for the inclusion of the phrase "or in the line of his profession" which reinstitutes the award for non-combat heroism which was not authorized by act of Congress of Feb. 4, 1919.[25]

The Army, Navy and Air Force administrative procedures and regulations governing the issuance of the award are quite similar. Recommendations are made in the field and sent, with the affidavits of eyewitnesses and a full exposition of the circumstances surrounding the act, through the chain of command to the respective headquarters in Washington. There they are referred to the appropriate Decorations Board and upon board approval are sent to the President for his signature. Time limits have been established for the recommendation and award of Medals of Honor. The recommendations for the Army award must be made within two years of the date of the heroic act, and the award must be made within three years of the act. Navy and Marine Medals of Honor must be recommended within three years of the act and awarded within five. The requirement for the Air Force award is that the recommendation be made within two years of the act. In 1950, Congress granted an additional period for the recommendations for World War II acts of heroism.[26] This period expired on May 2, 1951, as did the additional period for making awards on these recommendations on June 2, 1952. All provisions of law pertaining to extra pay for decorations, with minor exception, were repealed by act of Congress on Oct. 12, 1949.[27] Up to June 10, 1954, the Army had announced the award of 71 Medals of Honor, the Air Force four, the Navy six and the Marines 42 in the Korean fighting.

III. Special Legislation

From time to time Congress has seen fit to pass special acts authorizing the President to bestow a Medal of Honor on some specified individual. A few examples of those who received their Medals of Honor in this manner are Admiral Richard E. Byrd, Colonel Charles A. Lindbergh, our Unknown Soldier and those of some of our allies in World War I. This legislation has taken various forms, e.g., joint resolutions, private and public bills. The President authorized the award of Medals of Honor to the Belgian and Rumanian Unknown Soldiers by War Department General Order.

IV. Design

The Civil War Medals of Honor were designed by Christian Schussel and engraved by Anthony C. Paquet. "'…the star-shaped medal of bronze shows the figure of Minerva (the Union), wise in the industries of peace and war.' Encircled by the stars of the 34 States of 1861, she holds in her left hand the faces (badge of authority). The shield in her right hand is driving off the serpents held by the crouching figure of Discord, referred to in a letter of 6 May 1862, from the Director of the Mint as, 'the foul spirit of secession and rebellion.'"[28] The Army Medal was attached to a ribbon of 13 vertical stripes, alternating red and white, by a small metal American eagle standing on crossed cannons. The Navy Medal was attached to a similarly colored ribbon by a small anchor. The ribbon was changed in 1897 for both Medals of Honor and had progressively larger stripes from the center out of white, blue, and red.

In 1904 Congress authorized 3,000 Army Medals of Honor of a new design. In order that civilian organizations would not copy the new medal, a common practice in the past, General Gillespie, the designer, took out a patent on the medal and presented it to the Secretary of War. The medal is "a gold finished bronze star, one point down, 1-9/16 inches in diameter with rays terminating in trefoils, surrounded by a laurel wreath in green enamel, suspended by two links from a bar bearing the inscription 'VALOR' and surmounted by an eagle grasping laurel leaves in one claw and arrows in the other. In the center of the star is the head of Minerva surrounded by the inscription 'UNITED STATES OF AMERICA.' Each ray of the star bears an oak leaf in green enamel."[29] This medal is still used by the Army. The only change since 1904 has been in the ribbon. The ribbon of 1904 was blue with 13 white stars and was pinned on the chest. During the second World War, it was changed to a blue neck ribbon. "The medal is suspended by a hook to a ring fastened behind the eagle. The hood is attached to a light-blue moired silk neckband, 1-3/16 inches in width and 21-3/4 inches in length, behind a square pad in the center made of the ribbon with the corners turned in. On the ribbon bar are 13 white stars arranged in the form of a triple chevron, consisting of two chevrons of five stars and one chevron of three stars."[30] The Air Force uses the same type of ribbon and medal as the Army.

In 1913, the Navy changed to the blue ribbon with the stars, and in 1919 adopted a new medal. "It is in the form of a cross, and in the center of the obverse there is a small octagonal design bearing the Great Seal of the United States, surrounded by the words, 'UNITED STATES NAVY....1917-1918' and is in gold. The ribbon suspends form a bar with the word 'VALOR'.[31] In the second World War the Navy returned to the Civil War medal design and adopted the Army neck ribbon.

V. Presentation

"Wherever practicable, all person to whom the Medal of Honor has been awarded will be ordered or invited to Washington, D.C., where presentation will be made by the President."[32] The President, within his discretion, may personally present posthumous awards to the next of kin of the Medal of Honor winner. If the President does not make the presentation, it will be made by someone acting as his personal representative. The presentation ceremonies are to be simple and without troops or music.

Appendix A

Awards of the Medal of Honor

	Army	Navy	Marines	Coast Guard	Air Force*
Civil War	1,200	310	17		
1965-1898	416	117	8		
Spanish-American War	30	66	15		
1898-1917	75	136	62		
World War I	95	21	7		
1918-1941	8	15	5		
World War II	292	57	79	1	
Korea	71	6	42		4
As of June 10, 1954)					
Totals	2,187	728	235	1	4

*Medal winners in Air Corps in World War I and II are included in figures.

Sources: United States Department of the Army. The Medal of Honor of the United States Army. Washington, United States Government Print. Off., 1948 (Appendix II).

United States Navy Department Medal of Honor, 1861-1949. Washington ? 1950. p. 307

For Medal of Honor winners in Korean conflict: Awards Branch, Department of the Air Force; Medals and Award Division, Department of the Navy; Decorations and Awards Unit, Department of Army.

Appendix B

The following is the order of precedence for military decorations of the United States, based on degrees of valor and meritorious achievement, and the date each medal was established:

United States Army

1. Medal of Honor (1862)
2. Distinguished Service Cross (1918)
3. Distinguished Service Medal (1918)
4. Silver Star (1918)*
5. Legion of Merit (1942)
6. Distinguished Flying Cross (1926)
7. Soldier's Medal (1926)
8. Bronze Star (1942)
9. Air Medal (1942)
10. Commendation Ribbon (1945)

*The present medal itself was not authorized until 1942.

United States Navy, Marine Corps and Air Force

1. Medal of Honor (1861)
2. Navy Cross (1919)
3. Distinguished Service Medal (1919)
4. Silver Star (1942)
5. Legion of Merit (1942)
6. Distinguished Flying Cross (1926)
7. Navy and Marine Corps Medal (1942)
8. Bronze Star (1942)
9. Air Medal (1942)
10. Purple Heart (1782)

Medals for Civilians:

1. Medal for Merit (1942)
2. Medal of Freedom (1945)

Regulations

United States Department of Air Force. Air Force Regulation No. 30-14. Washington, 21 May 1953. (With AFR 30-14A 8 April 1954)

United States Department of Army. Army Regulations No. 600-45. Washington, United States Govt. Print. Off. 27 June 1950. p. 37.

United States Navy Department, Bureau of Naval Personnel. Decorations, medals, ribbons, and badges of the United States Navy, Marine Corps and Coast Guard, 1861-1945. Washington, 1950 ? 192 p. (MARCORPS DLD-298-vgp; BUPERS A10-3, Pers-B4b-MAC/rlo)

Frederick B. Arner
American Law Division
June 25, 1954

The Highest Award Bestowed Upon A Member of the United States Naval Service – Known

Chief Petty Officer Medal of Honor Recipients

1. Badders, William, chief machinist's mate, Interim 1920-0. Raising of USS *Squalus*, 23 May 1939.
2. Bennett, Floyd, machinist, Interim 1920-40. USS *Marblehead*, 11 May 1898.
3. Bennett, James H., chief boatswains mate, May 11, 1898. War with Spain. USS *Marblehead*.
4. Bonny, Robert Earl, chief watertender, Interim 1901-10. USS *Hopkins*, 14 February 1910.
5. Bradley, George, chief gunners mate, Mexican Campaign (Vera Cruz), USS *Utah* 1914.
6. Brady, George F., chief gunners mate, War with Spain. Torpedo Boat *Winslow* 11 May 1898.
7. Clancy, Joseph, chief boatswains mate, Boxer Rebellion, 13-22 June 1900.
8. Clausey, John J., chief gunners mate, Interim 1901-10. USS *Bennington* 21 July 1905.
9. Cooney, Thomas C., chief machinist, War with Spain. Torpedo Boat *Winslow* 11 May 1898.
10. Cox, Robert Edward, chief gunners mate, Interim 1901-10. USS *Missouri* 13 April 1904.

11. Crandall, Orson L., Chief Boatswain Mate, Interim 1920-40. Raising of USS *Squalus* 23 May 1939.

12. Crilley, Frank William, chief gunners mate, Interim 1915-16. USS F-4 17 April 1915.

13. Cronin, Comelius, chief quartermaster, Civil War. USS *Richmond*, 5 August 1864.

14. Densmore, William, chief boatswains mate, Civil War, USS *Richmond*, 5 August 1864.

15. Eadie, Thomas, chief gunners mate, Interim 1920-40. USS S-4, 18 December 1927.

16. Grace, Patrick H., chief quartermaster, Korean Campaign, 1871. USS *Benicia*, 10-11 June 1971.

17. Hamberger, William F., chief carpenters mate, Boxer Rebellion, 13-22 June 1900.

18. Hill, Edwin Joseph, chief boatswain, World War II, Pearl Harbor, USS *Nevada*, 7 December 1941.

19. Holtz, August, chief watertender, Interim 1901-10. USS *North Dakota*, 8 September 1910.

20. Itrich, Franz Anton, chief carpenters mate, War with Spain, USS *Petrel*, 1 May 1898.

21. Johannessen, Johannes J., chief watertender, Interim 1901-10. USS *Iowa*, 25 January 1905.

22. Johnsen, Hans, chief machinist, War with Spain. Torpedo Boat *Winslow*, 11 May 1898.

23. Jones, Andrew, chief boatswain's mate, Civil War. United States Ironclad *Chickasaw*, 5 August 1964.

24. Klein, Robert, chief carpenters mate, Interim 1901-10 USS *Raleigh*, 25 January 1904.

25. Mackenzie, John, chief boatswains mate, World War I. USS *Remlik*, 17 December 1917.

26. McCloy, John, chief boatswain mate, Second Award, Mexican Campaign, (Vera Cruz). Leading 3 Pickett launches, 22 April 1914.

27. McDonald, James Harper, chief metalsmith, Interim 1920-40. Raising of USS *Squalus*, 23 May 1939.

28. Monssen, Mons, chief gunners mate, Interim 1901-10. USS *Missouri*, 13 April 1904.

29. Montague, Daniel, chief master-at-arms, War with Spain. USS *Merrimac*, 2 June 1898.

30. Nordstrom, Isidor, chief boatswain, Interim 1901-10. USS *Kearsage*, 13 April 1906.

31. Ormsbee, Francis Edward Jr., chief machinists mate, World War I. NAS Pensacola, FL, 25 September 1918.

32. Peterson, Carl Emil, chief machinist, boxer Rebellion, Peking, China, 28 June—17 August 1900.

33. Peterson, Oscar Vemer, chief watertender, World War II USS *Neosho*, 7 May 1942.

34. Reid, Patrick, chief watertender, Interim 1901-10. USS *Dakota*, 8 September 1910.

35. Ross, Donald Kirby, machinist, World War II, Pearl Harbor, USS *Nevada*, 7 December 1941.

36. Rud, George William, chief machinist mate, Interim 1915-16. USS *Memphis*, 29 August 1916.

37. Schmidt, Oscar Jr., chief gunners mate, World War I. USS *Chestnut Hill*, at sea, 9 October 1918.

38. Semple, Robert, chief gunner, Mexican Campaign (Vera Cruz), USS *Florida*, 21 April 1914.

39. Shanahan, Patrick, chief boatswains mate, Philippine Insurrection, USS *Alliance*, 28 May 1899.

40. Smith, Eugene, chief watertender, Interim 1915-16. USS *Decatur*, 9 September 1815.

41. Snyder, William E., chief electrician, Interim 1901-10. USS *Birmingham*, 4 January 1910.

42. Stanton, Thomas, chief machinists mate, Interim 1901-10. USS *North Dakota*, 8 September 1910.

43. Stokes, John, chief master-at-arms, Philippine Insurrection. USS *New York*, 31 March 1899.

44. Summers, Robert, chief quartermaster, Civil War. USS *Ticonderoga* 13-15 January 1865.

45. Sunquist, Axel, Chief Carpenters Mate. War with Spain. USS *Marblehead*, 26-27 July 1898.

46. Romich, Peter, chief watertender, World War II, Pearl Harbor, USS *Utah*, 7 December 1941.

47. Tripp, Othniel, chief boatswains mate, Civil War. USS *Seneca*, 15 January 1865.

48. Troy, Jeremiah, chief boatswains mate, Interim 1871-98. United States Training Ship, New Hampshire, 21 April 1882.

49. Verney, James W., chief quartermaster, Civil War.

50. Walsh, Michael, chief machinist, Interim 1901-10. USS *Leyden*, 21 January 1903.

51. Westa, Karl, chief machinist mate, Interim 1901-10. USS *Dakota*, 8 September 1910.

Compiled by Dick Hilgendorf and Ray Emroy, The National Chief Petty Officers Association, December 20, 1990.

Above and Beyond

The headline in *The Batesville Guard* read *Navy to Name Center After Arkansas Hero*. The Associated Press story spoke of the heroism displayed 20-plus years ago during an attempt to put out a fire on an aircraft carrier in the Gulf of Tonkin off the coast of Vietnam.[33] *Life* magazine's Aug. 11, 1967, issue called it *Hell Aboard CVA-59*.

"In the Gulf of Tonkin, 150 miles off Vietnam, the inferno swept CVA-59 – The mighty USS *Forrestal*. Moments before, a skyhawk bomber, taking off on a mission against North Vietnam, had spewed flame in a "hot start," probably caused by excess fuel. The flame struck a missile aboard a F-4 phantom. The missile tore loose and struck the fuel tank of another plane. The fuel ignited and spilled over the deck, becoming a river of fire that vaporized steel and entered the bowel of the ship through holes ripped in the decks by explosions. With the carrier's fate at stake and more than 130 shipmates already doomed, sailors rushed to other rockets – which might explode any moment – to heave them over the side. That was the beginning of a day of horror, but what was to stand out in the minds of survivors was the memory of heroism – countless acts of bravery."[34]

One aviator, Lieutenant Commander John S. McClain III, who was seated in another plane directly behind a disabled plane, remarked, "that was the bravest thing I ever saw," speaking of the heroic act witnessed.[35] He was speaking of the desperate action attempted by Chief Aviation Boatswain's Mate Gerald F. Farrier of Batesville, Arkansas, who grabbed a fire extinguisher and walked directly into the flame, spraying a path ahead of him.[36] He was trying to reach a bomb at which a flame was already licking. It detonated right in front of him. "He died in a split second when a 1,000 bomb exploded, blasting a hole in the armored flight deck big enough to drop a truck through," a report said. An account of Chief Farrier's action on the *Forrestal* is now used in training sailors to fight fires at the Norfolk, VA, facility that bears his name. His heroism was filmed by three automatic cameras on the *Forrestal*.[37]

The Secretary of the Navy

Washington

The President of the United States takes pride in presenting the Navy and Marine Corps Medal posthumously to:

> Gerald W. Farrier
> Chief Aviation Boatswain's Mate
> United States Navy
> for service as set forth in the following

CITATION:

For heroism on 29 July 1967 while serving as petty officer in charge of the aircraft crash and salvage crew aboard USS *Forrestal* (CRA-59) in the Gulf of Tonkin. When a fire broke out on *Forrestal*'s flight deck and swept through the parked aircraft, Chief Petty Officer Farrier, armed with only a portable fire extinguisher, rushed toward the rapidly spreading fire and directed the contents of his fire extinguisher on the nearest burning aircraft and its bomb load in an effort to bring the configuration under control. Before he could effectively cool the flame-engulfed ordinance, one of the 1000-pound bombs detonated, taking his life. By his heroic actions and valiant leadership, Chief Petty Officer Farrier served to inspire the men of *Forrestal* to courageous endeavors in fighting the fol-

CQM John H. Lang, USN–View taken on 12 Dec. 1937, on the east bank of Yangtze River, after his ship, USS Panay *(PR-5) was sunk by Japanese planes that day between Nanking and Wuhu, China. Lang was wounded by the same bomb which also wounded the* Panay's *C.O., LCDR. James J. Hughes, USN, QN the ship's bridge. (Courtesy of U.S. Naval Historical Center)*

low-on explosions and fires. His selfless devotion to duty was in keeping with the finest traditions of the United State Naval Service.

For the President,

Paul R. Ignatius
Secretary of the Navy

Farrier Fire Fighting Dedication Ceremony

The Navy's largest Fire Fighting Training School was dedicated on 14 July 1987 to Chief Aviation Boatswain's Mate Gerald W. Farrier in a ceremony in Norfolk, Virginia, attended by his family. Farrier Fire Fighting Facilities Norfolk was named when a plaque was unveiled by Chief Farriers' three sons and only brother; Steve Farrier, Tim Farrier of Batesville, Arkansas, and David Farrier of Sensor, Oklahoma, along with his brother, Glenn Farrier, of Dayton, Ohio, who shared the honors for an assembled audience of 150 Fire School Instructors and honored guests.

The guest speaker Captain Lawrence G. Anderson, Chief of Staff to Commander Training Command, US Atlantic Fleet Training Center, US Atlantic Fleet. A Navy color guard performed to the inspiring march music provided by Commander In Chief, US Atlantic Fleet (CINCLANTFLT) Band.

After the brief ceremony family members and guests were permitted to view the Fire Schools 20-minute motivation movie entitled, *Trail by Fire*, which is an official documented visual account of the incident onboard USS *Forrestal* taken when it actually happened. The film is shown during the course of training instruct sailors, civilian employees, fire fighting civilians from the local area, as well as reserve personnel from all around the East coast and eastern part of the United States and vividly points out why fire fighting is considered such a serious business.

The permanently displayed plaque which is located in the Administrative building at Fire Fighting School sums up the sacrifice of

Chief Farrier and is meant to motivate all fire fighters everywhere. The plaque reads:

In Memory Of

Gerald W. Farrier
Chief Aviation Boatswain's Mate Handler
United States Navy

"Who gave his life in the service of his country while serving with the United States Navy. Chief Petty Officer Farrier was serving as crash and salvage chief aboard the USS *Forrestal* (CVA-59) on 29 July 1967 in the Gulf on Tonkin, when a fire broke out on *Forrestal's* flight deck and swept through the bomb laden parked aircraft. Chief Petty Officer Farrier, armed with only a portable fire extinguisher on the nearest burning aircraft and its bomb load in an effort to bring the conflagration under control. Before he could effectively cool the flame-engulfed ordinance, one of the 1000 pound bombs detonated, taking his life. By his heroic actions and valiant leadership, Chief Petty Officer Farrier served to inspire the men of the Forrestal to courageous endeavors in fighting the follow-on explosions and fires. His selfless devotion to duty was in keeping with the finest traditions of the United States Naval Service."

After the movie the president of the Tidewater Aviation Boatswain's Mate Association and a retired AB himself, Mr. Bill Sowers presented the family a plaque and for the school an identical one to hang in the Fire School entrance. The plaques are inscribed, "In memory of ABHC Gerald W. Farrier whose heroic actions on board USS *Forrestal* (CV 59) on 29 July 1967 exemplified the highest ideals of his chosen profession." Mr. Sowers' inspiring remarks were a fitting climax to an outstanding ceremony. It read:

"John Paul Jones once said that in defense of our nation he fully intended to go in harms way! We, the aviation boatswain mates have long recognized the commitment and dedication necessary when walking the dangerous path of our profession. Today we assemble to honor Gerald W. Farrier who in the Tonkin Gulf on 29 July 1967 gave his life leading his men in extinguishing a horrible fire aboard the USS *Forrestal*. On that tragic day many men freely went into "harms way" and many men made the ultimate sacrifice, their lives. Let us today measure ourselves as we dedicate this school in his name. Let us recall that others, like the chief, died for the flag and ideals that motivate us all. On behalf of all AB's everywhere, this small token of our respect is presented. Let the unselfish dedication of this gallant man and his shipmates serve as an inspiration to us all. And let us realize that the call to go in "harms way" waits in our future and may GOD help us when we face that call."

Endnotes:

[1] *12 Stat. 330.*

[2] *Public Resolution of July 12, 1862; 12 Stat. 623, 624.*

[3] *Act of March 3, 1863; 12 Stat. 744, 751.*

[4] *Including Medals of Honor stricken from Honor Roll in 1916.*

[5] *The wording of the Navy statute was modified slightly in 1901 by making the award available to an enlisted man "who shall have distinguished himself in battle or displayed extraordinary heroism in the line of his profession." Act of March 3, 1901 (31 Stat. 1099). In 1915, Congress made naval officers eligible for the Medal of Honor. Act of March 3, 1915 (38 Stat. 931).*

[6] *The Certificate of Merit, created in 1847, was technically a document rather than a decoration. By 1861 it had lost most of its meaning. See - US Dept of the Army. The Medal of Honor of the United States Army. Washington, US Govt. Print. Off., 1948. P. 5.*

[7] *US War Dept. Report of a Board of Officers to examine and report upon applications and recommendations for Medals of Honor and Certificates of Merit. Washington, US Govt. Print. Off., 1904, p. 17.*

[8] *Ibid., pp. 17-18.*

[9] *Act of April 27, 1916 (39 Stat. 53).*

[10] *Act of June 3, 1916; (39 Stat. 214).*

[11] *US War Dept. General Staff Corps and Medals of Honor. Washington, US Government Print Off., 1919. (66th Cong., 1st sess. senate. Document #58) pp. 108-478.*

[12] *40 Stat. 870.*

[13] *US Congress. Report to accompany S. 1720. Washington, US Govt. Print. Off., 1917. (65th Cong., 1st sess. Senate Report #73) p. 3.*

[14] *40 Stat. 1956.*

[15] *Act of March 3, 1901 - 31 Stat. 1099.*

[16] *Act of March 3, 1915 - 38 Stat. 931.*

[17] *Letter from Judge Advocate General to Chief of Bureau of Navigation 8 July 1921 - Citing decision of Comptroller Treasure 8 December 1919 - (26 Comp. December 464.)*

[18] *Act of August 7, 1942 (56 Stat. 743.)*

[19] *Act of April 7, 1922 - 42 Stat. 493.*

[20] *Act of May 26, 1928 - 45 Stat. 747*

[21] *US War Dept. Army Regulations 600-45. Personnel: Decorations. Sept. 22, 1943.*

[22] *US Navy Dept. General Instructions and Policies Relative to Decoration and Medals. Washington, December 18, 1944. n.p. mimeo*

[23] *40 Stat. 870.*

[24] *56 Stat. 743.*

[25] *40 Stat. 1056.*

[26] *Act of May 3, 1950 (64 Stat. 103.)*

[27] *63 Stat. 838-839*

[28] *US Navy Dept. Medal of Honor, 1861-1949. Washington ? 1950. p.2.*

[29] *US Dept. of Army. Army Regulations #600-45, Washington, US Govt. Print Off. 27 June 1950. p. 8.*

[30] *Ibid.*

[31] *Morgan, J. McDowell, Military Medals and Insignia of the US. Glendale, CA, Griffin-Patterson Publishing Co., 1941. p. 37.*

[32] *US Dept. of Army. Army Regulations #600-45. Washington, US Govt. Print. Off. 27 June 1950. p. 6.*

[33] *The Batesville, AR, Associated Press, 1987.*

[34] *Hal Wingo, "Hell Aboard CVA-59," Life magazine, Aug. 11, 1967, Vol. 63, p. 6.*

[35] *Ibid.*

[36] *(Wingo, p.6)*

[37] The Associated Press

Chapter 5

Anecdotes and Humor

The Chief's Basic Rules

Rule 1 – The chief is right.

Rule 2 – In the impossible hypothesis that a subordinate may be right, Rule 1 becomes immediately operative.

Rule 3 – The chief does not sleep; he/she rests.

Rule 4 – The chief is never late; he/she is delayed elsewhere.

Rule 5 – The chief never leaves his/her work: his/her presence is required elsewhere.

Rule 6 – The chief never reads the paper in his/her office; he/she studies.

Rule 7 – The chief never takes liberties with his secretary; he/she educates her/him.

Rule 8 – Whoever may enter the chief's office with an idea of his own must leave the office with his chief's ideas.

Rule 9 – The chief is always chief even in bathing togs.

and

The chief is always right

He may be misinformed, inexact, bullheaded, fickle, ignorant, even abnormally stupid, but never wrong.

Chiefs' Mates

The following poem was contributed by Rose Berlier for this issue, with the assistance of Jean Levicki, who is the wife of Charter/Life Member CSK Joe Levicki. Jean is a published author and her poem for the NCPOA is welcome.

> Chiefs' mates we are
> And proud to be
> Partners of those men
> Who went to sea
> And banded together
> In war and peace,
> Tragically seeing
> Their numbers decrease,
> As battles were fought
> And lives were lost.
> We question
> Was it worth the cost?
> Now for those who remember
> Many years have passed
> But they have memories
> Which will live and last.
> We as females,
> Partners, and special mates
> Can help to make NCPOA
> The pride of the States.
> With suggestions, actions
> And interest, too
> There is something
> Each one of us can do.
> We will build our
> Own memories
> As the years go by
> And together
> We'll "Reach for the Sky."
> *Jean Levicki*

Anecdotes and Humor

A group of "90 day wonders," AKA Officer Candidate students, were asked a question on their final examination – how to resolve a dilemma, namely that the base flag hoist was loose and stuck fast to the top of the flag pole – and given a multiple choice of the following possible answers:

1. Send for the cherry picker
2. Shoot the boatswain line
3. "just" allow the wind to blow free
4. None of the above.

The proctor, a former chief petty officer (CPO), was chagrined at the varied responses and told them they flunked the quiz. The obvious and correct choice, he stated, was "to send for the leading chief!"

Your book and endorsements sound great! Certainly my own four years as an officer were marked by CPOs saving my "you know what" many times. Good luck with the book. – The Reverend William B. Trimble Jr.

From my time at Naval Hospital, Memphis, I am pleased to include three funny incidents:

"A woman called the Information Desk and said several days ago she dropped her lipstick in the toilet. She fished it out and used it and now she had a sore on her lip. She wanted to know if dropping it in the toilet could have caused her problem. The CPO who answered the phone told her he doubted it, but perhaps she had a dental problem. She replied that she doubted that, as she only had one tooth!"

A CPO walked into the Personnel Office just as an officer spilled a cup of coffee on the floor. The chief remarked, "Well, I've always heard of the officer's mess, but this is the first time I've ever seen one."

The CPO's relief in our office was attempting to get acquainted after reporting. One asked the other, "You got any kids?" The response was, "Yes, five." The questioner though a minute and responded, "You're pretty good at begotting, ain't ya?" – George Ann Sanders, retired civil servant.

The late Captain George Grider from Memphis, an acknowledged submarine commander in World War II, was serving aboard the *Wahoo* in 1942. He received orders to participate in an extended submarine reconnaissance of the New Guinea Coast. The Japanese Navy bristled with strength at a remote place named Wewak. Captain Grider had informed his men that the United States Navy radiated technological expertise and no task was beyond the reach of the Navy. Accidentally, a CPO of the *Wahoo* had purchased a geography book in Brisbane before their departure. The *Wahoo* set sail and discovered as she approached the vast New Guinea coast that her charts offered little practical data. More and more, Wewak loomed as a mystery until the chief who had bought the geography book presented himself to Captain Grider and suggested reference to this publication. Captain Grider accepted the book and directed it to the navigator. Thereupon, Captain Grider recounted that this geography book played a significant role in the location of Wewak. Upon arrival in Wewak, the *Wahoo* bogged a Japanese destroyer – sunk by virtue of a direct bow shot at close quarters." – contributed by John R.S. Robilio.

Fiddler's Green is the traditional heaven of seagoing men, comparable to the Viking's Valhalla and the Indians' Happy Hunting Grounds. It's restricted to sailormen. Fiddler's Green is the only heaven claimed by an occupational group. According to legend, Fiddler's Green is well-supplied with joyous demoiselles, free drinks, and plenty of chow, and there are no regulations. Civilians ineligible for entrance. – anonymous.

Remember the adage, "There's more to knowing the ropes than knowing the ropes." When someone in the Navy knows the ropes, he is regarded as someone who knows his way around and is capable of handling most problems in a given area. Through the years, the meaning of the phrase "knows the ropes" has changed somewhat. Originally, the statement was printed on a seaman's discharge to indicate that he knew the names and primary uses of the main ropes aboard ship. In other words, "this man is a novice seaman and knows only the basics of seamanship." – anonymous.

Surely you will make it a point of "order" to single out the "to do" chiefs as able to leap tall buildings, solve the world's most difficult problems, and do so without spilling a drop of coffee" – Robert R. Mixon. HMC USN RET.

One new seaman reported to his chief that he was "ready for duty." The chief picked up a bucket and directed his new charge to "fetch a bucket of steam" from the engine room. Hours passed and this frazzled appearing seaman returned, almost in tears, to tell the chief, "there must be a hole in my bucket, 'cause it won't stay." – anonymous.

We had a group of CPOs and several chief warrant officers (CWO) in the Inspection Department at a North Island Naval Air Station. This one very senior CWO grew tired of hearing the chiefs reportedly state "the chiefs were the backbone of the Navy." One day, he went over to a group

of six or so and said, "OK, OK, I'll admit that you guys are the backbone of this spineless outfit," grinned, and went back to his desk. CWO Lester B. Tucker, USN, RET.

Meeting My First Chief Petty Officer

Early in 1960 I was assigned to the 1254th Air Transport Wing (special missions) at Andrews AFB, Maryland. This unit is now designated as the 89th Special Air Missions Wing. This unit is responsible for providing air transportation to the president and other important government officials. As a young Air Force captain, it was my job to obtain supplies and equipment necessary to support three VC-137 A/B aircraft. These were new jet aircraft similar to the Boeing 707-120 civilian models.

One day, my boss, Colonel Dick Yoder, called me in and gave me an assignment. He told me that he wanted me to get a full set of flags and flag plates to be used on the new aircraft. These items were displayed in the aircraft window and on the inside of the main aircraft entry door to indicate the ranking distinguished visitor (DV) on board.

These flags and flag plates are used by all cabinet officials as well as by many of their deputy secretaries, under-secretaries, and deputy under-secretaries, as well as all military general and flag officers.

In all there were dozens of the items to be obtained. My search took me to many levels throughout federal government. Before long I made steady progress in getting the items from all sources except those for Navy officials. After a number of attempts without progress, I decided out of frustration "to go to the top." I got out my handy Pentagon telephone directory and looked up the number for the Secretary of the Navy. It even listed his name – Secretary Franke. I dialed the number and it was answered by a very nice lady, who identified herself as Secretary Franke's secretary. I told her that I as "Captain" Freeman and would like to speak with Secretary Franke. What I didn't tell her was that I was an Air Force captain and not a Navy captain. She told me that the Secretary was not available at the moment, but asked if possibly she or someone else might be able to help me. I explained what I was looking for and she told me that she was sure that she could find someone who could solve my problem and also that she didn't think it would require Secretary Franke's personal time.

Almost before I had time to turn around my phone rang and it was an admiral from Secretary Franke's staff. He said that he had been instructed to give me whatever assistance I needed, and that he would get back with me in a couple of days to let me know when and where the flags and flag plates would be available.

As promised, he called back in a couple of days and said the items were now available. I told him that I would be glad to pick them up and he told me that if I would go to Anacostia Naval Air Station, to a specified aircraft maintenance shop and ask for Chief ? (I don't remember his name), he would give me what I was looking for.

I made my way to Anacostia and found the particular maintenance shop. As I was getting out of my Volkswagon, there were dozens of people in the area. They were all wearing either tan or blue work uniforms, except for one individual in a dark uniform, white shirt and tie and with gold braid that seemed to go from his shoulder down the entire length of his sleeve. He looked like he had just stepped out of a clothing store. He was immaculate and I just knew he must be some real high-ranking Navy officer. For at this time the only thing I understood about Navy ranks were that the more braid, the higher the rank.

Well, I walked up to him and came to attention and saluted him. I told him my name and asked if he could help me find Chief ?. After he saluted me back, he told me that he was Chief ? and that it was not necessary to come to attention and salute him first – that in the future it would be better to just return his salute. I knew then I had learned a lesson and I felt about one foot tall.

He was a very warm person and soon made me feel more at ease about my dumb mistake. But it was soon obvious that he was in charge in that shop. He gave a few instructions and before I knew it a couple of young seamen came back with the items I had come looking for.

He walked me back to my car and after I had gotten in, he came to attention and saluted me as I drove off. He smiled at me as I remembered to return his salute.

William E. Freeman Jr., COL USAF (RET)

Chapter 6

Biographical Sketches of Contributing Chief Petty Officers

Bits & Pieces
by Maurice L. Ayers, ATCM, USN Retired

We all know and recognize the individual that has the experience and "know how" to get the job done. When it is accompanied by a good personality and a positive "can-do" spirit it stands out in our memory even more. Chief Radioman Shoer was one of these.

Fresh out of boot camp in the spring of 1942, I was stationed at Fleet Air Wing 14 Radio School on North Island. Chief Shoer was the leading chief of the school. There were about 15 or 20 of us in this class and most of us were quite proud of the fact that we could copy code at three to four words a minute.

One afternoon we were all taking a break. Chief Shoer was in the code room and someone turned the radio receiver on to the "Fox Schedule." This was a plain language program with low priority traffic, repeated two or three times a day and transmitted in the blind. It was really moving along. I could only distinguish a little now and then. Chief Shoer and some other instructors said it was about 35 words per minute.

Chief Shoer put a clean sheet of paper in the "mill" (typewriter) and started copying. We were amazed. It was beautiful looking over his shoulder as he copied. We could also identify the letters coming over the speaker. After about three or four minutes, Chief Shoer stopped typing, took a pack of cigarettes out of his shirt pocket, removed one and lit it. He took a drag off the cigarette, laid it down in the ashtray and started copying again. He never missed a character. He had copied the message behind in his head.

Chief Shoer always stood tall in my mind from then on. I had admired him before mostly because of his clean-cut appearance. He was a credit to the Navy. Looked real good in his uniform. But his experience really made him stand out in my memory.

As my Navy career progressed, I encountered numerous chiefs. Some good, some not so good. But the squadron leading chief was always one that I had to reckon with. He controlled the liberty, the watches, etc. I frequently said if I ever got to be leading chief, things would be different. And they were. Heavy Attack Squadron 5 (VAH5), stationed at NAS Sanford, FL, attached to USS *Forrestall* (VA-59) gave me my chance to do it right over.

The leading chiefs desk is where "all the problems of the squadron meet." Anything relative to military duties personnel problems, berthing, messing, squadron movement transportation, maintenance of spaces, etc. In a squadron of 350 men and officers there will be many problems. I discovered that some things you can change, some you cannot, some should never be changed. But, every one should do his best to improve the operation of his organization. That's how Navy Es are won. Heavy Five made their mark.

A good leading chief will solve those problems and look out for the welfare and comfort of his men, always making sure that a high level of discipline is maintained. The squadron must also perform its assigned mission. The leading chief will be a primary link in the chain of success of any organization. If he's good, the squadron will function more smoothly, with high moral efficiency and pride.

When the executive or commanding officer put a plan into operation, the leading chief will be one of his primary tools. The degree to which he has successfully done his job determines in large measure the success of the squadron.

The Navy recognizes talent and ability to get the job done. That's why the chief is chief. He has had the experience, he has the knowledge, he has the authority and the officers and men of the organization know this and depend on it.

The story goes that a group of officers near graduation from a Maintenance Officers School were asked to explain how they would replace the lanyard of a flag hoist that had broken. After several explanations involving cranes, ladders, etc., the instructor gave the correct answer, "Chief get that flag hoist fixed."

Memoirs
by Norman D. Bender

My first encounter with a United States Navy Chief was aboard the USS *Nevada* as it rode at anchor in the harbor at San Pedro, California, in February 1934. The Navy had rejected me because of a slight malocclusion when I tried desperately to enlist upon high school graduation in May 1933. A few months later I managed to get in the CCC at Chicago and was shipped out to a camp in the mountains above Santa Barbara, California. The camp commander was an academy "jg" who had a choice of taking that duty or an indefinite furlough in a depression-raged Navy. I became his "striker" or "dogrobber," or whatever – anyway, I was his man-servant and earned a few extra bucks and enjoyed special privileges. He had come to the camp from duty on the *Nevada* and on a trip to Pedro one weekend he took me on board for a few hours visit. Boarding that ship was the opening of a new world. The jg asked a chief who came up to greet him if the CPO would give him a tour of the ship while he headed for the Ward Room. The chief was in dungarees and wearing his billed cap with a white cover. I didn't realize what good company I was in until I noted that everybody smiled and spoke nicely to him and to his waif he had in tow. Sailors were in action all around the decks. They were dressed in powder blue dungarees with bell bottoms and most had on gleaming white low cut "sneakers" which I later was advised were known as "boat shoes." Man! I wanted to be in that Navy so bad that I couldn't sleep for a week. But, back to the chiefs. Most of the chiefs that I saw were also in dungarees. The difference observed between the chiefs and sailors was in the caps and shoes they wore and the fact that the chiefs all seemed quite a few years older in that peace-time and slow-promotion Navy. Cruising back to the Pedro landing in the Nevada's motor launch, the jg explained the role of chief petty officer (CPO) in the United States Navy and it was then that I first heard the term, "backbone of the Navy." On several occasions during the following months, I had opportunity to visit the Navy boat landing at Pedro and watch the liberty boats unload. Happy and carefree sailors bounded from the boats and some rushed to link arms with waiting girl friends as they went merrily on their way. It was an intriguing and enviable scene to this shorebound soul. However, it was the sprinkling of Navy chiefs that really got my attention as they nimbly stepped to the floating dock and strolled away in a bit more dignified manner. The dress blue uniforms with white cap covers stood them out in high visibility, but it was the insignia and service stripes on their coat sleeves that really piqued my interest. Some sleeve decorations were on the right arm and some were worn on the left arm. Some were emblazoned with a sleeve full of gold while others were a less colorful red. It was a puzzlement of the moment but was lost as I literally drooled to get into the Navy and someday make chief. However, we soon left California and the Navy far behind as our camp was moved to summer location in Idaho. It would be almost seven years until the perplexity of gold/red and right arm/left arm, as it pertained to the United States Navy Chief Petty Officers, would be resolved.

Franklin D. in 1940 decreed that in late October males in my age bracket would register for a military draft. The Army Air Corps had provided opportunity to make Aviation Cadet but I just couldn't line up the two little pegs in that long trough and was rejected for bum depth perception. Dejectedly returning to Memphis from Maxwell Field in Montgomery, Alabama, I continued in my job working for Malcolm Fraser at Standard Parts – the Memphis NAPA warehouse.

The day that FDR invited me to Central High for inking the draft list, I dutifully signed and then strolled over to Madison and caught a streetcar – (yes, they were still running) – for downtown and reported in at the Navy Recruiting Station. A pharmacist's mate swirled a test tube of my specimen near a light and sang out with "Grade A!" Hypocritically, a medic peeked in my mouth and admired my perfect teeth, overbite and all. A chief pharmacist's mate handed me a fistful of papers and welcomed me into the United States Navy. I asked the chief where he was when I needed the Navy and they turned me away for being a physical wreck. He smiled as he replied, "The times they have changed Apprentice Seaman Bender. Enjoy your trip to Norfolk."

Arriving at Norfolk Recruit Training Station the first week of November 1940, I discovered that the Navy must have offered every retired Navy chief that could walk and talk an opportunity to get back in uniform. Never mind the age. If a chief thought he could stand the pace and had his hemorrhoids temporarily tucked away he was welcomed with open arms. I drew a retired chief fire controlman as a drill instructor. He was quiet but firm and truly Navy savvy. Our platoon was lucky indeed. Being significantly older at 26 than the 17-to-20-year-olds in the group, the chief appointed me "guidon" bearer. The job had some perks like carrying a little flag on a pole instead of a beat up old rifle without firing pin that the rest of the platoon drew. At every opportunity I asked the chief about the various rates and best opportunity in the Navy. He advised me to try to get into any aviation slot and I detected a bit of wistfulness in his manner. I had literally devoured that part of the *Bluejacket's Manual* that had to do with Navy tradition, procedures and ratings. I wasn't too interested in the knot-tying and seamanship sections of that informative little book. The chief, who was double my age, was a great help in those early weeks and unraveled the mystery of gold/red and right arm/left arm. He always wore his working blues with red insignia and two or three service stripes – on the right sleeve. It wasn't until graduation day and the dress parade that he arrived on the cold scene decked out with a right arm swatched in a gold insignia and eight gold "hash marks!" He was the epitome of a United States Navy Chief Petty Officer and it made a permanent impression on me. I set my goal right then to make chief as soon as I could. There had been some misgivings for my being a bit impetuous on enlisting as an apprentice seaman when groups of Naval Reserve "Feather Merchants" began arriving for "boot training" at Norfolk. All sorts of petty officers and even a sprinkling of "slick-arm chiefs" were in these reserve platoons. Too late, I discovered that with my professional experience and age I could have entered the Naval Reserve as a chief storekeeper (CSK). The regret lingered for a while, but I soon accepted the fact that I signed on for a six-year cruise in the United States "Regular" Navy and it was up to me to make the best of it.

The first two or three years of "wartime mobilization" there was a bit of resentment on the part of the "Regular Navy" to the Reserve or "Feather Merchant Navy." The resentment was caused by the Reserve Petty Officers arriving on the scene direct from civilian life with no actual Navy service. It was different for a first class "feather merchant" to be accepted by a second class or lesser Regular Navy Petty Officer who had spent long years in the peace-time Navy and now find rating opportunities usurped by the reserves. A "slick arm" Feather Merchant CPO (a chief who could not display at least one "hash mark" on his sleeve) had a particularly tough time. Disdain was not too well disguised by the "regulars," and even the regular chiefs who usually had an array of hash marks when they made the coveted CPO ranks didn't accept the "slick arms" with open fellowship. Although some of us "regulars" made chief comparatively early in the "duration," it was rare that a hash mark hadn't been earned in advance. However, I would have gladly accepted a "slick arm" appointment at any time and endured the insults of envy. In fact, my hash mark was barely a year old when I did make chief, but it may just as well have been of nearly eight years significance as far as prestige was involved.

It hadn't taken me very long to understand why a Navy chief was seldom "messed with" by either sailors or officers. I had experienced first hand that CPOs were the Navy!

An early incident had revealed to me the power and prestige of the Navy's CPOs Our platoon officer was a pompous little ensign who had apparently just drawn Norfolk duty out of the Academy Class of '40. Apparently he was part time as our guiding light for he only showed up for short stays two or three times a week. On one occasion while our chief had the platoon maneuvering, Mr. Ensign countermanded an instruction by the chief and proceeded to impart his version. The chief quickly, but calmly, gave out an "as you were" to the platoon. Up until then the chief had patiently tolerated a bit of upstarting by the smooth-faced one striper but this occasion apparently was too much. Chief walked up close to the ensign and spoke a few words, too soft to hear. The ensign snapped a salute to the chief, uttered a sharp "belay that order" to the platoon and returned from whence he had come. He did not show again until parade day and left immediately as he could – never saying another word, either to the chief or the platoon. The power, and wisdom, of Navy CPOs was manifested to me at that instant – and never forgotten. It stood me in good stead in the closing months of my enlistment.

Managing to get picked for Aviation Ordnance Service School at the end of boot camp, the instructor was another "retrad" CPO – a chief aviation ordnanceman. He was a great instructor and his profound knowledge of every item employed in aviation ordnance, plus his technical and theoretical knowledge was positively astounding. His authority in the class-

room was absolute. Occasionally he would allow a bit of horseplay as we dissected pistols, shotguns, machine guns, and other lethal weapons, but when that chief said, "OK, knock it off!" there was instantaneous silence and a return to serious work. Chiefs just seem to transmit a natural element of complete control in any situation.

Stationed at Jacksonville NAS in early Spring of 1941, the various uniforms that were required wardrobes for officers and chiefs were revealed. Although I had noticed a number of dark green uniforms walking around the Norfolk NAS with gold wings attached, I had erroneously assumed the wearers of those uniforms to be Marines. Naval aviators and aviation chiefs decked themselves out in those snazzy green "working" uniforms with smooth tan shoes and tan socks. I really yearned for the day that I could qualify. The dress blues and the dress whites with white shoes were standard. Then there were the universal khaki poplins. I also discovered that the billed caps could be disassembled and various colored cap covers installed to match the uniform. Apparently these uniform designs had been in service for a number of years prior to 1940. I deduced that from the fact that service stripes on CPO sleeves and rank stripes on officers sleeves often looked much newer than other uniform markings. Added "hitches" for the chiefs and promotion for the officers meant adding new stripes to present uniforms. (It is also remembered that with dungarees a chief could wear just about any cap cover that fit his fancy. Most of the older chiefs had accumulated several hats and a "salty" life was a symbol of seniority.)

About the time that I made chief in early 1945, or soon thereafter, some screwball in the Navy hierarchy, who apparently had little to do, sold top brass on authorizing a new "Air Force Gray" uniform design. No telling what it cost the Navy and the individuals to add this excess baggage to an already sufficient, if not overstuffed, wardrobe. I don't remember what it cost, but I added one to my blue and khaki threads. The one good thing with the gray version was the poplin work pants, shirts and hat covers that were quite practical. The gray may have disappeared after a few unpopular years.

The unique position that a United States Navy CPO commands in comparison to the Army and Marine Corps is established and perpetuated basically with their uniforms. At least it was until I think I recently observed some second and first class PO's wearing blue shirts and billed caps. Sergeants of the Army and Marines come in so many various denominations that the upper echelons of master sergeants and sergeants major, while enjoying authority within their own branch environment, sacrificed a bit of comparative clout against a distinctively uniformed CPO of The United States Navy.

CPOs are "appointed" to their positions in the United States Navy. The shedding of "Bluejacket" attire by a first class petty officer – which is the usual transition to CPO – creates an event that gets the attention of shipmates – sailors, officers, and fellow chiefs. The move into chief's quarters and joining the chief's mess is usually accompanied by a bit of hazing, for after all, the new chief is being accepted into a fraternity. He (remember, this reminiscence was situated prior to the advent of the WAVE contingent) is mantled with authority and dealt an aura of respect that he did not experience when additional chevrons were merely added to his petty officer insignia. (Army and Marine Corps must keep getting stripes and inserts also.) The newly appointed CPO becomes a member of management – "blue collar" perhaps, but solidly influential nonetheless.

Back in my day a chief was a chief. The fist year of "chiefdom" was noted on the payroll with a (T) for "Temporary Appointment." The pay was $96 per month along with a handy uniform allowance. At the end of a year as (T) the designation was changed to (PA) for "Permanent Appointment" and the pay was substantially bounced to $128 per month. It may have happened sometime in Naval annals, but it had to be rare that a (T) was automatically elevated to (PA). Nowadays there may be several grades of CPO as I have observed the rating of "master chief." Some prying busybody or infiltrator from the Army/Marine ranks must have felt that the CPO appointment needed dilution. The same has apparently happened to the once distinctive rank of warrant officer – but that is basis of another story.

Speaking of Warrant officer brings back to mind the puzzlement of why certain rates, including CPO, were tagged with "mate." My erstwhile drill instructor shed light on that question for me. Prior to WWII there were certain warrant officer positions which included boatswain, carpenter, pay clerk, machinist, gunner, and pharmacist. Certain petty officer ratings were "mates" to these warrants while others had "mateless" designation. Petty officers that made it to warrant boatswain usually transitioned from boatswain mate, quartermaster, and signalman ratings. Pay clerks usually rose from yeoman and storekeeper. Water tenders, electricians, metalsmith, and other "black gang" ratings usually ascended to warrant machinist. Carpenters and sailmakers could become warrant carpenters gunner's mates, fire controlmen, and perhaps others could make warrant gunner and so on. Pharmacist's mate could make warrant pharmacist. Pharmacist's mates served with combat Marines in the role of "corpsman" as the Marines did not have a separate medical unit, and probably none now. There was no "hospital corpsman" in that era of the Navy.

With the advent of WWII, and increasing technology, fertile ground for adding specialty ratings was cleared and many new ratings were added. One day I went from aerographer second class (aerog 2/c) to aerographer's mate second class (aerm 2/c) as the captain in charge of the aerology section wangled a position of warrant aerographer. A number of warrant positions or ranks followed and they provided more positive or less confusing identification in the warrant ranks – plus opportunity to reward more CPOs. The evolution of rating structures in the United States Navy would make interesting reading, and would be a good recruiting tool. It may already exist but no one has favored me with a complimentary copy.

Outfits that were staffed with multiple CPOs usually had a designated "leading chief" that acted as senior enlisted factotum. The position normally went to the chief with the longest tenure of service in rate. I enjoyed that status in my last duty assignment of leading chief of the Nanook Expedition in 1946 to establish a United States position at Thule in the upper region of Greenland. A photo of our aerological group is attached.

Although I can't remember ever entertaining serious desire for pursuing a career in the Navy, I did enjoy my last couple of years as a chief aerographer's mate. When the second A-Bomb landed on Japan the invasion expedition to which I had been assigned was canceled – thank Goodness. I became a "free agent" when my assignment to infiltrate Honshu evaporated. The Japanese surrender suddenly left the Navy with a wild surplus of ships and men. It was easy to get "lost in the shuffle" and I did just that. Riding the big carrier *Enterprise* back to New York as a passenger was fun. At New York the "Big E" was stripped and fitted with thousands of bunks and started ferrying victorious troops back to the USA. I wangled a transfer to the USS *Missouri* which was tied up at Pier 22 as an exhibition for hundreds of thousands of visitors to view the surrender plaque where MacArthur accepted the Samurai surrender sword. I had no assignment and merely used the "*Big Mo*" for a hotel as I kept a low profile and played around the Big Apple for a few months. The "*Mo*" had a luxurious chief's quarters and spacious chief's mess. A ship's cook and mess attendant were on hand at all hours to serve a la carte from the menu of desire. It was grand living and I considered it my due after sweating months in the South Pacific at battle stations. However, this good life got boring, and too expensive to maintain, so I checked in with Aerology Personnel in Washington to let them know where I was and to volunteer for a short assignment, as it was Spring 1946 and come November, I would be free to go back to standard parts. The guru in Washington gave me a choice of going to Bikini with the "A-bomb" nuts or accompanying the Nanook Expedition which was being formed to establish the USA at Thule, Greenland. I took Nanook and set out for Boston where it was to stage. Nobody there had heard of Nanook and I was the first guy that had mentioned it. So, I languished on Boston Common, Scolly Square, and saw a lot of Fenway Park. Finally, Nanook got put together and we sailed from Boston on the USS *Norton Sound* – a big Seaplane Tender. The chief's quarters were great and could accommodate at least a hundred CPOs. The 20 or so of us rattled around in luxury. CPOs of the United States Navy know how to make the best of a good situation!

Career Highlight

The opportunity to serve as the master chief petty officer of the Navy from 1979 to 1982 was the highlight of my 31-year Navy career. To function as an ombudsman and enlisted representative of the Navy enlisted community was demanding and rewarding at the same time. I was particularly fortunate to have served a leader, Admiral Thomas B. Hayward, that was openly supportive of the Navy family and concerned for the welfare and morale of the spouse as well as the service member. He supported the involvement of my wife as an equal partner in the function of the office of the MCPON and was the first to allow significant travel and official feedback to the naval leadership in Washington regarding the op-

eration of MWR, Navy Exchange and Commissary, Child Care Centers and the new, at that time, Family Services Centers. My wife and I will always remember our tour and the opportunity to travel and meet the sailors and their families that make up our great Navy. The byword of our duty with Admiral Hayward was pride and professionalism and one Navy. Caring for and supporting the whole team ashore and afloat.

Thomas S. Crow, MCPON
USN (Retired)

A Reflection by a Crusty Old Salt

"You will revert to your permanent rating of chief pharmacist's mate and be returned to inactive duty in the Fleet Naval Reserve" – these were my final words.

All of which means that I am now an old "has been." They told me that this job would not last when I first enlisted in 1914.

I sit in my home and let my mind wander back across the years to my alma mater – the old Hospital Corps School at Newport, Rhode Island. Dr. Kaufaman was the head of the school, assisted by hospital stewards, Bigelow and Dean, later by Maggie Benton.

Among the students who I recall, were Red Reed, Chester Fay, Eli Balrup, Baby Young, Dick Roberts, Dom Merry, Jack Davis, Ford Moe, Elzar Harwell, and Stritzsinger. A grand battle among us rebels. We avenged what our grandpappies took from the damyankees.

At last I am through school – can wear my blues and I'm ready for duty in the Brooklyn Naval Hospital as a hospital apprentice. My first "crow" – never was I as thrilled again in my entire naval career.

My first ship – the USS *New York* – queen of the seas – the newest ship in the Navy. I was still on the *New York*, after visiting in many ports, when we got in the first World War. But I was transferred to the destroyer base, Queenstown, Ireland, then to Aghada, Loch Foyle, Chardiff, Liverpool, and other ports for duty.

With the war over, I am a veteran at 21 – through with the Navy, I thought. I try to go back to school; I get a job; I am lonely; I drop into a recruiting station for a call; I am tempted; give in; I ship over – I know that I am a real Navy man.

Promotions came. On the *Hennie Maru*, on the trip to Alaska with President Harding. Europe again, the West Indies, Great Lakes, League Island – I finally make chief. Then, in order, Panama, the Special Service Squadron, Manilla, Chefoo, Tsingtao, Shanghai, Amoy, Hong Kong, the Yangtze Patrol, Nanking, Hankow, Changsha, Ichang, the Fourth Marines.

I have completed 20 years service. I am fat, middle-aged – the Navy has no further use for me – I thought. In the Fleet Reserve, back home in Memphis, the year, 1937. I buy a chicken farm in the manner of all good ex-sailors. I sell what was called a chicken farm.

Then in 1939 there came a wire from the Commandant, Eighth Naval District: "Will you return to active duty - If you do not, you can be ordered back."

Will I return" Boy, oh Boy, will I come back!

I burn the bearings out of my car getting back. Charleston, Parris Island – the fun of being back. Old shipmates – Captain Hale, Virgule Coulter, Bob Jones, Leon Merkle, Hal Swain, Red Neighbors.

We dug in and were ready for sea duty in the Fall of 1940 when all FNR's were called in. Orders to Quantico for duty in fitting out the dispensary in Antigua, British West Indies. Antigua – to set up a new base, with tents, land crabs and Negroes who spoke with a broad "A."

Old Sailors Never Die - They Just Sail Away
by Faure C. "Pappy" Frazer, CPhM, USN (FR)

Back to stateside and further Marine-ing at Quantico. Then Pearl Harbor Day and the hurried trip to San Diego, establishing the medical department of the first Marine Aircraft Wing. Ready to go to war, but too old and too fat. Then came "pharmacist" and duty as medical personnel officer, Marine Air, Pacific.

Orders again – what had happened? Can I read? "Report to the USNH, Memphis, Tennessee, for duty." My own home town. What is there?

I soon found out. What was a large cotton field is rapidly becoming a large hospital. Our colors are run up and we are commissioned. I find that all hands know me only as "Pappy," and I am the Hospital Corps education and detail officer. My swimming classes and the standing dare to "try and duck" 'Old Pappy.' What is the Navy coming to? WAVES!

Off again to Great Lakes.

My old shipmate, John McCormack, who has been a "Pappy" to me all these years. Glenview, then back South to Pensacola. Returning trains filled with wounded, to be unloaded and received at the hospital. V-E Day, and business as usual.

Orders again, back to my home town, Memphis, and the usual work. V-J Day, and your job slowly folds up under you. A separation center has been set up. Sailors become ex-sailors.

The phone rings in my office. "Mr. Frazer, your orders are here. You are going to be released from active duty – at once. Terminal leave will start immediately. The final parting from my boys and girls.

As one of the old sailors, I will say we leave the Navy to you young folks. But if you ever get it in a mess again, please remember to call on us, and we will leave our wheel chairs at Gray's Ferry Road or the sunny beaches of Florida and California, throw away our canes, and come back to the world's greatest institution – the United States Navy. Just remember: Old Sailors Never Die – They Just Sail Away.

David Calvin Graham
Senior Chief Signalman
United States Navy (Retired)

David C. Graham was born in Atlanta, Georgia, on September 10, 1924, son of David and Lena (Harris) Graham. At an early age his family moved too Lynchburg, Virginia, where he attended both elementary and junior high school before entering the United States Navy on December 13, 1941. He received his recruit training at the United States Naval Training Station in Norfolk, Virginia.

Upon graduation from recruit training, he was assigned to the battleship USS *Idaho* (BB-42), where he served from January 1942 until July 1945. After making the rate of signalman second class in 1944, he was assigned to the Flag Allowance of Commander Battleship Division Three. His second naval assignment was in the USS *Peter H. Burnett* (IX-103), a station supply ship operating in the Philippines. In October 1945, he was honorably discharged from the United States Navy as a signalman second class.

He returned to his hometown, Lynchburg, but civilian life was not to his liking, so he re-enlisted in the Navy in November 1945. After a short stay at the Recruiting Station in Pittsburgh, Pennsylvania, awaiting assignment, Graham was ordered to the Fargo Building in Boston, Massachusetts. There he awaited assignment to the Enlisted Allowance, Commander Destroyer Division 42, embarked in the destroyer USS *Vogelgesang* (DD-862), at Norfolk, Virginia. Graham reported to this duty in April 1946. His outstanding performance as both a signalman and a petty officer convinced his Commodore that he should be advanced to signalman first class. He was so promoted on May 1, 1946.

Graham departed the Staff of ComDesDiv 42 in April 1947 to exchange signal billets with another SMIc at the Signal Tower in Guantanamo Bay, Cuba. After a six-month tour there, he was transferred to the Naval Air Station, Jacksonville, FLorida, for discharge for the second time.

Following re-enlistment in Jacksonville, he was sent to the USS *Canisteo* (AO-99), operating out of Norfolk, Virginia. Graham became a quartermaster first class on April 2, 1948, when the enlisted rating of signalman was incorporated with that of quartermaster. In May 1948 he was ordered to the USS *Amphion* (AR-13). The following month Graham went aboard the USS LCI-545 at Norfolk.

His first tour of shore duty was in the First Naval District in Boston, Massachusetts, in September 1948. The first year of his two-year tour was spent as a stationkeeper aboard the Reserve Training Ship USS LCI-1093.

His second year was spent in the Signal Tower at the United States Naval Base, Newport, Rhode Island.

In October 1949, Graham was discharged for the third time and was re-enlisted for the third time, this cruise being for six years. During the month of September 1950, he received orders to the USS *Siboney* (CVE-112) which was to be recommissioned at the Naval Shipyard, Philadelphia, Pennsylvania, in November 1950. He remained in *Siboney* until February 1951, at which time he joined the Flag Allowance, Commander Carrier Division 18. He donned the hard hat of a chief petty officer on June 16, 1952. It was during this sea duty tour that Graham officially recommended to the Secretary of the Navy that the rating of signalman be restored in the United States Navy.

His next tour of duty came over four years later when he was ordered detached from ComCarDiv-18 Flag Allowance, to report to Naval Schools Command, Norfolk, Virginia, for duty at the Class C-1 Instructor's School. After serving in the capacity as an enlisted instructor from June 1955 until July 1958, he was ordered to the Enlisted Allowance of Commander Destroyer Squadron 22, at that time embarked in the new Sherman Class Destroyer, USS *Du Pont* (DD-941). Meanwhile, Graham had re-enlisted for the fourth time in July 1955 to serve for another six years. He departed from his assignment with ComDesRon-22 in November 1959, and joined the Flag Allowance of Commander Destroyer *Flotilla Four*, who at the time was embarked in the USS *Sierra* (AD-18) at Norfolk, Virginia. After eight-and-one-half years of waiting, Graham was cited by the Secretary of the Navy for having been primarily responsible for the restoration of the rating of signalman in November 1956.

Graham was discharged from the naval service in May 1961 while he was serving with ComDesFlot Four, and re-enlisted for another six years the same month. In January 1963 he was ordered detached and because of his previous experience as an instructor, he was sent to the Instructor's School at San Diego, California, for a normal tour of shore duty. While at the school, Graham taught in the areas of career counseling, leadership and instructor training.

In March 1966, he reported aboard the USS *Providence* (CLG-6) home port at San Diego. His tour in *Providence* was brief as he was ordered to the USS *Oklahoma City* (CLG-5) in November of the same year. Graham's enlistment expired in 1967 and in February of that year, he re-enlisted aboard *Oklahoma City* for another six years. May 1967 saw Graham packing his bags again and this time it was for duty with the Enlisted Allowance, Commander Amphibious Squadron Three, embarked in the USS *George Clymer* (APA-27) at San Diego, California.

May 1969, Graham was ordered detached from ComPhibRon Three, as squadron signalman, and reported to the commanding officer, Fleet Training Center, San Diego, for duty as an instructor at the Fleet Signal School. This was Graham's last active duty assignment as he retired from this duty station with 30 years of active duty.

Graham is a recipient of the Secretary of the Navy's Achievement Medal with the Combat "V" Meritorious Unit Citation, Good Conduct Medal, (Sixth Award), American Theater Medal, Asiatic-Pacific Theater Medal with seven stars, World War II Victory Medal, Navy Occupation Medal with the European Clasp, National Defense Medal with star, Vietnam Service Medal, Vietnamese Campaign Medal and the Philippine Liberation Ribbon.

Senior Chief Signalman Graham (having been advanced to this pay grade on Dec. 16, 1961, while serving with ComDesFlot Four), has been cited for outstanding performance of duty numerous times during his lengthy span of 30 years of naval service. His record reflects this information in great detail.

Graham resides in San Diego, California, with his wife, Margaret. Their six children, all grown, some of them married, are living away from the home. He was placed on the retired list to date from May 1, 1971.

Graham is founder and chairman of both the American Battleship Association and the USS *Idaho* (BB-42) Association. His hobbies are stamps, old coins, boxing, and personalized auto plate inscription collecting. He wrote the book, *Those Crazy License Plates*, Price/Stern/Sloan/Publishers, Los Angeles.

Keeton Brothers Make Tar History

When the results of this year's Senior and Master Chief Petty Officer Selection Board were announced in NAVOP 041/85, a dream came true for two brothers. Aviation Ordnance Master Chief David R. Keeton, USNR-R(TAR), called his brother Yeoman Master Chief Robert Keeton, USNR-R (TAR), who was conducting an administrative inspection at Naval Reserve Center, South Charleston, West Virginia, to inform him of his selection for master chief.

Both brothers are in the Training and Administration of Reserves (TAR) Program. David's selection to master chief is a first for the TARs – having two brothers, both E-9, on active duty at the same time.

Robert is the command master chief for the Commander, Readiness Command Region Six, headquartered at the historical Washington Navy Yard, Washington, D.C. Robert has performed tours aboard USS *Constellation* (CV-64) and USS *Forrestal* (CV-59). As a TAR, he has had tours at Naval Air Reserve Unit Memphis, Tennessee; Naval Air Station Glenview,

Illinois; Recruit Training Department, Memphis, Tennessee; Chief of Naval Reserve, New Orleans, and was the first director of the Naval Reserve Management School, New Orleans.

David serves as the maintenance coordinator for Commander, Fleet Logistics Support Wing in New Orleans, Louisiana. He is responsible for the overall maintenance supervision of 11 Fleet Logistics Support Squadrons (VR) that transport military personnel and cargo world-wide, two Fleet Composite Squadrons (VC), which act as adversary support for our Active and Reserve squadrons, and one detachment that is responsible for airlift requirements of the Secretary of the Navy and the chief of Naval Operations.

Robert encouraged and recruited his younger brother David into the Navy in 1971 after David completed four years with the United States Air Force, helping to strengthen the family's Navy tradition.

A third, and the youngest brother, James T. Keeton, a park manager for the Tennessee State Parks Division, also was recruited into the Navy by Robert under the 2x6 program. James served aboard Robert's former duty station, the USS *Forrestal* (CV-59) during a tour supporting operations in Vietnam. They are the sons of Mr. and Mrs. Robert F. Keeton of Decaturville, Tennessee.

King Neptune – The Patriotic Pig

King Neptune is often thought of as the ancient god of the sea, identified with the Greek God Poseidon, but this testimonial is about people pulling together for a common effort.

"People were intensely fixed with patriotism – win the war, great days then."

"This stirring statement described the fervor of excitement created by CPO Don Lingle and King Neptune everywhere they went in Illinois." *(George E. Parks, Union County Illinois History, 1986-87, Vol. 1-3, pp. 1540.)* Don Lingle was his name, a union country boy, friend of many people who grew up in the neighborhood of the campground. His family was well known and were respected citizens. When World War I was on, Don was enrolled in the Student Army Training Corps at Milliken University, Decatur, Illinois. Perhaps this was just long enough to whet his military desires. When World War I came on, Don grew and achieved a business success in association with Charles A. Dean and Company of Anna where he learned the business of meeting people and selling his wares. He married early to a neighborhood girl from Saratoga and had one daughter, Betty.

As World War II arrived, Don enlisted in the Navy early, hoping for sea duty and to see the world. The Navy looked at his resume and saw the bit about his years of salesmanship so they decided he could best serve the Navy by selling the Navy to young men of Southern Illinois. He was stationed at Marion, Illinois.

One day, at West Frankfort, Don was watching a crowd at a sale whereby the school boys and girls were selling their victory pigs, the idea being that the school children raised their pigs and sold the hog later to buy United States War Bonds. Standing in the crowd with local men, Don expressed the desire to own that pig and said he could sell him for one million dollars. The men promptly bought the pig and presented it to Don and he promptly named the 265 pound porker King Neptune. Four years later this pig had made his name known to the middle west of the United States. Don Lingle had sold $19,000,000 in War Bonds by the system of auctioning the pig off. The proceeds of the auction would then be used to buy War Bonds. The pig, in turn, was given back to the committee to be auctioned off again for the same purpose. In fact, this King Neptune became the most expensive piece of pork known in history. He was also toasted on Navy ships, given a Navy citation, and acclaimed in many huge newspaper spreads announcing the War Bond rallies within different communities.

Both Lingle and King Neptune became famous and, at War Bond rallies, this pig was auctioned off again and again. The people at Herrin and Colonel L. Oard Sitter, local popular auctioneer, conceived the idea of selling the pig as whole or piecemeal. There, the side brought $175, his feet $250, and backbone $5,000, his ears brought $3,000 – even his grunt sold for $75. The top sale was at Anna for $1,0000,000. The resultant success of this sale was communicated by the gift of a treasury award to a local druggist who happened to be Bond Committee Chairman for that period of the war. The Treasury Department Silver Medallion is yet around, although it has been carried by the recipient for so many years in his pants pocket that it is now worn smooth, and the fine figures and print upon the silver piece have long since been erased.

As the war came to a close, the sales were drawing enormous crowds and the activity was great. King Neptune lived for these occasions and, for each auction, he was shampooed and groomed until he glistened. The King mounted the auction block in regal fashion, wearing a royal Navy blue robe and a golden crown fashioned by a company of West Frankfort sorority women. He was adorned with silver earrings, a tribute from one of his loyal subjects, Norman Wahl, an Anna Jeweler (now deceased). When the war was over, Don brought his pig back to Anna, where it was given royal attention at the form of the Ralph Goddard Seed Company of Anna, and the Corno Manufacturing Company had given a lifetime supply of oats for his sustenance. The pig lived out five more years of his life before succumbing to pneumonia. He was buried at the site of Route 146, East of Anna, and there was an adequate bronze monument erected by Don Lingle and friends. The monument became a roadside park and is existing as such at the present date.

Enlisted Prize Essay, 1948

The Post-war Chief Petty Officer:
A Closer Look
by Chief Machinist's Mate Richard M. McKenna, USN

"The 'eathen in 'is blindness must end where 'e began"
(Source: US Naval Institute Proceedings, 1948)

The chief was not happy. With both an elusive feeling of shame and a real and immediate feeling of distress, he was being driven to acknowledge a fact which he found to be as unpleasant as it was inescapable. Service in the post-war Navy was not the same as before. It was not very satisfying. And – admit it, old-timer – the only real reason he had for hanging on was to protect his investment of 17 year's service.

The prospective retirement benefits, while they had exerted considerable weight, had not been the primary factor in shaping his decision to make a career in the United States Navy. For many years he had been pleasantly conscious of being well paid and provided for in return for following a mode of life which he had found to be attractive in itself. He had frequently advised uncertain youngsters not to choose a career in the Naval Service solely on the basis of the remuneration involved; he had told them that, if they did not have an affinity for the life, no reasonable amount of money could make it worth their while. That advice had been given sincerely and he believed it now more strongly than ever. It was a shock to find himself in much the same position as that against which he had advised. If he were 10 years younger, he thought, he would call his service a bad investment and go on out; and having his share of human frailty, he would probably carry with him a sense of injury and become one more nameless citizen with an emotional bias against the "military." Certainly a disquieting number of the younger men were doing so.

The old chief did not feel that he could afford that gesture; nor, indeed, did he wish to. No man who had spent his most productive years in the United States Navy could ever willingly condemn it in general terms. Sentiment aside, he would seem too great a fool. The chief sourly regarded the three year period still separating him from his retirement date, and those three years shaped up as a sort of endurance contest. He felt depressed and somehow at fault. It was a sad prospect.

Consider this chief in his collective thousands. Are his troubles inconsequential – or do they exist at all? Can their basic nature be established? What is their relation to the larger picture of the naval service generally? Finally, can practical measures be devised to minimize them?

Such an inquiry must be conducted warily. The field is that of human relations and behavior with its inherent pitfalls of subjective prejudice, over-simplification, and too easy generalization. Standard procedure would require a survey of conditions and attitudes to be made over a wide enough range and over a long enough period of time to avoid being misled by the local and temporary. All contributory factors should be considered and should be assigned their proper values. This paper can claim no such authority. Preliminary research was confined to general conversation over the coffee cups in a half-dozen CPO messes during the past two years (1947-1948). Where unqualified statements are made and where positive conclusions are seemingly jumped at, it should be remembered that this report is hardly more than an expanded case history; that all conclusions are extremely tentative; and that suggested remedies are scarcely more than hopes.

In view of these qualifications restricting the general validity of this investigation, it will be well at the outset to fix upon which, if any, of the various annoyances involved are real beyond question. A diligent clearing away of the trivial and of the illusory reveals two vexations which stubbornly refuse to vanish. These may be defined as (1) the subjection to an apparently endless series of transfers, and (2) the difficulty of obtaining a duty assignment commensurate with training and ability.

The primary cause of these afflictions is obvious: we are engaged in operating four ships with three crews, so to speak. In the present state of the planet no Navy man will question the wisdom of that policy. It is a fact. The chief petty officer is especially subject to the consequent frantic shuttling about because, in many ratings, he is vastly in excess of allowance. Here at last is the crux of the matter. Here, too, is a fact.

Facts cannot be shrugged off. The enlisted structure heads up in the chief petty officer whose function, within that structure, is traditionally and correctly regarded as being of critical importance. It is reasonable to infer that whatever contributes to loss of moral effectiveness among the chief petty officers will have secondary ill effects extending through the entire enlisted body. It is the central purpose of this inquiry to determine whether or not the facts established above have brought about a condition where such an adverse influence is demonstrably at work.

It is advisable to take up separately the logical consequences of the numerical disproportion of chief petty officers. One of these is the overcrowding of available living space, which occasionally produces a congestion surpassing that prevailing in time of war. This consequence, being physical and of minor importance, may be dismissed. Of greater concern is the insecurity of duty which now obtains in the service. Too often a chief is not given time to develop a sense of personal identity with his ship or with his station in that ship before being ordered to another. He may be granted ample scope to exercise his technical proficiency, but he has small chance to exert that moral guidance by precept and example which he knows to be the real reason for the preferential status attached to his rating. The exercise of leadership, in the best sense of that word, demands sustained effort extending over an appreciable period of time. On the short term it is as futile as is its counterpart in some civilian activities. Official recognition has long been accorded to the desirability of relatively long tours of duty. In pre-war days a man was required to complete a year of service in his ship before he was permitted to request a transfer. Such requests, moreover, were frequently disapproved by the commanding officer "in the interests of permanency of personnel aboard this vessel." The disregarding of those interests tends to create, among the men who are subject to continual short terms of duty, an attitude comparable to that of passengers on a transport. Any ship seems to be just another Chaumont bound for, but never arriving on, a China Station that is lost somewhere over the rainbow.

Of final and most pointed concern is the differentiation of status among chief petty officers which becomes unavoidable when they are in considerable excess of allowance aboard a given ship. Where five chiefs are assigned to a single station, one of them must be in charge. Whether he is selected on the basis of seniority in service, of seniority on the station, of professional competence, or for any other reason, the result is necessarily unsatisfactory. The man who has been given charge has the duty of impressing his will on the junior chiefs to an extent that neither he nor they can relish. These latter, relegated to duties more properly pertaining to petty officers first or second class, become victims of something resembling frustration which may make itself evident in any of several ways. Where one may accept his lot cheerfully and make the best of it, another may suffer a general loss of interest. Where one may constitute himself the champion of the other enlisted men and lead them in a chorus of complaints and destructive criticism, another may relieve his feelings by the converse course of pushing around the non-rated men. In most cases these junior chiefs' attitudes will consist of a fluid and unpredictable mixture of those mentioned. The human spirit will not yield a precise analysis.

The over-all result, however, is found to be anything but indeterminate. Influences radiating from this group of men, that is, the chief petty officers who are by force of circumstance excluded from the duties and responsibilities suited to their training, tend to lower the standards of all chief petty officers. This tendency, accentuated by the transient nature of duty now current, is reflected in a wider tendency toward increased discontent and slack discipline noticeable in all levels of the enlisted order.

The corps spirit of the chief petty officers is waning. The once well defined boundary between the CPO and the other enlisted men is becoming less definite. The outer semblance, in the form of a distinctive uniform and separate living and messing arrangements, is unchanged; the inner substance, a particular way of living and thinking, is wearing perilously thin. What is being lost is an intangible something compounded of attitudes and relationships and conditioned responses, something probably ultimately undefinable in exact terms. It is no less real, for all that. Its inherent sources are an inward awareness of being a necessary and important part of a larger whole and an outward prestige that is the general acknowledgment of that necessity and importance. Neither of these attributes can exist without the other. There is a difference of spirit between a chief whose position affords him full realization of what has been called "the instinct of workmanship," and a chief who knows that he is "spare gear." Any seaman apprentice can sense that difference. To say that pay and privilege is all-important in respect to maintenance of morale is to show a lack of understanding of the ideals of military organization. It must be emphasized that in these circumstances the chief petty officer loses, in the same measure, both the will and the ability to perform his traditional function.

Visible evidence of deteriorating principle is plentiful. Bearing and behavior are not what they were. Chiefs show up poorly turned out for personnel inspection. Chiefs are summoned to Captain's Mast and muster with the restricted men. Chiefs sit in their quarters while their work suffers until, to the embarrassment of all concerned, they must be driven to their stations.

These statements do not refer to any separable group but are true to some extent of all chief petty officers. The casual conditions, although at a given time they may apply with varying force to different individuals, nevertheless affect them all. The constant pressure of these conditions has led to a competitive striving for advantage that brings out some of the least admirable qualities of the men engaged in it. This struggle for preferment has brought an addition to the United States Navy's unofficial vocabulary, the word "wheel," which in its connotations is an ugly word.

Weakness in any part may logically be expected to make itself felt through the entire organism. On taking a more general view, one of the first points to claim attention is the comparatively large proportion, in many ship's companies, of youngsters with less than two years of service. These men are plastic material for the impression of whatever mood and spirit is prevalent. Too often the mood is one of fretful discontent and the spirit is one of adolescent rebellion. Not infrequently a situation is encountered reminiscent of Kipling's:

"We was rotten 'fore we started – we was never disciplined;
We made it out a favour if an order was obeyed.
Yes, every little drummer 'ad 'is rights an' wrongs to mind,
So we had to pay for teachin'."

The Theme Needs No Elaboration

A competent chief petty officer, confronted with such an attitude among his men, should have little difficulty in correcting it. Primarily, he must impress them with his own devotion to the ideals of which they are indifferently aware and which they hold in low esteem. Secondly, he must win their respect for himself which will tend to include respect for his attitude. These objectives must be achieved by dealing with the men on an individual basis, exercising strict impartiality, constraint of praise and blame, maintenance of a just and consistent control, and all the other ethical principles known in their sum as leadership. Space does not permit a detailed account of how a small group of men may in this way be brought gradually to a mood of quiet satisfaction in their work and to a spirit of loyalty to the ideals of the service. The story is well known and is as old as military life. Equally well known is the fact that in general only the chief petty officer or his equivalent is concerned with small enough groups of men to be enabled to exert the moral force necessary to accomplish it. The acknowledgment of this fact is implicit in his traditional appellation, "backbone of the Navy."

Note well that the chief petty officer must feel honestly the devotion which it is his duty to instill in others, and that he must gain the respect of those others for himself if his example is not to be wasted and the process to be a failure. These requirements are most easily met when the CPO, considered symbolically and collectively, maintains his prescribed status and material condition above reproach. That this is not the case, due to the present excess of CPO ratings, has been established. The individual chief, under this handicap, frequently encounters difficulties that are insuperable. The fallen estate of the chief petty officer is indeed a part of the larger picture and it is a prior part.

Here perhaps is a reason for the conceded failure to achieve satisfactory indoctrination of the men on two year enlistments. If so, it augurs ill for the success of any scheme of universal military training in which the United States Navy may have a part. It is easy to understand that a promising petty officer first class may choose to leave the Navy because he feels that there is no hope of his making chief for many years to come. It is not so easy, but possibly more important, to understand that he may not feel sufficient attraction toward making chief to make it worth the waiting for.

To sum up: the disproportional number of chief petty officers has an adverse effect on their morale and on their prestige; this effect is, at least in the negative sense, a causal element in the unwholesome attitude toward military life displayed by a disturbing number of the first cruise men. A problem is posed.

Possible solutions suggest themselves at once. The simplest is to do nothing. In time, as the older men continue to transfer to the Fleet Reserve and the younger men continue to leave the service, the number of chief petty officers will dwindle until it comes into the correct ratio to the total enlisted strength. This solution, perhaps because of its obvious appeal, makes two unwarranted assumptions. The first is that there is no qualitative difference between the younger chiefs who leave the service and those who elect to "stick it out." A little reflection will make it clear that the men who are not content to continue indefinitely at work demanding less than all of their ability and experience are the very ones whose retention in the service is most desirable. The second assumption is that the sorry condition remarked on earlier, which may well intensify before real recovery begins, is not a matter of serious concern. This belief is at least questionable.

Why not sever the Gordian knot with a selective elimination designed to bring all ratings within allowances and at the same time to retain the most desirable men? To this proposal both legal and moral objections may be made of such cogency that it must be considered impractical.

Perhaps there is no perfect solution to this problem. Nonetheless, there may be ways to minimize its clearly undesirable effects. It is essentially a problem of finding suitable employment for a reasonably small number of men. Congressional criticism has been heard recently in reference to the rising ratio of civilian employees of the Armed Forces to the men in uniform. Are there not instances where such civilian employees might be replaced by these excess and unwanted chief petty officers?

At first glance it would seem that such a program embodies the ancient fallacy of robbing Peter to pay Paul. The lesser evil of impermanence of duty springs from the general lack of sufficient personnel to man all the ships in service. While the greater evil of the invidious distinction among CPOs will be abated by employing some of the surplus ratings as proposed, there is the corresponding certainty of intensifying the lesser evil. However, the favorable effect will be measured by the ratio of the number of CPOs employed to the total number of excess CPOs; the correlative unfavorable effect will be measured by the ratio of the number of CPOs so employed to the total number of petty officers in the ratings in which CPOs are in excess of allowance. It is readily seen that the first ratio must be considerably the greater of the two. Hence there must be an initial favorable effect to be derived from such a course. Whether it would obtain until all surplus CPOs were placed, or whether it would increase to a maximum at a prior point, must be determined largely by trial and error.

Turning to possible conflict with established policy, it must be admitted that the suggested disposition of personnel might encroach on spheres set up by long usage if pursued in an indiscriminate way. Much of this danger would be avoided if the area of employment were restricted to overseas bases and occupational groups where political considerations would not figure greatly. If in addition it were presented as a strictly temporary expedient, designed to cope with a situation of limited duration, no really significant opposition should be aroused. As the attrition of time made a place for them, the men involved could be re-absorbed in the Fleets and the entire episode could be written off as post-war re-adjustment.

Precedent is not lacking. At one time the administrative structure of Guam was staffed largely with naval personnel. There was even an experimental farm managed by a warrant machinist assisted by a chief boatswain's mate. Although this and other less extreme instances may

have seemed incongruous to a naval traditionalist, it is not on record that the personnel assigned to such extra-military activities failed in the discharge of their duties. There is not reason to suppose that United States Navy men are any less versatile at present.

It would be no great task to institute parallel surveys, one to determine where the surplus CPO's might be usefully employed, and the other to discover such special skills and abilities as exist among these men which might fit them for such employment. Whether or not the outcome provided adequate relief, the knowledge that such surveys were being made, or that active consideration of any kind was being given to this problem, would administer a much needed tonic to the presently ailing spirits of the chief petty officer.

Harold F. Mull

The Bill of Sale from the initial uniform purchase when Chief Harold F. Mull was promoted to CPO is a postscript to this biographical sketch he provided. He explains that three distinctive uniform styles were then authorized, in addition to the traditional white, khaki and dress blue. They were working dungarees, aviation green and gray. The gray was believed to be unusual and short-lived mostly worn during the early to late years of World War II. One size cap device was worn on the overseas cap and a larger device on the "Bill" style cap that has since evolved to the chiefs hat. Chief Mull further states he believes a book as written concerning the Gray uniform but that cannot be validated by this writer. Chief Mull is "giving back" to the Navy he loves by spending several hours of volunteer service as a docent at the Navy Memorial in Washington, D.C. Chief Mull, indeed, continues to serve as a role model to all whose presence graces the hallowed halls of countless memories this grand edifice represents.

C.C. Baker Co.

Mens Wear

403 Granby Street – Phone 25488

Norfolk, March 2, 1944

Sold to Chief H.F. Mull
Bus Address: HedRon t-1, F.A.W.
Home Address: Elizabeth City Detachment, Elizabeth City, North Carolina.

1.	1 Grey Insignia	1.00
2.	1 Blue Insignia	2.50
3.	1 Green Insignia	2.50
4.	3 Shirts @ 3.50	10.50
5.	2 White Shirts @ 2.25	4.50
6.	3 Pr. Sox	1.10
7.	1 Cap Device	3.00
8.	1 Cap Device	1.00
9.	1 Cap Device	2.00
10.	1 Blue Uniform	45.00
11.	1 Rate	4.50
12.	Hash Mark	.10
13.	Buttons	1.00
14.	1 Top Coat	50.00
15.	1 Suit Greens	55.00
16.	Rate	.75
17.	Hash Mark	.10
18.	1 Grey Suit	15.38
19.	Rate	.75
20.	Hash Mark	.10
21.	Buttons	1.00
22.	2 Caps @ 4.25	8.50
23.	1 Blue Cover	2.50
24.	1 Grey Cover	1.25
25.	2 Fin. @ 1.00	2.00
26.	1 Belt	1.50
27.	1 Belt	.55
		223.33

The Memoirs of John A. Nelson, CM

Served Aboard, March 1942 – June 1945

(Source: "Memoirs" of the Crew of the Battleship USS *Maryland* (BB46), Fred R. Vreeken, Editor

I came aboard the battleship USS *Maryland* during March of 1942, as a hospital apprentice 2nd class, assigned to the "H" Division, and left the Maryland

in June of 1945 as chief pharmacist's mate. With the exception of Pearl Harbor, I've participated in every action involving the *Maryland* during her illustrious World War II tour of duty. The experience I wish to share, has to do with the perplexing mystery of fate. It focuses on a series of incidents for which there is no explanation as to why they happened as they did?

For a period of two years, my battle station had been the area occupied by the Ship Repair Shop. Then just after getting underway for the invasion of the Philippines, my battle station was changed to the boat deck. I objected to this change, but to no avail. Up I went topside, to an entirely different kind of environment. My specific battle station was in the tailor shop.

As those who were there know so well, during the daylight hours we were under constant threat of Kamikaze attacks. The last attack of the day, usually always occurred at sunset. (A definite advantage to a diving enemy aircraft bent on total destruction, because they become more difficult to see at that particular time of day.) Once this attack had subsided we would be secured from our battle stations, and allowed to return to our duty area. For me, this was Sick Bay.

On the evening of November 29, 1944, we were secured from our battle stations, and I returned to Sick Bay to prepare for the evening sick call. Preparations were just underway, when General Quarters was sounded. My immediate reaction was to ignore it, stay in Sick Bay, and continue to prepare for sick call. I reasoned that I would no sooner arrive at my battle station on the boat deck, and surely the all clear would be sounded. If I remained in Sick Bay, I could get a running start on sick call patients. Because of the constant threat of kamikaze attacks, coupled with our daily firing assignments during the invasion, fatigue and stress had substantially increased the numbers of the crew requiring medical attention. So if I stayed in Sick Bay, I would be ready for them immediately after the "All Clear!" Hopefully, sick call would then be over in time for me to get caught up on some of my personal needs. However, an inner voice spoke to me, which urged me to drop everything and get to my battle station.

I no sooner arrived topside, when I saw the Japanese kamikaze diving on the *Maryland*. It's line of flight made it appear as if it were going to crash on the boat deck! I knew all along I never wanted to be there yet there I was! I just stood there and watched. The rest is history. The forward gun crews shot off the diving kamikaze's left wing, causing it to crash at the forward part of the ship. The armor piercing bomb it carried, found it's way between 16-inch gun turrets one and two, crashing through

to the armor deck, (three decks below), where it exploded. The explosive impact traveled forward, aft, and upwards. It killed most everyone in the Sick Bay area, which I had just vacated. Everyone in the Ship Repair Shop area my former battle station, were also killed! Fate was on my side. If I had been in either place, chances are I would not be writing these memoirs!

USS *Maryland*

April 18, 1945

CONFIDENTIAL

From: Commanding Officer.

To: Commander in Chief, United States Pacific Fleet.

Subject: NELSON, John Alfred, 620-24-02 PhM1/c
           V-6, USNR-Request authority for advancement
           in rating for meritorious conduct in case of:

Reference: (a) AlNav 163-41

1. In accordance with reference (a), authority is requested to advance the subject man to the rating of chief pharmacist's mate (Acting Appointment) for meritorious conduct as set forth in the following paragraphs:

(a) On the night of 7 April 1945, this vessel was struck by a Japanese plane and it's bomb off Okinawa. Nelson immediately ran to the scene of the damage. The area was littered with debris and casualties. Great flames were leaping form the scene and 20mm ammunition was constantly exploding among the wounded from the heat of the fires. Nelson, without regard to his own personal safety, coolly dashed among the flames and the exploding 20mm ammunition and rescued casualties, which otherwise undoubtedly would not have lived. His gallant action in the face of great danger is in keeping with the highest tradition of United States Naval service.

(b) Amid the tumult and excitement, Nelson immediately organized the corpsmen with him, and personally supervised the control of hemorrhage, the administration of morphine, the application of splints to fractures and organized stretcher bearers for evacuation of the casualties to the main deck, collecting station. His cool courageous leadership was greatly responsible for the prompt and adequate first aid treatment the casualties received at the scene of action. Following the removal of casualties from the damaged scene to the collection station, Nelson brought his hospital corpsmen to the main deck collecting station and supervised their activities with further treatment. In the meantime, he personally administered plasma infusions and applied some of the more difficult dressings about the face and eyes. In moving patients from the collecting station to the sick bay, NELSON again demonstrated his usual ability in

The USS Maryland *(BB-46)*

taking charge of the corpsmen and getting the job done quickly and carefully. In the sick bay, he gave blood transfusions and assisted with the re-dressing and operations. He worked cheerfully and untiringly throughout the emergency.

(c) Nelson was recommended for the Navy and Marine Corps Medal for heroic conduct as hospital corpsmen aboard a United States Battleship during and following the action when a Japanese Plane and bomb struck the ship on 29 November 1944. He immediately ran forward to a burning area, administering treatment and evacuating seriously wounded personnel. Nelson then went below, into a burning and smoke filled compartment, treating and rescuing additional wounded. A third time he had himself lowered through a damaged hatch into a burning and smoke-filled compartment, where he treated and helped supervise evacuation of the wounded. With untiring effort and without gas mask, and without regard to his own safety, he undoubtedly helped save lives that might otherwise have been lost.

(d) After the sick bay was demolished on 20 November 1944, a temporary sick bay was established in the junior officer and warrant officer mess. Nelson greatly assisted in establishing this temporary sick bay and also in setting up a clerical office. His assistance was invaluable in completing various clerical forms, casualty reports and official correspondence. Nelson greatly assisted in the surveying of all damaged and missing items of supplies and equipment. His excellent work in requisitioning supplies and equipment for the new sick bay and assistance in outfitting and getting the new sick bay in operation were invaluable.

(2) From the time of the explosion on April 7, 1945, until the transfer of the injured and dead, Nelson worked without rest or sleep, assisting the medical officers, administering himself or directing other corpsmen. It is believed that his recommendation for advancement to the rating of chief Pharmacist's Mate (Acting Appointment) is a fitting reward in recognition of his capability and efficiency and the cheerfulness and willingness with which he successfully completed every arduous tasks, and is well within the purview of reference (a).

(3) Nelson enlisted in the United States Navy in January 1942 and has been on continuous active duty since that date. He was advanced to the rating of pharmacist's mate first class on 1 April, 1944, and completed a Navy Training Course for chief pharmacist's mate on 9 September 1944 with a final mark of 3.8.

ROBERT B. GOLDMAN

An Old Salt's Old Salt

Millie Tamberg

Problems?

Everyone has their share, but most people don't try to take on someone else's – except for Thomas C. Oneyear.

Why?

Because that's his job.

As command master chief for Assault craft Unit (ACU) Two at Little Creek Naval Amphibious Base in Norfolk, Virginia, Oneyear helps solve day-to-day problems for the more than 300 men and women assigned to the unit.

"I'm the waterfront supervisor for all the craftmasters and coxswains (men in charge of small craft)," says the boatswain's mate.

Although Oneyear is assigned to shore duty, he works amid a fleet of mechanized landing craft (LCMs or Mike-eights) and utility landing craft (LCUs). The 74-foot-long Mike-eights and the 134-foot-long LCUs support the unit's dual mission of transporting troops and equipment to and from amphibious assaults and training boat crews.

Because these crews operate at sea and on the beach, they are known as the amphibious Navy and dubbed "Gators."

"Gators are a unique part of the Navy," says Oneyear. "We're not fancy

The USS Iowa *(BB-61)*

cruiser or battleship sailors – like I used to be – but we get an important job done. If you look back in history you'll read about the operations they've done – D-Day and Iwo Jima, for example – and you'll find that the unsung heroes were the boat coxswains."

Those assigned to the Norfolk-based unit who deploy to the Atlantic, Mediterranean and Caribbean coasts to conduct periodic amphibious exercises often return to Norfolk with their own branch of problems for their command master chief.

"Sometimes when the crews come back from a deployment they come to me because they're not being paid correctly for their living quarters or food. I also deal with the average problems that other command master chiefs do, and those are marital problems and money management problems," he explains. "I deal with everything a sailor can get into while he's a young person maturing.

"I'm also the command master chief – the senior enlisted advisor and direct liaison between the commanding officer and all the enlisted personnel at this command. In other words, I take care of all their problems."

What are the prerequisites for the job?

There was no formal schooling, according to Oneyear, just the combination of assignments and working with people throughout his Navy career.

"I've been to the 'school of hard knocks' – 25 years in the Navy, 12 ships, two tours in Vietnam, and two tours of 'pushing boots,'" says the former recruit company commander in a raspy baritone voice.

"I've commissioned the USS *Dubuque* (an amphibious transport ship), and recommissioned the battleship *Iowa*," boasts the Dubuque, Iowa, native. "I've recommissioned a few too."

Oneyear's well-rounded career is evidenced by three different breast insignia indicating special qualifications, and a colorful array of ribbons – some awarded two and three times. And nautical tattoos on the fingers of both hands, wrists and ear lobes, not to mention the gold inlayed crossed anchors in two front teeth – each holding a sea story of its own – are outdone only be those his uniform covers.

Besides two tours in Vietnam, the 42-year-old sea service veteran has made two around-the-world cruises, nine to the Caribbean, eight to the Western Pacific, seven to the Mediterranean and five to the North Atlantic.

Because Oneyear is a husband and father of two, he appreciates the types of problems Navy men and women with families face, and takes a different approach with their problems.

"We have an Ombudsman (wife of a unit member who assists other wives with their problems) for our command who is a very young lady, but if I have someone who is a little older and needs to talk to someone with a little more experience, then I'll let them talk with 'ma' – my wife Cheryl," he says. "It's just a nice touch, and it does work. The job can get nerve-racking at times, but the satisfaction comes with knowing that people are being taken care of."

William G. Ray

Chief William G. Ray enlisted in the United States Navy, September 23, 1939, in Macon, Georgia, for a six-year term of enlistment. Roy was processed in the sub-recruiting station, Jacksonville, Florida. Because of a delay in getting his birth certificate, the term of enlistment had changed from four years to a six-year enlistment. The war in Europe was underway, with Poland conquered and the low countries taken by Hitler, war clouds loomed on the horizon.

Norfolk, Virginia, was the East Coast boot training for sailors and this is where he went. After boot camp came by bus home to Florida, in his ill fitting uniforms (they were not tailored in those days), Roy went aboard the USS *Arkansas*, BB33.

The battleships: *Texas, New York, Arkansas* and *Wyoming* made up Battleship Division 5. The *Wyoming*, of course, had been reduced in guns by the 1926, disarmament.

The *Arkansas* made midshipmen cruises to Panama and South America, The Caribbean Islands, Puerto, Cuba, Virgin Islands; and East Cost ports of Boston, New York and Norfolk, which was home port. We made an amphibious landing in the islands, where both the First Army Division and the First Marine Division made an assault. I was a boat engineer on a 50-foot launch with a cargo net over the bow. You may recall the First Army Division and the First Marine Division, both were later in the South Pacific and doing the amphibious landings there.

I served aboard the *Arkansas* as a fireman in the fire room in the engine room as a messenger, in the machine shop and lastly in the boat crew. This was the most exciting of all. I was in the 26-foot lifeboat as the engineer when a sailor fell overboard and we were launched in the Atlantic to recover him. The other boat got the sailor and we were to recover the Franklin life rings.

The flares went off each time as they were brought into the life boat. For recovery the captain had lost time and was in a hurry and had no patience to get our painter line on and recovered before there was screaming at the officer of the deck to get underway and the boat was beat up fairly good. One time we hit a coral reef in Cuba and bent the shaft and busted the propeller. Another time in New York, with a big liberty party we hit the dock (Camel) and busted the keel and were taking on water

The USS Arkansas *(BB-33)*

back to the ship. We requested to be hoisted aboard and the officer of the deck did not understand we were sinking since we had power to come alongside. We lost one boat in the Atlantic repairing a target float during 12-inch gunnery practice. We saved the crew.

I received several letters of commendation both as a landing party member, when I served as a machine gunner with a Lewis Machine Gun. Nobody knew anything about the Lewis gun but I had served in the Florida National Guard and qualified as an expert machine gunner on the 1918, 39 Cal. water cooled gun during three summer camps. Also, with the 45 cal. pistol and hand grenades. My division officer looked good with his landing party so we got letters. In the machine shop we worked 70 hours straight on a fresh water feed pump, that provided water to the boilers, while out in the Atlantic. The battleship would stop without that operating.

It was while serving on board the *Arkansas* that I fist saw the prestige and respect that a chief received from officers and other enlisted alike. The boatswain's mates and gunners mates, they were the leaders that stood out and their world was like God. No one dared to even think to speak let alone talk back to them. It was a little sad though to learn that two of our first class petty officers had been chief's and had been busted, supposedly for getting drunk and missing ship. One chief had been busted two times and he is the one who would hang around the passageway outside the chief's quarter; and get a cup of coffee from some kind chief who knew him.

The chief master at arms, Chief Sprayberry, put the fear of God in all when he came down the starboard side deck and yelled at sailors to get out of the way…Get out of sight…the captain is coming…and the old man neither looked, spoke or certainly smiled and acknowledged anyone. It took Sprayberry over 12 years in the Regular Navy to make chief and everybody was pleased, and surprised that he had made it. Of course the big war was on the horizon and promotions were starting to move. I never dreamed that I would make chief. That was something that I never contemplated, because it was "way out there, almost beyond reach." There were many first class sailors who had 12 to 17 years in the service at that time.

I also served as the official ships photographer for the USS *Arkansas*, because I had worked for a commercial photographer while in high school. The ships store concession paid extra money and I could use it. We were paid two times a month. The 1st and 15th and until I got aboard the *Arkansas*, I was paid $10 one pay and $11 the next…that is 75 cents a day…but the Navy did not deduct any federal tax or social security.

I was studying at night on my professional and military with the course books, since there was not time during the day, because at sea the watches were four on and four off, remember we were short crew members and could only operate the battleship when the money for Bunker "B" fuel oil had accumulated to allow this.

We did the hull and ships painting and values at Portsmouth, Virginia Navy Yard. On one occasion we had just come out of the navy yard and were coming up the long island sound to our anchorage on the Hudson about 135th Street when a freighter and the *Arkansas* collided near Jones Point and the freighter grounded and we had all the news people come out to report on the ship. That was big news. I did go to the New York World's Fair several times.

I reported to Naval Air Station, Pensacola, Florida, in July 1941, in full dress blues, sea bag and hammock, sea going fashion. After riding the L&N train (late and nasty) no air and lots of cinders and soot. Took the bus to the air station and walked from the gate about two miles to building 45 to check in for Photo School. Dress blues are hot in July and especially with a sea bag and hammock, which was everything that I owned.

At Pensacola things were gearing up for the war to come and there I learned to take photos from airplanes, mapping and motion pictures and laboratory work. We had perhaps the last small class before the on rush of the war and hundreds of students. I had received my private pilots license in Norfolk at Glenn Rock Airport, Virginia… This was a grass trip landing area and lessons were $5 dual or solo…a real bargain today. I never thought about trying to be a Navy pilot, again this was beyond my wildest dream, such as ever being a chief in the Navy. The opportunity then did not seem to present itself and there was no encouragement or programs as such.

After Photo School I was assigned to the USS *Yorktown*, CV-5, there I participated in the island bombing of Marshall-Gilberts, and targets all the way down the Pacific to Tulagi. In the Battle of the Coral Sea I filmed the last of the USS *Lexington*, CV-2, into the late evening to show the fires and explosions... I filmed this battle from the bridge on the Yorktown alongside Captain Elliott Buckmaster... He and I were missed when the Japanese fighters strafed the bridge, but the chief was hit. From the bridge the torpedoes were very clear in the water and we were able to turn into them and watch them go by on either side at times. Water from bomb explosions alongside showered the bridge at times. Captain Buckmaster fought the battle from the bridge, not the conning tower. After the battle I make both still and motion picture, 35mm movies of the damage. We had sustained a bomb burst on the flight deck and also down in a repair party. It was very sad that evening to photograph the burial at sea as the ship sped through the night out of the battle area. The destroyer *Simms* was sunk and the fleet oiler that had just provided us with crucial fuel, USS *Neoshio* was also sunk. We took on some of the *Lexington's* planes and left the area. We did have some Japanese planes try to land on our ship since their carriers were sunk, but we fired our 50 caliber guns at them and they left for a watery grave.

At Pearl Harbor we worked around the clock to repair the *Yorktown* and in 72 hours had made emergency repairs and took on provisions to get underway for Midway. I was on the work party to bring crates of can goods and stuff on board, all hands turned to.

Once at sea we brought our air group aboard. One of my jobs was the flight deck movie camera, a 35mm black and white film camera. The next to last aircraft, just at dusk, to land came in high and floated near the island, when the right wing struck the island and the propeller went into the executive officers' aircraft who had just landed and taxied forward. I filmed the aircraft with the officer in it, covered with canvas, as it was pushed overboard that night. We were at sea and heading for midway.

The Battle of Midway on June 4, 1942, was a beautiful clear, blue sky, white cloud day. I selected a battle station on the signal bridge and manned a motion picture camera and a still camera to record the battle. This provided a better all around visibility than on the bridge. During the battle, some of the Japanese torpedo aircraft were so close that almost the entire 35mm movie frame was filled with the plane. The sky was black with five-inch shell bursts and red tracers. The bombing came in several waves and then the torpedo planes came in and I could see the wake of the two torpedoes that struck the ship. After the battle I made the damage photos with movie camera and still photos. When the order was given to abandon ship it was a very sad moment for us all. I saved a mess cook who could not swim and then helped to save others. Finally we got away from the *Yorktown* and the oil and out to a destroyer and there volunteered to be a member of the salvage party to try to save the *Yorktown*. About 125 of us went back as a salvage party.

I worked on the fire party to try to put out the fires in the forward locker, way below decks. The ship was listing about 21 degrees and that made it very bad to walk on the decks, let along carry large fire extinguishers, which we popped and threw into the burning holds. Then I worked with a pharmacist mate to bury the dead and take their identification. Also, such as cutting away the port side five-inch guns to remove the weight to reduce the list.

It was at this time the Japanese submarine fired four torpedoes at the *Yorktown* and sank the USS *Hammann*, destroyer that had returned us to the *Yorktown*. I made the last photo of the *Hammond* as it was going down then the depth charges blew up.

At Midway, I had saved 100-foot rolls of movie film, taped up and in my shirt. Now I was able to get off onto a sea going tug and remain relatively dry. This was a very sad and forlorn cruise back to Pearl, on a destroyer, having seen the *Yorktown* roll over and go down stern first at 7:03 a.m. the next day, June 6, 1942.

Back at Pearl I was at a place in the hills for a few days with no clothes, no money, no records and no one that I knew.

I worked at Pearl on burial parties and as a longshoreman unloading ships and moving stuff on the docks. Orders were then cut for Treasure Island, California. There things were getting organized, such as a partial clothing issue and some records were made up. I still had to get money from home to survive. I got a 30-day survivor leave home to Florida to see my family, but no one knew the *Yorktown* had been sunk or anything about it.

Upon my return to San Francisco I was transferred to San Diego Naval Air Station to a replacement squadron. Ensign George Gay was also in this squadron. I then received orders to NAS Pensacola to be an instructor at the Photo School, because they wanted someone with combat experi-

ence. There I was chief Instructor of the motion picture class and later the aerial camera class.

Here I was promoted to first class photographers mate and then in May 1943 took the examination for chief photographer's mate. During the war every one was on a fast track and I always tried to study hard and be ready for the next promotion. I was told that my grade 3.85 on the chief's exam was the highest ever made at that time at Pensacola. I remember at sea after the Battle of Midway, when I was on on the destroyer and passed my movie films over the cruiser, where they had a movie film processor, that it would be an honor if I could ever be a Navy CPO. I was second class photo mate at that time, but my dream was coming into focus.

At Pensacola, I passed the chief's exam in May but I did not have enough time in rate. In August 1943, I had the time and went to SAX Fifth Ave. in Pensacola and bought my chief's uniforms. I was then given orders to be the chief master at arms of the Navy Photographic School, some 1,100 personnel, three barracks, two hangers and two school buildings. There I was a slick arm, Regular Navy chief, until September when I could wear a hash mark. I was up for ensign when the war was over and I was asked to stay in, but I got out, and at first regretted doing so, because the Navy had given me so much opportunity. When I finished my career I had gone to over 17 schools and courses.

I joined the Navy Reserve at Naval Air Station, Jacksonville, Florida, and was the photo chief in Arm Wing Staff.

In 1953 I received a direct commission as a senior lieutenant and was the photo officer for the Air Wing Staff. In this capacity I make the 16,000 mile flight VP-741, in the P2V2's, a first ever for a reserve squadron to go half-way around the world.

I transferred to Naval Air Station, Anacostia, D.C., and there served as the photo officer, Air Wing Staff and as an intelligence officer. Subsequent promotions brought me to retirement an space a travel.

As a lawyer I still keep busy here in beautiful Naples and Barbara and I love to travel to places like China, Australia, Poland, Africa, and of course Germany, where we have been many times. We love to play golf and enjoy our social life and family.

Of all the many positions I have held, such as corporate attorney in Washington, D.C., manager/director of space programs; Mace, Matador, Gemini, Titan III, for Martin-Marietta, and the Apollo Moon Program, Engineering support for NASA, by Dow, the one that I cherish the most and hold so fondly is being "chief," United States Navy.

(Source: A Rogers, Ret. CWO-2 1962 on 1 Jan)

Briefly – I was working in a sole leather factory in Haverhill, Massachusetts, when the war broke out in 1941. Went down to the PO – saw the recruiter and told him I wanted to ship for six with the intent of doing 20. After I spent the next two weeks getting and paying for own dental work, the recruiter shipped me off for NavTraCen Newport, Rhode Island. Quickly out of there on February 12, 1942, had vol for submarines (passed all prelim) sub school at New London was loaded so shipped out for Sub Base St. Thomas VI via Norfolk. Left the sub base and went to USMCAS St. Thomas in base radio, did some flying with VMS-3 in OS2U Kingfishers on sub patrol. Back to States in 43 on leave, stopped at New London, still no sub school. Hung around Pier 92 New York for awhile and was "offered" a spot with Communications Liaison Group 3, sailing aboard British Pacific Fleet as Communicator Advisor/Crypto, sailed on HMS *Empire Battleaxe* APA type, then MHS *Quilliam* DD, short stint on HMS *Black Prince* cruiser, HMS *Chaser* CVE, on and off several CV types. Plenty of action around Taiwan (Formosa) and Okinawa, kamikaze fun time! Back to USN PCE-903 and PCE-880 – then *Essex* CVA-9 then *Bon Homme Richard* CVA-31, short cruises on *Boxer* CVA 21, Pine Island AV12-MAAG Taiwan advisor to Chinese – PacResFlt *Astoria* – Put *Iwo Jima* in commission as 1st Electronics Officer, two years at IT Sch San Diego, finished up 20 years and two days at NRF San Diego as ship supt...then into the civilian type shipbuilding, Dir for a large Washington School District (retired again) chief examiner for Sheriffs Dept. and then hung it up for good in 1982...finally there's some interesting stories with the British as I served as Rum Bosn'n aboard the *Quilliam*, some job!!

(No Source Named)

A CPO gives of himself many times during a varied career. Many times he gives of himself by making the supreme sacrifice. The giving, indeed, is the greatest gift we can share with our shipmates. The unusual, if not unique talents and abilities incumbent to the CPO relates to who we are and what we can do. As the CPO develops in wisdom and maturity, the evolution and development process provides the sense of well being and positive influence with those whom we serve and share. The leadership example reflected in this progression enables the CPO to be more creative and visionary in his role toward those he interfaces with in achieving mission objectives.

Master chief Equipment Operator Herman C. Hart Assumes Duties as Seabee Force Master Chief

Master Chief Equipment Operator Herman C. Hart, USN, is the new master chief of the Seabees and senior enlisted advisor to the Commander, Naval Facilities Engineering Command.

He assumed duties as master chief at a ceremony on June 29, 1990 at the Construction Battalion Center, Port Hueneme, California.

Master Chief Utilities Man William R. (Billy) Brower, the former master chief, will retire later this summer after more than 29 years of distinguished naval service and three years of exceptional performance as master chief of the Seabees.

Master Chief Hart, who has 23 years of service, most recently was master chief of the Pacific Fleet Seabees, Commander, Naval Construction Battalions, United States Pacific Fleet, Pearl Harbor, Hawaii. His career has included duty with both Pacific and Atlantic Fleet Naval Construction Force Units (NMCB 10, NMCB 11, NMCB 40, 20th NCR, 30th NCR, 31st NCR, and the CBPAC staff), and the Naval Construction Training Centers at Gulfport, MS, and Port Hueneme. He is a highly decorated Vietnam combat veteran.

His awards include the Defense Meritorious Service Medal, Navy Commendation Medal, Navy Achievement Medal with Combat "V," Navy Combat Action Ribbon, and Navy Unit Commendation Ribbon.

Congratulations to Master Chief Hart and best wishes to Master Chief Brower!

Bibliography

Master chief Hospital Corpsman Orten C. Skinner, USN (Ret)

Master Chief Skinner enlisted in the United States Navy on his 17th birthday and served on active duty for 24 years, retiring on June 1, 1970. After boot camp and Hospital Corps School in San Diego, Master Chief Skinner had his first duty at the USNH Long Beach, California.

He graduated form the Naval Medical School Laboratory course in 1950. He then served aboard USS *Yellowstone* (AD27). His first tours of independent duty were aboard USS *Remey* (DD-688) and USS *Thaddeus Parker* (DE-369), all home ported in Newport, Rhode Island.

He then reported to Washington as the research assistant to the attending physician, United States Congress, at the United States Capitol, for a four-year tour. He was then assigned to Mobile Construction Battalion Nine on Midway Island.

From Midway he reported to USS *Worcester* (Cl-144). He served on *Worcester* for two years decommissioning her at Mare Island as the Senior Medical Department Representative. He then served four years as chief-in-Charge of Laboratory services at USNH Charleston, SC.

As a senior chief hospital corpsman, Skinner commissioned USS *Semmes* (DDG-18), spending time on the pre-commissioning detail at Newport and the building site at Westwego, Louisiana. After three years, Skinner was promoted to master chief and became the administrative assistant to the force medical officer, Cruiser-Destroyer Force, United States Atlantic Fleet in Newport.

During this tour he initiated a newsletter, *The Transfusion* which was sent to all ships in the force with the latest information. After four years he was transferred to EPDOCONUS as the detailer for medical and dental personnel at Bainbridge, MD. He also served as master chief of the command for the Bainbridge Training Center.

The Commanding Officer at EPDOCONUS was nominating Master Chief Skinner to be the second master chief of the Navy when he decided to retire and become the administrator of the Cumberland Technical Center in Cookeville, Tennessee, a school for allied health personnel.

In 1974, Mr. Skinner became the executive vice president of International Clinical Laboratories and in 1977 assumed his present role as the administrative director of Pathology and Laboratory Medicine at St. Jude Children's Research Hospital in Memphis.

Skinner graduated from George Washington University and did graduate work at the University of Maryland. He served as cChairman of the Tennessee Laboratory Licensure Board for a number of years and served on the national board of the American Medical Technologists for nine years and served as National President for two terms 1978-80.

In addition to his administrative duties at St. Jude, he serves as an MC for radiothons and telethons and traveled with Danny Thomas before his death. Mr. Skinner is recognized as an outstanding public speaker and does seminars and workshops throughout North America. He is the immediate past president of the Tennessee Organization for Public Speakers.

He has received every award from both state and national levels for his efforts on behalf of medical technology. He has published over 150 articles and four books.

During his career he received numerous awards and decorations and still feels his Navy career was most rewarding. He states, "I owe everything to the United States Navy. They educated me and prepared me for civilian life. I feel very fortunate to have had two very successful careers in my lifetime."

My Friend Lester B. Tucker Who Helped Immensely with "History of a CPO"

Born February 22, 1921, El Paso, Texas, of Thelma E. (Young) Tucker and Joseph B. Tucker. Raised and attended schools in Long Beach-Bellflower areas of California, graduating from Excelsior Union High School in Norwalk, on June 2, 1939. Enlisted in Civilian Conservation Corps July 2, 1939, until joining the United States Navy on Nov. 9, 1939. Underwent recruit training at San Diego (Co. 39-37). Stationed at USN Hospital, PH, TH, and then assigned to the USS *Memphis* CL-13 in July 1940. Served aboard making seaman first class and during neutrality patrol in the Caribbean blockading German and Italian ships in violation of the neutrality. Transferred to the USS *North Carolina* BB55 in Spring 1941 and promoted to gunner's mate 3/c. Transferred to the Aviation Unit (OS2U Kingfishers) changing rating to aviation ordnance man. Was elevated to AOM2c and First Class prior to transfer in August 1943. Saw action at Guadalcanal and Solomon Islands prior to transfer. Selected for flight training attending same in January-July 1944 at which time was released there with 7,000 other aviation pilot and aviation cadet trainees. Attended Combat Aircrew School (PBY Catalinas) and Air Bombers School in Florida. Was transferred to NAAS Oceana, Virginia and promoted to aviation chief ordnance man August 1945. Served in VP-98 (PBM Mariner) refresher training and assigned to VP-26 on the China and Japan stations. Upon decommissioning of VP-26 was assigned to VPHL (PB4Y2 Privateer) at NAS Kaneohe, TH. Upon decommissioning of that squadron was assigned for a short time to VP-42 (MS-2, PBM's) then was ordered to recruiting duty at Modesto, California, remaining there 1947-1949. July 1949 to January 1951 was attached to VP-46 (PBM's) serving in San Diego and then the Far East from June 1950 flying patrols during the Korean War. In January 1951 was selected for warrant officer (Gunner Aviation) effective Aug. 1, 1950. Filled the billet of ordnance officer at NAS Atsugi, Japan 1951-1953. Transferred to Bureau of Ordnance (Ma8a1) Aircraft Machine Guns (and other aviation ordnance equipment) from July 1952-1953. Served a split sea duty tour aboard the USS *Oriskany* CVA-34 as V-5 division officer and the USS *Salisbury Sound* AV-13 as aviation ordnance section and ship's gunner until 1958. Attended Armed Forces Special Weapons School, Sandia Base, Albuquerque, New Mexico, and attached to Special Weapons Unit Pacific, retitled Nuclear Weapons Training Center, until 1960. That duty was followed by a tour on ComFair Hawaii until 1963. Transferred to Naval Missile Facility, Pt. Arguello, California, and served as ground clearance and safety officer for a year before that activity was taken over by the Air Force. Returned to NWTCP in 1964 to end 27 1/2 years service and retiring on February 1, 1967. While in service was awarded the Distinguished Flying Cross, Air Medal (4), Good Conduct (7), National Defense Medal with (A), American Theater Campaign, Asiatic-Pacific Campaign with three stars, WWII Victory Medal, China Service (extended), Navy Occupation Medal (Japan), National Defense Medal (2), Korean Service Medal, United Nations Service (Korea), Korean Presidential Citation.

Tucker holds a commercial pilots license and has flown as a glider tow-plane pilot. He held residence in Victor, Montana 1972 -1990, although was in Washington, D.C. 1974-1981 conducting research for his books, *Enlisted Ratings of the United States Navy 1774 to Present*. He is author of *The History of the CPO Grade*, presently being published in NAS Whidbey Island "Crosswind" in four parts. (Probably will be in *Navy Times* March 1993). He has written other articles published in *Trading Post* on several Navy Ratings. Tucker is Charter Member #11 of the National CPO's Association. He has urged that Active Duty and Retired CPS's celebrate the centennial birthday of the founding of the CPO grade April 11, 1993. Tucker is a 47-year continuous member of the American Legion. He also is a member (life) of TROA, and belongs to Company of Military Historians and American Society of Military Insignia Collectors, and Benevolent and Protective Order of Elks among other organizations.

Lester is married to the former Helen O. (Crow) Yates whom he met at BuPers while doing his research. He has, by a previous marriage, three daughters and one son. His father served as a Signal Quartermaster Seaman First Class in WWI and a chief ship fitter in the 40th CB's in WWII.

One Time CPO Remolded Navy with an Idea and a Philosophy Acquired Coming Up Through the Ranks:

Forty-one years ago, a New Hampshire man helped refine Navy Personnel policies –

As he watched his shipmates stand in line for movie stars' autographs on the warships' rolling flight deck, the sailor from New Hampshire had a revelation.

It was the early years of World War II and Van Watts was stationed aboard the aircraft carrier, USS *Enterprise*. He was a "plank owner" of that majestic ship, meaning he was a member of the carrier's original crew in 1938.

Well, here we are out here in the Pacific, he mused, – actually not that far out, just a ways off the coast of Hollywood, California, – on a ship that was serving as a floating movie set for the Warner Bros. film, *Dive Bomber*. Glamour boys, Erroll Flynn, Fred MacMurray and Ralph Bellamy were playing Navy heroes. These were the VIPs whose autographs were in demand.

One of those guys, for instance, standing up there waiting for an autograph, had jumped into the sea to save the life of a pilot whose aircraft had missed the carrier deck and crashed.

"We've got plenty of enlisted men out here doing heroic things. We ought to be recognizing them and telling the folks back home about them," Watts whose home was in Chesterfield, New Hampshire.

It would be more than a decade before Watts' revelation was realized. World War II got in the way.

Watts, who retired to North Hollywood, California, in 1972 after a quarter century in the Navy, is the founder of a worldwide Navy recognition program that brought praise from admirals, accolades to seamen, and improved relations between the United States Navy the civilian communities.

The Sailor of the Week – Sailor of the Month – Sailor of the Year program turns 40 on June 23, four decades after Watts organized the first such project at Norfolk.

During his boyhood on the family's 150-acre Chesterfield farm, Watts read books about the sea. He was troubled by the oft-times punitive attitudes demonstrated by officers toward seamen in the past.

Though Richard Henry Dana's book about the sea, *Two Years Before the Mast*, generated reforms in the treatment of sailors in the mid-19th century, the Navy still could be demeaning when Watts joined in 1937.

"It could still be a pretty rough place. Some of the officers swore at and used demeaning language toward the men. If I had attacked this problem frontally, I myself would have been discharged from the Navy, so I decided that some day I would do something to create a different atmosphere in the Navy.

"As a result of such programs the enlisted man, once celebrated by the Navy, would be less likely to be demeaned," he said. Besides, it would be good for everybody's morale and provide an opportunity for ships or Naval stations to link up with the local community as well as the home town of the honored sailor.

In 1952, Watts requested – to everybody's amazement – that he be transferred to Norfolk, a community that was notorious for strained relations with the Navy, he recalls. He was planning his pilot project and found Norfolk ripe for the kind of reform he had in mind.

"I arrived two days before Norfolk was to become the international sea capital of the North Atlantic Treaty Organization (NATO). If Norfolk can't get along with one Navy how can it get along with 13 navies? I asked myself. I thought my program would do more good there (than elsewhere) because the Navy morale was so low there."

On June 23, 1952, Watts produced his first Sailor of the Week, a program designed to honor an outstanding seaman. "My program sparked a goodwill drive in the city, it was placed under the civic auspices of the mayor, endorsed by the Chamber of Commerce and the 150-church ministerial association," he recalled in an interview from North Hollywood this month.

"The 'Sailor of the Week' was a show, a big 'welcome to the town' on television. At the end of the week there was a Bon Voyage on the radio."

The program expanded to include a Sailor of the Month and a Sailor of the Year. The recipient was feted not only on his base and in the community, but in his home town, where he was given a special leave that was not counted toward his regular leave time. Over the years it spread to more than 2,500 ships and stations. The program is a Navy fixture today.

Watts has received acclaim over the years for the program he generated.

(From the Manchester, NH, *New Hampshire Sunday News*, June 7, 1992.)

A Tale of the South Pacific
by Van Watts

Up front with James Michener

We would take him on what may have been, I guess, the ride of his life, picking him up and leaving him, two weeks later, at our rear base, the New Hebrides Island of Espiritu Santo. It was there, during the next 18 months, he would write his first novel. In it, he would have his narrator say in his only partly fictionalized work, "I never saw a battleship except from a distance. I never visited a carrier, or a cruiser, or a destroyer. I never saw a submarine."

But – old shipmates remind me – Michener did sail on at least one naval vessel in the first of her narrow escapes from the Solomons – the little seaplane tender we called the *Mighty Mac*. And when we didn't call her that, we called her by the number on her bow – the *Lucky Thirteen*! She was lucky all right! Lucky enough to survive, with Michener aboard, "the most advanced post" in World War II's famous "First Offensive."

Strange – but not so strange – that modest fellow never seems to have mentioned it. But if you ever wondered where he was the day the action began, he was with us – up front – way up front – at an island shown on military maps behind enemy lines – refueling planes under the very noses of the enemy.

For, on August 7, 1942, the day of the landings on Guadalcanal, 60 miles north of there, at Malaita, the USS *Mackinac* would be refueling the planes that scouted the enemy. The success of one of the war's most daring exploits would depend completely on lack of detection. But, charged off in advance as part of the price of Guadalcanal, there was no question of her presence eventually being discovered. The only question was who got to her first, enemy subs or planes or, from the landward side, their demolition experts.

An over-aged destroyer converted into a seaplane tender and sent into the area to refuel scouting planes in advance of our landings in the southern Solomons had been gambled and lost. And no one could have honestly expected the *Mackinac* to return. Within four days all but one of the nine big PBYs in her first assigned squadron had been shot down by the huge Japanese fleet which, history says, "roamed the area at random."

But, on the fifth day, with still a bare chance of her own survival, the *Mackinac* would be ordered to evacuate "the most advanced post." As we upped anchor and sailed, the last of her big PBYs flew over to warn that 40 Japanese bombers had been sighted headed her way. And either they were after bigger game – or didn't see the speck below that was us.

For tailing far behind a departing fleet that had deposited the Marines on Guadalcanal and, in the face of superior enemy forces, retired as fast as it could go, the *Mighty* – and mighty lucky little – *Mac* escaped with Michener and company!

Eight days later we would head north to set up another base in Ndeni, leaving him at our rear base, Espiritu Santo. We never saw him again but, years later, would read in the press that he "retreated to a jungle shack and began writing the stories that were to appear in *Tales of the South Pacific*."

(From the National CPOs Association Newsletter *The Chiefs*, Spring 1992.)

Chapter 7

Chief Petty Officer's Creed

New Chief Petty Officers' Creed

During the course of this day you have been caused to humbly accept challenge and face adversity. This you have accomplished with rare good grace. Pointless as some of these challenges may have seemed, there were valid, time-honored reasons behind each pointed barb. It was necessary to meet these hurdles with blind faith in the fellowship of Chief Petty Officers (CPOs). The goal was to instill in you that thrust is inherent with the donning of the uniform of a chief.

It was our intent to impress upon you that challenge is good; a great and necessary reality which cannot mar you—which, in fact, strengthens you. In your future as a CPO, you will be forced to endure adversity far beyond that imposed upon you today. You must face each challenge and adversity with the same dignity and good grace you demonstrated today.

By experience, by performance and by testing, you have been this day advanced to CPO. In the United States Navy (only in the United States Navy) the rank of E7 carries unique responsibilities. No other armed force in the world grants the responsibilities, nor the privileges, to enlisted personnel comparable to the responsibilities and privileges you are now bound to observe and are expected to fulfill.

Your entire way of life is now changed. More will be expected of you; more will be demanded of you. Not because you are an E7, but because you are now a CPO. You have not merely been promoted one pay grade, you have joined an exclusive and, as in all fellowships, you have a special responsibility to your comrades, even as they have a special responsibility to you. This is why we in the United States Navy may maintain with pride our feelings of accomplishment once we have attained the position of CPO.

Your new responsibilities and privileges do not appear in print. They have no official standing; they cannot be referred to by name, number nor file. They exist because for over 100 years, chiefs before you have freely accepted responsibility beyond the call of printed assignment. Their actions and their performance demanded the respect of their seniors as well as their juniors.

It is not required that you be the fountain of wisdom, the ambassador of good will, the authority in personal relations as well as in technical applications. "Ask the chief" is a household phrase in and out of the Navy. You are now that chief. The exalted posi-

*A most senior chief: Retired senior chief Rhea Rohn appears at the April 1 ceremony of the "Chiefs Bell" presentation in Annapolis. Rohn, who is 82 years old, joined the Navy in 1929 and first made chief in 1949. (*Times *photo by Steve Elfers)*

tion you have now achieved—and the word "exalted" is used advisedly—exists because of the attitude and performance of the chiefs before you. It shall exist only as long as you and your fellow chiefs maintain these standards.

It was our intention that you never forget this day. It was our intention to test you, to try you, and to accept you. Your performance has assured us that you will wear "the hat" with the same pride as your comrades in arms have before you.

We take a deep and sincere pleasure in clasping your hand and accepting you as CPO in the United States Navy.

Permission from *Navy Times* to reprint. John Burlage is a retired CPO.

Incoming CPOs Can Expect New Creed, Initiation Scutiny

by John Burlage, Times Staff Writer (By permission: Navy Times)

Something will be missing from the time-honored creed that will be read September 16 as the last event of initiations intended to help 3,600 sailors make the transformation to Navy CPO.

Gone from a new version are words telling selectees it was necessary to suffer humiliation during the initiation in order to appreciate fully what it means to be a chief. New words instead stress the need to overcome "hurdles" and the importance of fellowship.

Some top chiefs argue the new creed represents just one more nail in a coffin that soon will contain the remains of another fine Navy tradition. But others say the rewrite, distribution fleet in mid-August, is another important step in a continuing effort to replace excesses of past initiations with a meaningful ceremony.

Most chiefs consider reading of the creed the most poignant moment in the initiation that's meant to assure acceptance of new CPOs by those already "wearing the hat."

But like many other aspects of the initiation, say the Navy's most senior chiefs, the creed must change to meet new realities.

Fleet and force master chiefs determined actions or words that support humiliation and harassment no longer can be tolerated. Nor can the notion that Navy chiefs are somehow superior to noncommissioned officers of other services, a contention included in the most popular version of the creed read at initiations in recent years.

Words thought offensive have been stricken from the new creed as a result. It was sent to the fleet with a strongly worded suggestion from the outgoing and incoming master chief petty officers of the Navy that it replace any other version.

Inserted in place of offending words are phases referring to "total trust in our fellow chiefs" missing from earlier versions, said Duane Busahey and John Hagen in a joint letter. Bushey was relieved by Hagan as MCPON August 28.

Now, Hagen said September 8, the creed "tries to summarize some of the emotions and capture some of the accomplishments going with this huge step forward" of advancement to chief without offending modern sensibilities.

Issuing a new creed is the latest move in a continuing effort to clean up initiations, long contentious because too many have turned into drunken melees that left new chiefs humiliated or even hurt.

Changing the thrust of the rituals was given impetus by a 1989 *Navy Times* survey that asked whether they should be continued. Most responding said yes. And hundreds were like Senior Chief Radioman James R. Dodge, who said the creed helps new chiefs realize they "belong to one of the strongest and tightest social and professional organizations in the world."

Public scrutiny of this years' initiations will be the closest ever. In the wake of sexual harassment and assault episodes like Tailhook, which have knocked the Navy's reputation for a loop, the most senior leaders declare this year's rites must be squeaky-clean.

"The CPO initiation is an important part of Navy tradition, and I endorse initiations," said Admiral Frank Kelso II, chief of naval operations. "But they must be conducted in such a way that people feel comfortable bringing their husbands or wives, parents or friends and they must be done in a way we would not be embarrassed to see on local television."

"The command master chiefs all are talking to their communities, putting out the word in strong language not to cross the line" into excess, said a retired master chief in San Diego until active in CPO affairs.

Chapter 8

Evolution and Development

Aviation Pilot: Chief

1924 - 1933
1927 - 1933
1942 - 1948
1942 - 1948
1942 - 1942
1st Class
2nd Class
3rd Class

This mark was originally specified to be yellow on all rating badges, giving the general appearance of the naval aviator's wings. The Specialty mark was changed to white on the blue rating badge and blue on all other rating badges in BPCL 199-44 dated 12 July 1944. The mark with straight wings more nearly like those of the aviator's wings appeared during the World War II period. Both styles of mark exist on rating badges of the World War II type. Some speculation has been advanced that the curved wing style after 1942 was used by the Coast Guard and the straight wing style by the Navy, but nothing has been found to substantiate this distinction.

Distinguishing Mark: Naval Aviation Pilots 1933 - 1935 "...men designated as aviation pilots will wear the present specialty mark of aviation pilot midway between the shoulder and elbow of the left sleeve for men of the seaman branch, and on the right sleeve for others."

Chapter 9

Getting There - Alpha to Omega

Evolution

The Recruits Handy Book[1] reflects an era of evolution important to one's understanding of the important aspects of basic management essential to the success of a sailor.

Pay, Promotion and Rewards in the Navy

Recruits do not always realize what the Navy offers them in the way of pay and promotion. Of course a man must not expect promotion at once. He must be willing to begin at the bottom and work up. In civil life he must "learn his trade," or prove this ability and faithfulness before he can rise. The same is required in the Navy; but it may be said that a man who deserves promotion is more certain to get it in the Navy than in any civil trade. He need never lose his job, but is always sure of his pay, and is taken care of in case he is sick or disabled. The following tables were then in effect:

1. Petty officers of the Navy, performing duty which deprives them of quarters, and their rations or commutations thereof, shall receive $9 per month in addition to the pay of their rating.

2. Subsistence furnished to enlisted persons attached to ships of the Navy, when unavoidably detained on shore under orders, or absent, by authority, from the ship to which attached, must be charged to appropriate "Provisions, Navy." During the time of such subsistence their rations shall be stopped on board ship, and no credit for commutation therefore shall be given.

3. Enlisted men attached to permanent recruiting offices will hereafter be allowed necessary expenses for their maintenance, not exceeding $1 per day, chargeable to appropriation "Recruiting, Navigation."

4. Men that have successfully completed a prescribed course of instruction for seaman gunners or petty officers may be given, by the Bureau of Navigation, a certificate to that effect, which shall entitle them to receive $2.20 per month in addition to the pay of the rating in which they are serving; such certificates to continue in force only during the enlistment in which the men were respectively graduated, unless renewed by re-enlistment for four years within four months from date of honorable discharge.

5. Each enlisted man of the Navy shall receive 82 cents per month, in addition to the pay of his rating, for each good-conduct medal, pin, or bar which he may heretofore have been, or shall hereafter be, awarded. On and after September 5, 1904, the date of the award of a good-conduct medal, pin, or bar shall be the date of the holder's discharge by reason of the expiration of the enlistment for which the medal, pin, or bar is given, the allowance of 82 cents per month to be reckoned from said date of award: Provided, that nothing herein contained shall be construed to authorize any change in the date of award of any good-conduct medal, pin, or bar heretofore awarded, or to grant any arrears of allowances on account thereof.

A chief aviation pilot prepares to board an SNJ aircraft for a flight during World War II. He bears the uniform insignia of the aviation machinist's mate rate. He wears the traditional scarf, an inflatable life vest and a parachute.

6. Coxswains detailed as coxswains of boats propelled by machinery, or as coxswains to commander-in-chief, shall receive $5 per month in addition to their pay.

7. All enlisted men of the Navy shall receive $5 per month in addition to their pay while serving on board of submarine vessels of the Navy.

8. Seamen in charge of holds shall receive $5 per month in addition to their pay.

9. Ordinary seamen detailed as Jacks-of-the-Dust, or as Lamplighters, shall receive $5 per month in addition to their pay.

10. Enlisted men detailed as crew messmen shall, while so acting, except when assigned as reliefs during the temporary absence of the regular crew messmen, receive extra compensation at the rate of $5 per month.

11. Enlisted men detained beyond their regular term of enlistment until the return to the United States of the vessel to which they belong shall receive for the time during which they are so detained an addition of one-fourth of their former pay, "computed on the total pay which they are entitled to receive."

12. Seamen and ordinary seamen detailed for duty as firemen or coal passers shall receive in addition to the pay of their rating extra pay at the rate of 36 cents per day for the time so employed.

13. Enlisted men of the navel service regularly detailed as signalmen shall receive the following extra compensation in addition to the monthly pay of their rating: signalmen, first class, $3; signalmen, second class, $2; signalmen, third class, $1. (General Orders, No. 110.)

14. All chief petty officers of the Navy whose pay is not fixed by law, including chief water tenders, who, on or after July 1, 1903, shall receive permanent appointments after qualifying therefor, shall be paid at the rate of $77 a month; those who serve under permanent appointments issued prior to said date, or under acting appointments, shall be paid at the rates now in force. The pay of chief water tenders who hold acting appointments shall be $55 a month. (General Orders, No. 134.)

15. After October 1, 1903, enlisted men of the Navy, after having qualified as gun pointers, and who are regularly detailed as gun pointers by the commanding officer of the vessel, shall receive monthly, in addition to the pay of their respective ratings, extra pay as follows:

Heavy gun pointers: First Class...$10, Second Class...$6.
Intermediate gun pointers: First Class...$8, Second Class...$4.
Secondary gun pointers: First Class...$4, Second Class...$2.

16. Enlisted men of the Navy regularly detailed by the commanding officer of a vessel as gun captains, except at secondary battery guns, shall receive, in addition to the pay of their respective ratings, $5 per month, which, in the case of men holding certificates as gun captions, or of graduation from the gun-captain class, petty officer's school, shall include the

$2 per month to which such certificates entitle them. (General Orders No. 137.)

17. From and after July 1, 1905, any enlisted man of the Navy detailed to perform the duties of "ship tailor" on board of a vessel having a complement of 600 men or more, exclusive of Marines, shall receive $20 per month in addition to the monthly pay of his rating; on a vessel having a complement of from 300 to 600 men, exclusive of Marines, $15 per month in addition to the monthly pay of his rating; on a vessel having a complement of less than 300 men, exclusive of Marines, $10 per month in addition to the monthly pay of his rating. Any enlisted man of the Navy detailed as "tailor's helper" on board a vessel having a complement of 600 men or more, exclusive of Marines, shall receive $10 per month in addition to the monthly pay of his rating: Provided, that the total pay of an enlisted man detailed to perform the duties of a "ships tailor" shall not exceed $50 per month, and of "tailor's helper" shall not exceed $40 per month. (General Orders 186).

18. To provide adequate compensation for trained men, the pay now prescribed by Executive Order for each rating in the Navy is hereby increased to $5.50 per month during the second period of service and a further sum of $3.30 per month during each and every subsequent period of service: *Provided* - That only enlisted men who are citizens of the United States, and whose second and subsequent periods of service each follow next after service in the Navy, that was terminated by reason of expiration of enlistment, shall receive the benefits of the increased pay named herein: *Provided further* - That in the cases of men who are or were finally discharged from the Navy by reason of expiration of enlistment, the first enlistment on or after the date of this order shall be considered the second period of service which shall carry with it the increased pay provided by this order; except that men discharged on recommendations of boards of medical survey, shall, if they re-enter the service, be given credit for any previous periods of service in the Navy which were terminated by reason of expiration of enlistment.

19. Chief petty officers detailed as instructors of apprentice seamen at naval stations who qualify as instructors by examination shall receive hereafter in addition to their pay the sum of ten dollars per month while so detailed, such pay to be considered extra pay for special duty.

20. Apprentice seamen detailed as apprentice chief petty officers, apprentice petty officers, first, second, or third class, in connection with the instruction of apprentice seamen at naval stations, shall receive hereafter in addition to their pay the sum of two dollars and fifty cents, and one dollar each per month, respectively, while so detailed, such pay to be considered, extra pay for special duty.

21. An outfit of clothing not exceeding in value the sum of $60 shall be furnished, on first enlistment, to all enlisted men of the Navy.

22. Any man who has received an honorable discharge from this last term of enlistment, or who has received a recommendation for re-enlistment upon the expiration of his last term of service of not less than three years, who re-enlists for a term of four years within four months from the date of his discharge, shall receive an increase of $1.50 per month to the pay prescribed for the rating in which he serves for each consecutive re-enlistment.

23. Seamen distinguishing themselves in battle, or by extraordinary heroism in the line of their profession, may be promoted to warrant officers, if found fitted, upon the recommendation of their commanding officer, approved by the flag officer and the Secretary of the Navy. And upon such recommendation they shall receive a gratuity of $100 and a medal of honor prepared under the direction of the Secretary of the Navy.

24. Men must remember that their pay, except what is necessary to clothe them, is nearly all clear money. They have many "allowances" in addition to their pay. Their "ration" is quite sufficient to feed them. They have no board-bill to pay! If they are sick they get medical attendance *free*. In case of serious illness they are sent to a hospital and *cared for by the Government*. If they are injured in "line of duty" they get a pension. They would get no such allowances in civil life. The great rewards that come to a man from long and faithful service are referred to below under the heads of "Retirement," "Honorable Discharge," and "Continuous Service."

25. It will be noted above that there are certain rewards in the way of pay, etc., for men who graduate from the different classes in the petty officers' school of instruction. Proficiency in ordnance, torpedoes, engineering and electricity will ensure promotion; and men who have a special taste for clerical work are sent to the Yeoman's School. *And for the man who strives to be a good seaman, there is always promotion.* In fact, there are many different "trades" in the Navy, and a man is usually assigned to the work for which he is best fitted.

26. Recruits should carefully consider the great rewards which the Navy offers them. They should not be discouraged in the beginning, nor leave the service before they know what they are doing. They should settle down to work and remember that good conduct and continuous service will always earn promotion, and that life in the Navy has many bright sides. A man who is easily discouraged, or who will not work or "learn a trade' will never succeed at anything.

Source: *Blue Jackets Manual, Naval Institute Press*, p. 20

Ratings and Rates

A rating is a Navy job—a duty calling for certain skills and aptitudes. The rating of engineman, for example, calls for persons who are good with their hands and are mechanically inclined. A paygrade (such as E-4, E-5, E-6) within a rating is called a rate. Thus an engineman third class (EN3) would have a rating of engineman and rate of third class petty officer.

The term petty officer (PO) applies to anyone in paygrades E-4 through E-9. E-1s through E-3s are called non-rated personnel. Personnel in general apprenticeships are identified as recruit (E-1), apprentice (E-2) or at the E-3 level by the apprenticeship field, such as seaman, fireman, airman, constructionman, dentalman, or hospitalman. A person training for a specific job in paygrades E-1 through E-3 is called a striker - one who has been authorized to "strike" or train for a particular job.

Enlisted seniority is determined by time in rate and time in the Navy. If two POs are in the same paygrade, the one in that grade the longest is considered senior. In the case of two POs in the same paygrade with the same amount of time in grade and in the Navy, the one having the most time in the next lower paygrade is senior.

Ratings

These are divided into three categories: General, service and emergency.

General ratings are broad occupational fields for paygrades E-4 through E-9. Each general rating has a distinctive badge. General ratings are sometimes combined at the E-8 or E-9 level, when the work is similar, to form even broader occupational fields. For example, senior chief instrumentman (IMCS) and senior chief opticalman (OMCS) can be combined to form the rating of master chief precision instrumentman (PICM). Some general ratings include service ratings; others do not.

Service ratings are subdivisions of a general rating that require specialized training. There are service ratings at any PO level; however, they are most common with E-4s and E-5s. In the higher paygrades, service ratings merge into a general rating, usually at the E-8 level. For example, a chief fire-control technician C (gunfire control) is an FTGC until the E-8 level and then becomes an FTCS. An FTCS must know both guns and missiles.

Emergency ratings are used to identify civilian occupational fields that are only used in time of war, for example, stevedore, transportationman, and welfare and recreation leader. There are no emergency ratings in use today.

Rates

A rate identifies the level of your rating. For example, the yeoman rating is broken down into rates E-1 through E-9. General rates (not to be confused with general ratings) are the general apprenticeships that identify enlisted personnel in grades E-1, E-2, and E-3. Within these apprenticeships, enlisted personnel receive their recruit training and initial technical training, as preparation for advancement to PO or a service rating. E-1 is generally where recruits start. E-2s (apprentices) perform the routine duties of their occupational groups, but they also perform duties with more responsibility. General rates are identified by various colored stripes.

Title	Color of Stripe
Seaman (SN)	White
Hospitalman (HN)	White
Dentalman (DN)	White
Fireman (FN)	Red
Constructionman (CN)	Blue
Airman (AN)	Green

Stripes are navy blue on a white uniform, except those for FN, CN and AN, which are the same color as described above on all uniforms.

The following is a basic description of the duties on E-3s:

Seaman (SN): Keeps compartments, lines rigging, decks and deck machinery shipshape. Acts as a lookout, member of a gun crew, helmsman, and security and fire sentry.

Hospitalman (HN): Arranges dressing carriages with sterile instruments, dressings, bandages, and medicines. Applies dressings. Gives morning and evening care to patients. Keeps medical records.

Dentalman (DN): Assists dental officers in the treatment of patients. Renders first aid. Cleans and services dental equipment. Keeps dental records.

10. Sensor operations—EW, OT, St
11. Weapons system support—(none currently assigned)
12. Data systems—DP, DS
13. Construction—BU, CE, CM, EA, EO, SW, UT
14. Health care—DT, HM
15. Administration—LN, NC, PN, PC, YN, RP
16. Logistics—AK, DK, MS, SH, SK
17. Media—DM, JO, LI, PH
18. Musician—MU
19. Master-at-Arms—MA
20. Cryptology—CT
21. Communications—RM
22. Intelligence—IS
23. Meteorology—AG
24. Aviation sensor operations—AW

Here's how the general rates of E-2 and E-3 fit into the occupational fields:

SA, SN—1, 2, 4, 8, 9, 10, 12, 15, 16, 17, 18, 20, 21, 22
FA, FN—3, 4; CA, CN—13
AA, AN—5, 6, 7, 16, 17, 23, 24
HA, HN—14
DA, DN—14

Members of the following ratings who are in advanced electronics require a six-year enlistment: AT, AV, CT, DS, ET, DW, FC, FT, OT, RM and ST.

Members of the following ratings who are in an advanced technical field require a six-year enlistment: BT, GS, HM, HT, IC and MM.

Enlisted Ratings[2, 3]

Aviation Boatswain's Mate: ABs operate, maintain, and repair aircraft catapults, arresting gear and barricades. They operate and maintain fuel-and lube-oil transfer systems. ABs direct aircraft on the flight deck and in hangar bays before launch and after recovery. They use tow tractors to position planes and operate support equipment used to start aircraft.

Air Traffic Controller: ACs assist in the essential safe, orderly and speedy flow of air traffic by directing and controlling aircraft. They operate field lighting systems, communicate with aircraft, furnish pilots with information regarding traffic, navigation and weather conditions. They operate and adjust GCA (ground-controlled approach) systems. They interpret targets on radar screens and plot aircraft positions. This is a five-year enlistment.

Aviation Machinist's Mate: ADs usually maintain jet aircraft engines and associated equipment, or engage in any one of several types of aircraft maintenance activities. ADs maintain, service, adjust and replace aircraft engines and accessories, as well as perform the duties of flight engineers.

Aviation Electrician's Mate: AEs maintain, adjust and repair electrical-power generating and converting systems in aircraft, lighting, and control and indicating systems. They also install and maintain wiring and flight and engine instrument systems.

Aerographer's Mate: The Navy has its own weather forecasters, AGs, who are trained in meteorology and the use of aerological instruments that monitor such weather characteristics as air pressure, temperature, humidity, wind speed and wind direction. They prepare weather maps and forecasts, analyze atmospheric conditions to determine the best flight levels for aircraft, and measure wind and air density to increase the accuracy of antiaircraft firing, shore bombardment and delivery of weapons by aircraft.

Aviation Storekeeper: AKs ensure that the materials and equipment needed by naval aviation activities are available and in good order. They take inventory, estimate future needs, and make purchases. AKs store and issue flight clothing, aeronautical materials and spare parts, ordnance and electronic, structural and engineering equipment.

Aviation Structural Mechanic: The maintenance and repair or aircraft parts (wings, fuselage, tail, control surfaces, landing gear, and attending mechanisms) are performed by AMs working with met-

Navy bandsman in blue dress uniform, 1886.

als, alloys, and plastics. AMs maintain and repair safety equipment and hydraulic systems.

Aviation Ordnanceman: Navy planes carry guns, bombs, torpedoes, rockets and missiles to attack the enemy on the sea, under the sea, in the air, and on land. AOs are responsible for maintaining, repairing, installing, operating and handling aviation ordnance equipment; their duties also include the handling, stowing, issuing, and loading of munitions and small arms.

Aviation Support Equipment Technician: ASs perform intermediate maintenance on "yellow" (aviation accessory) equipment at naval air stations and aboard carriers. They maintain gasoline and diesel engines, hydraulic and pneumatic systems, liquid and gaseous oxygen and nitrogen systems, gas turbine compressor units and electrical systems.

Aviation Electronics Technician: ATs perform intermediate-level preventive and corrective maintenance on aviation electronic components supported by conventional and automatic test equipment. They repair weapons-replaceable assemblies (WRAs), shop replaceable assemblies (SRAs), and microminiature (2M) components, and perform test-equipment qualification and associated test-bench preventive and corrective maintenance.

Aviation Maintenance Technician: AVs perform organizational-level preventive and corrective maintenance on aviation electronics systems, including equipment used in communications, radar, navigation, antisubmarine warfare, electronic warfare data, fire control and tactical display.

Aviation Antisubmarine Warfare Operator: AWs operate airborne radar and electronic equipment used in detecting, locating, and tracking submarines. AWs also operate radar to provide information for aircraft and surface navigation. They perform helicopter-rescue duties and serve as part of the flight crew on long-range and intermediate-range aircraft. This is a five-year enlistment.

Aviation Maintenance Administrationman: The many clerical, administrative and managerial duties necessary to keep aircraft maintenance activities running smoothly are handled by AZs. They plan, schedule and coordinate the maintenance workload, including inspections and modifications to aircraft and equipment.

Boatswain's Mate:[2] BMs are expert seamen who maintain the ship, serve as steersmen, take command of tugs and other small craft, serve as gun captains, look after rigging, paint, handle and care for deck equipment, and serve on working parties and damage-control teams. BMs in upper grades train and supervise others in caring for and handling deck equipment and small boats. There are no Navy technical schools for this rating.

Boiler Technician: Because the propelling agent of our large naval ships in steam, the Navy relies on BTs to keep its ships moving. BTs operate and repair Marine boilers and fireroom machinery, and they transfer, test, and inventory fuels and water.

Builder: Navy BUs are like civilian construction workers. They may be skilled carpenters, plasterers, roofers, cement finishers, asphalt workers, masons, painters, bricklayers, sawmill operators, or cabinet-makers. BUs build and repair all types of structures, including piers, bridges, towers, underwater installations, schools, offices, houses and other buildings. This is a five-year enlistment.

Construction Electrician: CEs are responsible for the power production and electrical work required to build and operate airfields, roads, barracks, hospitals, shops, and warehouses. The work of Navy CEs is like that of civilian construction electricians, powerhouse electricians, telephone and electrical repairmen, substation operators, linemen, and others. This is a five-year enlistment.

Construction Mechanic: CMs maintain heavy construction and automotive equipment (buses, dump trucks, bulldozers, rollers, cranes, backhoes, and pile drivers) as well as other construction equipment. They service vehicles and work on gasoline and diesel engines, ignition and fuel systems, transmissions, electrical systems and hydraulic, pneumatic and steering systems. This is a five-year enlistment.

Cryptologic Technician: CTs control the flow of messages and information. Their work depends on their special career area: **administration** (CTA), administrative and clerical duties that control access to classified material; **interpretive** (CTI), radiotelephone communications and foreign-language translation; **maintenance** (CTM), the installation servicing, and repair of electronic and electromechanical equipment; **collection** (CTR), Morse Code communications and operations of radio direction-finding equipment; and **technical** (CTT), communications by means other than Morse Code and electronic countermeasures.

Damage Controlman: DCs perform the work necessary for damage control, ship stability, firefighting, and chemical, biological and radiological (CBR) warfare defense. They instruct personnel in damage control and CBR defense, and repair damage-control equipment and systems.

Disbursing Clerk: DKs maintain the financial records of Navy personnel. They prepare payrolls, determine transportation entitlements, compute travel allowances, and process claims for reimbursement of travel expenses. DKs also process vouchers for receiving and spending public money and make sure accounting data are accurate. They maintain fiscal records and prepare financial reports and returns.

Illustrator-Draftsman[2]**:** DMs prepare mechanical drawings, blueprints, charts and illustrations needed for construction projects and other naval activities. They specialize in a number of areas, among them graphics, structural drafting, graphic arts mechanics and illustrating.

Data-Processing Technician: The Navy needs an extensive accounting system to maintain personnel records, to keep tabs on the receipt and transfer of supplies and the disbursement of money, and to inventory Navy equipment. DPs operate and maintain transceivers, sorters, collators, reproducers, interpreters, alphabetic accounting machines, and digital electronic data-processing (EDP) machines for these purposes. This is a five-year enlistment.

Data-Systems Technician: DSs are electronics technicians who specialize in computer systems, including digital computers, video processors, tape units, buffers, key sets, digital display equipment, data-link terminal sets, and related equipment. They clean, maintain, lubricate, calibrate and adjust equipment. They run operational tests, diagnose problems, make routine repairs, and evaluate newly installed parts and systems units.

Dental Technician: Navy dentists, like many civilian ones, are assisted by dental technicians. DTs have a variety of "chairside," laboratory, and administrative duties. Some are qualified in dental prosthetics (making and fitting artificial teeth), dental X-ray techniques, clinical laboratory procedures, pharmacy and chemistry, or maintenance and repair of dental equipment. This is a five-year enlistment.

Engineering Aide: EAs provide construction engineers with the information needed to develop final construction plans. EAs conduct surveys for roads, airfields, buildings, waterfront structures, pipelines, ditches, and drainage systems. They perform soil tests, prepare topographic and hydrographic maps, and survey for sewers, water lines, drainage systems and underwater excavations. This is a five-year enlistment.

Electrician's Mate: The operation and repair of a ship's or station's electrical power plant and electrical equipment is the responsibility of EMs. They also maintain and repair power and lighting circuits, distribution switchboards, generators, motors, and other electrical equipment.

Coxswain in dress whites, 1905.

Boatswain's Mate, First Class, in blue dress uniform, 1905.

Engineman: Internal-combustion engines, either diesel or gasoline, must be kept in good order; this is the responsibility of ENs. They are also responsible for the maintenance of refrigeration, air-conditioning, and distilling-plant engines and compressors.

Equipment Operator: EOs work with heavy machinery such as bulldozers, power shovels, pile drivers, rollers and graders, etc. EOs use this machinery to dig ditches and excavate for building foundations, to break up old concrete or asphalt paving and pour new paving, to loosen soil and grade it, to dig out tree trunks and rocks, to remove debris from construction sites, to raise girders, and to move and set in place other pieces of equipment or materials needed for a job. This is a five-year enlistment.

Electronics Technician: ETs are responsible for electronic equipment used to send and receive messages, detect enemy planes and ships, and determine target distance. They must maintain, repair, calibrate, tune, and adjust all electronic equipment used for communications, detection and tracking, recognition and identification, navigation, and electronic countermeasures.

Electronics Warfare Technician: EWs operate and maintain electronic equipment used in navigation, target detection and location, and the prevention of electronic spying by enemies. They interpret incoming electronic signals to determine their source. EWs are advanced electronic technicians who do wiring and circuit testing and repair. They determine performance levels of electronic equipment, install new components, modify existing equipment, and test, adjust, and repair cooling systems.

Fire Controlman:[3] FCs maintain the control mechanism used in weapons systems on combat ships. Complex electronic, electrical, and hydraulic equipment is required to ensure the accuracy of guided-missile and surface gunfire-control systems. FCs are responsible for the operation, routine care, and repair of this equipment, which includes radar, computers, weapons-direction equipment, target-designation systems, gyroscopes, and rangefinders. FCs are in the advanced electronics field, which requires a six-year enlistment.

Fire Control Technician:[3] FTs maintain advanced electronic equipment used in submarine weapons systems. Complex electronic, electrical, and mechanical equipment is required to ensure the accuracy of guided-missile systems and underwater weapons. FTs are responsible for the operation, routine care and repair of this equipment. They are in the advanced electronics field, which requires a six-year enlistment.

Gunner's Mate: Navy GMs operate, maintain and repair all gunnery equipment, guided-missile launching systems, rocket launchers, guns, gun mounts, turrets, projectors and associated equipment. They also make detailed casualty analyses and repairs of electrical, electronic, hydraulic, and mechanical systems. They test and inspect ammunition and missiles and their ordnance components, and train and supervise personnel in the handling and storage of ammunition, missiles, and assigned ordnance equipment.

Gas Turbine System Technician: GSs operate, repair, and maintain gas-turbine engines, main propulsion machinery (including gears, shafting and controllable-pitch propellers), assigned auxiliary equipment, propulsion-control systems, electrical and electronic circuitry up to printed circuit modules, and alarm and warning circuitry. They perform administrative tasks related to gas-turbine propulsion-system operation and maintenance.

Hospital Corpsman: HMs assist medical professionals in providing health care to service people and their families. They act as pharmacists, medical technicians, food-service personnel, nurses' aids, physicians' or dentists' assistants, battlefield medics, X-ray technicians, and more. Their work falls into several categories: first aid and minor surgery, patient transportation, patient care, prescriptions and laboratory work, food service inspections, and clerical duties.

Hull Maintenance Technician: HTs are responsible for maintaining ships' hulls, fittings, piping systems, and machinery. They install and maintain shipboard and shore-based plumbing and piping systems. They also look after a vessel's safety and survival equipment and perform many tasks related to damage control.

Interior Communications Electrician: ICs operate and repair electronic devices used in a ship's interior communications systems—SITE TV systems, public-address systems, electronic megaphones, and other announcing equipment—as well as gyrocompass systems.

Instrumentman: The Navy uses many meters, gauges, watches and clocks, typewriters, adding machines, and other office machines. Repairing, adjusting, and reconditioning them is the IM's job. An IM also repairs mechanical parts of electronic instruments, and is often called upon to manufacture parts such as bushings, stems, jewel settings, mainsprings and spring hooks.

Intelligence Specialist: Military information, especially secret information about enemies or potential enemies, is called intelligence. The IS is one of the people involved in collecting and interpreting intelligence data. An IS analyzes photographs and prepares charts, maps, and reports that describe in detail the strategic situation all over the world.

Journalist: JOs are the navy's information specialists. They write press releases, news stories, features, and articles for Navy newspapers, bulletins, and magazines. They perform a variety of public relations jobs. Some write scripts and announcements for radio and TV; others are photographers for radio and television broadcasters and producers.

Master at Arms in blue dress uniform, 1886.

The photo work of JOs ranges from administrative and clerical tasks to film processing. This is a five-year enlistment.

Lithographer:[2] LIs run Navy print shops and are responsible for producing printed material used in naval activities. LIs print service magazines, newspapers and bulletins, training materials, official policy manuals, etc. They operate printing presses, do layout and design, and collate and bind printed pages. The usual specialties are cameraman, pressman, and binderyman.

Legalman:[2] Navy LNs are aides trained in the field of law. They work in Navy legal offices performing administrative and clerical tasks necessary to process claims, to conduct court and administrative hearings, and to maintain records, documents, and legal-reference libraries. They give advice on tax returns, voter-registration regulations, procedures, and immigration and customs regulations governing Social Security and veterans' benefits, and perform many duties related to courts-martial and nonjudicial hearings.

Master-at-Arms:[2] Members of this rating help keep law and order aboard ship and at shore stations. They report to the executive officer, help maintain discipline, and assist in security matters. They enforce regulations, conduct investigations, take part in correctional and rehabilitative programs, and organize and train sailors assigned to police duty. In civilian life, they would be detectives and policemen.

Molder: MLs make molds, cores, and rig flasks. They make castings of ferrous and nonferrous metals, alloys, and plastics for the repair of ships, guns, and other machined equipment. MLs identify metals and alloys, heat-treat them, and test them for hardness. They operate furnaces used to melt metals for castings, and they use a variety of special hand and power tools.

Machinist's Mate: Continuous operation of the many engines, compressors and gears, refrigeration, air-conditioning, gas-operated equipment and other types of machinery afloat and ashore is the job of the MM. In particular, MMs are responsible for a ship's steam propulsion and auxiliary equipment and the outside (deck) machinery. MMs may also perform duties in the manufacture, storage, and transfer of some industrial gases.

Mineman: MNs test, maintain, repair, and overhaul mines and their components. They are responsible for assembling, handling, issuing, and delivering mines to the planting agent and for maintaining mine-handling and minelaying equipment.

Machinery Repairman: MRs are skilled machine-tool operators. They make replacement parts and repair or overhaul a ship engine's auxiliary equipment, such as evaporators, air compressors, and pumps. They repair deck equipment, including winches and hoists, condensers, and heat-exchange devices. Shipboard MRs frequently operate main propulsion machinery in addition to performing machine-shop and repair duties.

Mess Management Specialist: MSs operate and manage Navy dining facilities and bachelor enlisted quarters. They are cooks and bakers in Navy dining facilities ashore and afloat, ordering, inspecting, and stowing food. They maintain food-service and preparation spaces and equipment, and keep records of transactions and budgets for the food service in living quarters ashore.

Missile Technician:[2,3] MTs assemble, maintain, and repair missiles carried by submarines. They maintain the specialized equipment used in missile handling. Although missile components and related testing and handling equipment are primarily electrical and electronic, MTs must also work with mechanical, hydraulic, and pneumatic units in launcher, fire-control, and missile flight-control systems.

Musician: MUs play in official Navy bands and in special groups such as jazz bands, dance bands, and small ensembles. They give concerts and provide music for military ceremonies, religious services, parades, receptions, and dances. Official unit bands usually do not include stringed instruments, but each MU must be able to play at least one brass, woodwind, or percussion instrument. Persons are selected for this rating through auditions.

Navy Counselor:[2] NCs offer vocational guidance on an individual and group basis to Navy personnel aboard ships and at shore facilities, and to civilian personnel considering enlistment in the Navy. They assess the interests, aptitudes, abilities, and personalities of individuals.

Opticalman: OMs perform organizational - and intermediate - level maintenance on small navigational instruments, binoculars, night-vision sights, rangefinders, turret and submarine periscopes, and other optical instruments. OMs must be highly mechanical, with the ability to perform close, exact, and painstaking work.

Operations Specialist: OSs operate radar, navigation, and communications equipment in a ship's combat-information center or on the bridge. They detect and track ships, planes and missiles. They operate and maintain IFF (identification friend or foe) systems, ECM (electronic countermeasures) equipment and radiotelephones. OSs also work with search-and-rescue teams.

Ocean Systems Technician: OTs operate special electronic equipment used to interpret and document oceanographic data, such as the depth and composition of the ocean floor and how sound travels through water. They operate tape recorders and related equipment, prepare reports and visual displays, and convert analyzed data for use in statistical studies.

Postal Clerk: The Navy operates a large postal system manned by Navy PCs, who have much the same duties as their civilian counterparts. PCs collect postage-due mail, prepare customs declarations, collect outgoing mail, cancel stamps, and send the mail on its way. They also perform a variety of record-keeping and reporting duties, including maintenance of an up-to-date directory service and locator file.

Photographer's Mate: PHs photograph actual and simulated battle operations as well as documentary and newsworthy events. They expose and process light-sensitive negatives and positives; maintain cameras, related equipment, photo files, and records; and perform other photographic services for the Navy. This is a five-year enlistment.

Patternmaker: In a Navy foundry, PMs are the important link between the draftsman (DMs), who make the drawings, and the molders (MLs), who produce the castings. PMs make patterns in wood, plaster, or metal from which castings are made. PMs use drafting, carpentry, metalworking skills, and shop mathematics to create their patterns.

Personnelman: PNs provide enlisted personnel with information and counseling about Navy jobs, opportunities for general education and training, promotion requirements, and rights and benefits. In hardship situations, they also assist enlisted persons' families with legal aid or reassignments. PNs keep records up to date, prepare reports, type letters, and maintain files.

Aircrew Survival Equipmentman: Parachutes are the lifesaving equipment of aircrewmen when they have to bail out. In time of disaster, a parachute may also be the only means of delivering badly needed medicines, goods, and other supplies to isolated victims. PRs pack and care for parachutes as well as service, maintain, and repair flight clothing, rubber life rafts, life jackets, oxygen-breathing equipment, protective clothing, and air-sea rescue equipment.

Quartermaster: QMs are responsible for ship safety, skillful navigation, and reliable communications with other vessels and shore stations. In addition, they maintain charts, navigational aids, and records for the ship's log. They steer the ship, take radar bearings and ranges, make depth soundings and celestial observations, plot courses, and command small craft. QMs stand watches and assist the navigator and officer of the deck (OOD).

Radioman: Naval activities often involve people working at many different locations on land and at sea, and RMs operate the radio communications systems that make such complex teamwork possible. RMs operate radiotelephones and radioteletypes, prepare messages for international and domestic commercial telegraph, and send and receive messages via the Navy system.

Religious Program Specialist: RPs assist Navy chaplains with administrative and budgetary tasks. They serve as custodians of chapel funds, keep religious documents, and maintain contact with religious and community agencies. They also prepare devotional and religious educational materials, set up volunteer programs, operate shipboard libraries, supervise chaplains' offices, and perform administrative, clerical, and secretarial duties. They train personnel in religious programs and publicize religious activities.

Ship's Serviceman: Both ashore and afloat, SHs manage barbershops, tailor shops, ships' uniform stores, laundries, dry-cleaning plants, and cobbler shops. They serve as clerks in exchanges, soda fountains, gas stations, warehouses, and commissary stores. Some SHs function as Navy club managers.

Storekeeper: SKs are Navy's supply clerks . They see that needed supplies are available, everything from clothing and machine parts to forms and food. SKs have duties as civilian warehousemen, purchasing agents, stock clerks and supervisors, retail sales clerks, store managers, inventory clerks, buyers, parts clerks, bookkeepers, and even fork-lift operators.

Signalman: SMs serve as lookouts and, using visual signals and voice radios, alert their ship of possible dangers. They send and receive messages by flag signals or flashing lights. They stand watches on the signal bridge, encode and decode messages, honor passing vessels, and maintain signaling equipment. SMs must have good vision and hearing.

Sonar Technician: STs are responsible for underwater surveillance as well as assistance in safe navigation and search, rescue, and attack operations. They operate and repair sonar equipment and jam enemy sonar. They track underwater objects and repair ASW fire-control equipment and underwater radiotelephones.

Steelworker: SWs rig and operate all special equipment used to move or hoist structural steel, structural shapes, and similar material. They erect or dismantle steel bridges, piers, buildings, tanks, towers, and other structures. They place, fit, weld, cut, bolt, and rivet steel shapes, plates, and built-up sections used in the construction of overseas facilities. This is a five-year enlistment.

Torpedoman's Mate: TMs maintain underwater explosive missiles, such as torpedoes and rockets, that are launched from surface ships, submarines, and aircraft. TMs also maintain launching systems for underwater explosives. They are responsible for the shipping and storage of all torpedoes and rockets.

Utilitiesman: UTs plan, supervise, and perform tasks involved in the installation, operation, maintenance, and repair of plumbing, heating, steam, compressed-air systems, fuel storage and distribution systems, water treatment and distribution systems, air-conditioning and refrigeration equipment, and sewage-collecting and disposal facilities.

Weapons Technician: WTs maintain, store, inspect, test, adjust, repair, and package nuclear-weapon components and associated equipment for surface ships. They also assemble and disassemble nuclear weapons, warheads, and/or components.

Yeoman: YNs perform secretarial and clerical work. They greet visitors, answer telephone calls, and receive incoming mail. YNs organize files and operate duplicating equipment, and they order and distribute supplies. They write and type business and social letters, notices, directives, forms, and reports. They maintain files and service records.

Enlisted Service Record

Your service record contains all the papers and records concerning your Navy career. It is the Navy's official file on you.

Actually, you have two service records, one in the personnel office of your ship or station, which goes with you when you are transferred, and another in the Naval Military Personnel Command (NMPC). At the end of your naval service, the two records are combined and sent to a records-storage center.

Airman apprenticeship: Airman and airman apprentices wear emerald-green stripes on blue and white uniforms.

Hospital apprenticeship and dental apprenticeship: Hospitalmen and hospital apprentices, dentalman and dental apprentices, all wear white stripes on blue uniforms and navy-blue stripes on white uniforms. They also wear specialty marks that indicate their particular apprenticeship and distinguish them from seamen apprentices.

Rating Badges

Rating badges, worn on the left sleeve, consist of an eagle (called a crow), chevrons indicating the wearer's rate, and a specialty mark indicating rating.

Once you reach paygrade E-7 (CPO), a rocker, or arch, is added to the rating badge. The specialty mark is centered in the space between the eagle and the upper chevron. Senior chief petty officers (SCPOs) (E-8) also have a single silver star, centered above the eagle's head, while MCPOs (E-9) have two silver stars arranged horizontally above the eagle's wingtips. The rating badge for command MCPOs is the same as that for MCPOs except that an inverted five-point silver star takes the place of the specialty mark. The badge for fleet/force MCPOs is the same as that for command MCPOs except that all the stars are gold. The MCPON has three gold stars arranged in a horizontal line above the eagle's head.

Enlisted people in all services wear chevrons that indicate their paygrade. Personnel in the Coast Guard, for the most part, wear badges identical to those of the Navy.

Service Stripes

Service Stripes, or "hashmarks," are worn on the left sleeve below the rating badge and indicate length of service. Each stripe signifies completion of four full years of active or reserve duty (or any combination thereof) in any of the armed forces. Scarlet stripes are worn on blue uniforms, navy-blue stripes on forest-green uniforms (worn by aviation personnel).

Navy enlisted rates and paygrades are compared with those of the other services. The Coast Guard's badges are the same as the Navy's. Chevrons are red on blues with a white eagle, blue on whites with a blue eagle. CPO chevrons are red or gold as appropriate. Specialty marks are the same color as the eagle. Badges worn on dungaree shirts have dark blue chevrons but no specialty marks.

Decorations and Awards

Awards include any decoration, medal, badge, ribbon, or attachment thereof bestowed on an individual or a unit. A decoration is awarded to an individual for an act of gallantry or meritorious service. A unit award is presented to an operating unit and can be worn only by members who participated in the action cited. A service award is made to those who have participated in designated wars, campaigns, and expeditions, or who have fulfilled a specified service requirement. The Navy Cross, Bronze Star Medal, and Purple Heart are examples of decorations; the Presidential Unit Citation (PUC) and Meritorious Unit Commendation (MUC) are examples of unit awards. Service awards, often called campaign or theater awards, include the Good Conduct Medal and the Vietnam Service Medal.

The Navy recognizes 19 military decorations, 5 unit awards, 23 non-military decorations, 40 campaign and service awards, and a number of others by foreign governments. Many foreign awards may be accepted, but cannot be worn. There are more than 150 awards, including those bestowed by military societies and other organizations.

Military decorations and unit awards may be given at any time. They

Sitting: Joe B. Havens, HMC, USN, reenlisting LCDR H. L. Brown, CO, officiates in 1963, Port Arthur, Texas.

are listed below in order of precedence, with the four unit awards last. All other decorations are worn below these. Ribbons for decorations and awards are worn on the left breast. One, two, or three ribbons are worn in a single row, centered above the pocket. When more than three ribbons are authorized, they are worn in horizontal rows of three each. If not in multiples of three, the uppermost row contains the lesser number, with the ribbon(s) centered over the row beneath.

Medal of Honor
Navy Cross
Defense Distinguished Service Medal*
Distinguished Service Medal
Silver Star Medal
Defense Superior Service Medal*
Legion of Merit
Distinguished Flying Cross
Navy and Marine Corps Medal
Bronze Star Medal
Purple Heart
Defense Meritorious Service Medal*
Meritorious Service Medal
Air Medal
Joint Service Commendation Medal*
Navy Commendation Medal
Joint Service Achievement Medal*
Navy Achievement Medal
Combat Action Ribbon
Presidential Unit Citation Ribbon
Joint Meritorious Unit Award*
Navy Unit Commendation Ribbon
Navy "È"
*Not Navy decoration—listed for precedence only.

See USN Uniform Regulations, Chapter 10, for the order of precedence and proper wear of all awards.

Ownership Markings

Articles of clothing are legibly marked with the owner's name and the last four digits of the owner's social security number in black marking fluid for white clothes and chambray shirts. White marking fluid is used for blue clothes and dungaree trousers, or indelible ink when labels are provided for the purpose. All markings other than those on labels should be made with a half-inch stencil or stamp.

Detailed ownership marking instructions are furnished to recruits when clothing is issued to them. These instructions must be followed explicitly. As a general rule, instructions for marking clothes, as laid down in the uniform regulations, should be followed when additional uniform articles are obtained.

The word right or left means the owner's right or left when the article is worn. On towels, it means the owner's right or left when standing behind the article laid out for inspection. Markings on all articles, properly rolled or laid out from bag inspection, will appear right side up to the inspecting officer and upside down to the person standing behind them.

Divine Services

When divine services are held on board, the church pennant is flown, and word is passed that services are being held in a certain space of the ship and to maintain quiet about the decks. A person entering the area where services are held uncovers, even if the person is on watch and wearing a duty belt and sidearm. There is one exception: remain covered for a Jewish ceremony.

Sick Bay

In the days of sailing ships, it was customary to uncover when entering sick bay, out of respect to the dying and dead. Though modern medicine has transformed the sick bay into a place where people are usually healed and cured, the custom remains. As in any hospital, silence is maintained. Smoking is usually not permitted in sick bay, partly because the oxygen used for medical purposes is a fire hazard.

Officers' and CPO Country

Officers' country includes all staterooms and the wardroom. CPO (chief petty officer) country includes CPOs living spaces and mess. Do not enter these areas except on business, and do not use their passageways as thoroughfares or shortcuts. When entering the wardroom, or any compartment or office in officer or CPO Country, uncover. Watchstanders wearing a duty belt or sidearm remain covered, unless a meal is in progress. Always knock before entering any officer's or CPO's room.

Enlisted Mess Deck

The mess deck for enlisted personnel is treated with the same courtesy as the wardroom. Always uncover when on or crossing mess decks, even if you are on watch and wearing the duty belt.

Sentry Duties and Recruit Drills

One of the first military duties a recruit will perform is a sentry or security watch. Security means protecting a ship or station against damage by storm or fire, and guarding against theft, sabotage, and other subversive activities. Chapter 7 discusses security in greater detail.

Security involves sentry duty, guard duty, fire watches, and barracks watches. Sentry duty is formal military duty governed by specific orders. Guard duty may be the same as sentry duty, or a guard may be permitted to relax military bearing, so long as he or she is on the job and ready to act. A fire watch may mean covering an assigned area on foot or in a vehicle, or it may mean assignment to a certain place for a specified period. A barracks watch may mean standing sentry duty, or merely being available to answer a phone, check people in and out, turn lights off and on, and preserve order and cleanliness.

Requirements for standing sentry duty are the same as those for all watches: keep alert, attend to duty, report all violations, preserve order, and remain on watch until properly relieved. The basic rules or orders for sentries are the same for all security watches.

Being detailed to a sentry watch involves two sets of orders: special orders and general orders. Special orders apply to a specific type of watch. They will be passed on and explained to you by the petty officer of the watch or the petty officer of the guard. General orders never change. You will—on any watch or duty, now and in the future—be responsible for carrying them out, even if no one has explained them to you or reminded you of them. The 11 general orders, with a brief explanation of each, follow. Memorize them and be ready to recite them whenever called on to do so.

The General Orders

1. To take charge of this post and all government property in view.
2. To walk my post in a military manner, keeping always on the alert and observing everything that takes place within sight or hearing.
3. To report all violations of orders I am instructed to enforce.
4. To repeat all calls from posts more distant from the guardhouse than my own.
5. To quit my post only when properly relieved.
6. To receive, obey, and pass on to the sentry who relieves me all orders from the commanding officer, command duty officer, officer of the deck, and officers and petty officers of the watch.
7. To talk to no one except in the line of duty.
8. To give the alarm in case of fire or disorder.
9. To call the officer of the deck in any case not covered by instructions.
10. To salute all officers, and all colors and standards not cased.
11. To be especially watchful at night, and during the time for challenging, to challenge all persons on or near my post, and to allow no one to pass without proper authority.

Orders 1, 2 and 3 mean that all persons in the service, whatever their ranks, are required to respect you in the performance of your duties as a sentinel and a member of the guard.

Report immediately, by telephone or other means, every unusual or suspicious event.

Apprehend and turn over to proper authority all suspicious persons involved in a disorder on or near your post, and anyone who tries to enter your post without authority.

Report violations of orders when you are inspected or relieved. If it is urgent and necessary, apprehend the offender and call the petty officer of the guard.

Order 4 means that you "pass the word" by calling "Petty officer of the guard, no. __," giving him or her the number of your post, when you need that person for any purpose other than relief, fire, or disorder.

Order 5 means that if you become sick or for any reason must leave your post, you call "Petty officer of the guard, no.__, relief." Do not leave your post for meals or other reasons unless properly relieved. If your relief is late, telephone or call the petty officer, but do not leave your post.

Order 6 names the officers whose orders you must obey. However, any officer can investigate apparent violations of regulations when he or she observes them.

Give up possession of your rifle only on receiving a direct order to do so from a person authorized to give you orders while you are on your post. No other person may require a sentinel to hand over his or her rifle or even require it to be inspected.

Order 7 is self-explanatory. When challenging or holding conversations with any person, take the position of "port arms" if you are armed with a rifle, the position of "raise pistol" if you are armed with a pistol.

Order 8 means that if fire is discovered, you must immediately call, "Fire, no. __," then turn in the alarm or make sure it has been turned in. If possible, put out the fire.

Order 10 covers saluting. (For more details on saluting, see p. 91). A sentry salutes as follows: If walking post, he or she halts. If armed with a rifle, he or she salutes by presenting arms. If otherwise armed, he or she renders the hand salute. On patrol duty, he or she does not halt, unless spoken to, but renders the hand salute. In a sentry box, he or she stands at attention in the doorway upon the approach of the person or party involved and renders the hand salute (or, if armed with a rifle, presents arms).

When required to challenge, a sentry salutes an officer as soon as the officer is recognized.

A sentry salutes an officer as he or she comes on the post. When an officer stops to talk, the sentry assumes the position of port arms if armed with a rifle, or the position of attention throughout the conversation, and salutes again when the officer leaves.

When talking to an officer, the sentry does not interrupt to salute another officer unless the officer being addressed salutes. Then the sentry follows his or her example.

When the flag is raised at morning colors or lowered at evening colors, the sentry stands at attention at the first note of the national anthem or the call to colors and salutes. A sentry engaged in a duty that would be hampered doesn't have to salute. The sentry should face the flag while saluting, but if duty requires he or she may face in another direction.

Security of Information

The word security, as it is used in the Navy, can mean many things, but its most common usage refers to the safeguarding of classified information. Security can also mean the protection of ships and stations or property, which will be discussed later in relation to external security.

Because the safety of the United States in general, and of naval operations in particular, depends greatly on the protection of classified information, it is important that you understand what classified information is, who may have access to it, some rules for safeguarding it, and the penalties for security violations.

Security Classification

Information is classified when the interests of national security are at risk. It is assigned a classification designation, which tells you how much protection it requires. There are three classification designations—top secret, secret, and confidential— to indicate the anticipated degree of damage to national security that could result from unauthorized disclosure. The expected impact for each designation is as follows: top secret—exceptionally grave damage; secret—serious damage; and confidential—damage. Regardless of the level, all classified information must be protected against unauthorized disclosure. Unauthorized disclosure, or "compromise," means that classified information becomes available to a person not authorized to have it.

There is another category of information, for official use only (FOUO). This is not classified information (it does not involve national security), but it cannot be divulged to everyone. Results of investigations, examination questions, bids on contracts, etc., are "privileged information," kept from general knowledge under the designation FOUO.

Senior Enlisted Grades

Until 1958 a person who had advanced to chief petty officer (CPO, E-7) had gone as high as possible in the enlisted rating structure. Then the grades of senior chief petty officer (SCPO, E-8) and master chief petty officer (MCPO, E-9), were established to give additional recognition to people with outstanding technical, leadership and supervisory abilities.

While advancement to E-7 is considered normal for a 20-year career, the E-8 and E-9 paygrades are regarded as 30-year career plans. Candidates for E-8 and E-9, having met all the requirements, have their service records closely screened by a selection board at NMPC. This is where a good record becomes extremely important—those with the best records are selected over those with records that may be good, but not good enough.

Master Chief Petty Officer of the Navy

The master chief petty officer of the Navy (MCPON) is the Navy's senior enlisted member. Assigned to the Office of the Chief of Naval Operations for a 3-year tour of duty, the MCPON serves as senior enlisted representative of the Navy and as senior enlisted adviser to the chief of naval operations and the chief of naval personnel in all matters pertaining to enlisted personnel.

The MCPON also serves as an advisor to many boards dealing with enlisted personnel; accompanies the chief of naval operations on some trips; serves as the enlisted representative of the Department of the Navy at special events, celebrations, and ceremonies; and maintains a liaison with the Navy Wives' Club of America.

Minimum requirements for advancement to specific paygrades, including professional knowledge and practical work, are part of the military standards listed in the PQS. Occupational standards—specific practical skills and knowledge you'll need for advancement in a general rate or rating—are also listed. The career-pattern section of the manual outlines the path of advancement you will normally take. It also tells you some of the schools you must attend.

Service Schools

Navy service schools are located at the three training centers and at Memphis, Tennessee; Gulfport and Meridian, Mississippi; Norfolk, Virginia; and Port Hueneme (pronounced Y-neemee), California, among other places. For some ratings, graduation from a particular service school is necessary for advancement; the PQS Manual will tell you which ones. Selection for a service school depends on your rate, time in service, current duty assignment, school quotas, and the operational schedule of your unit. Although you can attend a service school on a temporary additional duty (TAD) basis from your current duty station, most school assignments are made with a permanent change of station (PCS).

The five types of enlisted service schools are:

Class A: Provides the basic technical knowledge required for mob performance and later, specialized training. A Navy enlisted classification (NEC) code may be awarded to identify the skill.

Class C: Advanced skills and techniques needed to perform a particular job are taught. This category includes schools and courses previously identified as Class B. An NEC code may also be awarded to identify the level of skill.

Class E: Designed for professional education leading to an academic degree.

Class F: Trains fleet personnel who are en route to, or are members of, ships' companies. Also provides individual training such as refresher, operator, maintenance, or technical training of less than 13 calendar days. An NEC code is not awarded.

Class R: This is the basic school that provides initial training after enlistment. It prepares the recruit for early adjustment to military life by inculcating basic skills and knowledge about military subjects. Class R schooling does not include apprenticeships.

Living Aboard Ship

All ships have standard routines for in port and at sea. The routine varies on different ships; a submarine on extended patrol will run on a schedule different from that of an aircraft carrier on around-the-clock flight operations. Below is a sample standard routine. Departures from the normal routine are published in the plan of the day (POD).

Plan of the Day

This is "the word," the daily schedule of events, prepared and issued by the executive officer. It will name duty officers, assign various watches, and include any changes or additions to the normal routine and orders of the day—drills, training schedule, duty section, liberty section and hours, working parties, movies, examinations, or inspections.

The POD, distributed to the officer of the deck (OOD), all offices, officers, and division bulletin boards, is carried out by the OOD and all division officers.

Daily Routine at Sea

Note: When time is left blank, it is specified in the POD. Standard reports appear in quotation marks.

Sundays, Weekdays and Holidays

		Routine
0030	0030	JOOW (junior officer of the watch) inspects the lower decks; hourly thereafter until sunrise.
0030	0030	Call the morning watch. Call galley force.
0400	0400	Reveille for duty cooks (cooks may specify time to be called by signing up in the wake-up log kept by the boatswain's mate of the watch).
Sunrise	Sunrise	Turn off running lights. If the ship is darkened, light ship. Hoist pennants and flags as necessary.
0530	0600	Call masters-at-arms (MAA), division police petty officers, and mess cooks. Early mess for designated persons.
0600	0630	"Up all idlers" (non-watchstanders).
0605	0635	Announce weather.
0615	0645	"Turn to. Scrub down weather decks. Sweep down all compartments. Empty all trashcans." "Lay below to the MAA office for muster, all restricted persons." Pipe sweepers, MAA report to OOD, "Idlers turned out." Clean boats and fuel boats as necessary.
0655	0655	"Clear mess decks." Mess call.
0700	0700	Pipe to breakfast. "Testing general alarm." On completion: "Test of general alarm completed." "Uniform of the day__," or "Uniform for captain's inspection is __."
0720	0720	"Relieve the watch, on deck the __ section. Lifeboat crew of the watch to muster."
0755	0755	OOD reports "8 o'clock: request permission to strike eight bells" to the captain.
0800		Officers' call.
0805		Assembly.
0815	0815	Sick call.
0815		Rig for church. "Knock off ship's work; shift into uniform for inspection." Uniform for inspection is __."
0915		Officers' call. "All hands to quarters for inspection."
0930		"Knock off work. Shift into clean uniform of the day." Church call. "Maintain quiet about the decks during divine service." Hoist the church pennant. At the end of service, pennant is hauled down. Commence holiday routine.
1115	1115	Pipe sweepers. "Sweepers, man your brooms. Make a clean sweep down fore and aft. Empty all trashcans." Early mess for cooks, mess cooks, and MAA.
1145		"Knock off all ship's work."
1150	1150	Quartermaster reports "Chronometers wound and compared" to OOD. OOD reports "12 o'clock" to the admiral if embarked, and "12 o'clock, chronometers wound and compared; request permission to strike eight bells" to the captain.
1200	1200	Pipe to dinner.
1220	1220	"Relieve the watch, on deck the __ section. Lifeboat crew of the watch to muster."
1300		"Turn to." At this time, extra duty persons muster to MAA office.
1300	1300	Pipe sweepers. "Sweepers, man your brooms. Make a clean sweep down fore and aft. Empty all trashcans."
1300		Friday, or when ordered, inspections call, "Stand by for inspection on lower decks."
1545		"Relieve the watch, on deck the __ section. Lifeboat crew of the watch to muster."
1600	1600	Pipe sweepers. "Sweepers man your brooms. Make a clean sweep down fore and aft."
1600		"Knock off all ship's work."
1600	1600	Test running lights and emergency identification signals, report their readiness to the OOD (or at least one hour before sunset).
1630	1630	Early mess for mess-deck MAA and mess cooks.
1645		"Observe sunset." Set the prescribed material condition. Division damage-control petty officers report closures to damage-control central or sign the closure log maintained by the OOD on the bridge.
1655	1655	Mess call. "Clear the mess decks."

Sunset	Sunset	If the ship is to be darkened: "Darken ship. The smoking lamp is out on all weather decks." If ship is not to be darkened, turn on running lights and haul down colors following motion of senior officer present afloat (SOPA). Lookouts report running lights bright; lifebuoy watch on fantail reports stern light and lifebuoy light bright (and on the half hour thereafter until sunrise). Boatswain inspect weather decks and boats; when secured, report to the OOD.
1700	1700	Pipe to supper. Close watertight doors, etc.: "Set material condition __ throughout the ship."
1720	1720	"Relieve the watch. On deck, the __ section. Lifeboat crew of the watch to muster."
1730	1730	Security patrols make reports to the OOD, and hourly thereafter. Coxswain of life boat reports lifeboat crew mustered and boat ready for lowering (every half hour) and engine tested (once each watch). Corporal of the guard reports police conditions (every half hour thereafter).
1745	1745	Sick call.
1800	1800	Pipe sweepers. "Sweepers, man your brooms. Make a clean sweep down on lower decks and ladders. Empty all trashcans." "Lay below to the MAA office for muster, all restricted persons." Rig for movies. Movie call. Time and place as designated in POD.
1930	1930	"Now lay before the mast all 8 o'clock reports."
1945	1945	OOD reports "8 o'clock" to the admiral if embarked, and "8 o'clock, lights out and galley ranges secured; prisoners and lower decks secured; request permission to strike 8 bells," to the captain.
Dark	Dark	Dump trash and garbage. Pump bilges (Oil Pollution Act permitting). Blow tubes if wind favorable and plant requires it.
2000	2000	Hammocks. "Out lights, and silence in all berthing spaces."
2100	2100	MAA reports to the OOD "9 o'clock, lights out."
2155	2155	Tattoo.
2200	2200	MAA reports to the OOD "10 o'clock, lights out."
2330	2330	Call the watch.
2345	2345	Relieve the watch, on deck, the __ section. Lifeboat crew of the watch to muster.

Daily Routine in Port

0030	0030	JOOW inspects lower deck and boats in water (and hourly thereafter until reveille).
0330	0330	Call the watch.
0350	0350	Relieve the watch.
Daylight		Call galley force. Turn off all unnecessary lights.
Sunrise		Turn off anchor, boom, and gangway lights. Hoist guard flags, absentee pennants as necessary. If darkened, light ship; the smoking lamp is lighted on the top side.
0540	0600	Call MAA, division police petty officer, mess cooks, and boat crews. Early mess for designated personnel.
0600	0600	Reveille. "Reveille, all hands, heave out and trice up."
0605		Announce weather to the crew.
0615		Pipe sweepers. "Turn to. Scrub down weather decks, sweep down compartments, empty all trashcans." Division police petty officers report to the duty MAA that persons of their divisions are turned out. MAA reports to OOD "Crew turned out." Fuel all boats and test engines.
0655	0715	Mess call. "Clear all mess decks."
0700	0720	Pipe for breakfast, "Uniform of the day is__," "or "Uniform for inspection is __," Meal pennant is hoisted. Announce weather over officer's circuit.
0720	0740	Relieve the watch.
0750	0750	Guard of the day.
0755	0755	First Call. Hoist PREP, signifying prepare for colors. OOD reports "8 o'clock" to the admiral if embarked, and "8 o'clock, request permission to strike 8 bells" to the captain.

0800	0800	Morning colors. Meal pennant hauled down.
0800		All hands to quarters for muster. Officers' call. Assembly.
0815		Pipe retreat. "Turn to, commence ship's work."
0800		"Turn to, sweep and clamp down weather decks and living spaces."
0815	0815	Sick call.
0815		Rig for church. "Knock off work. Shift into uniform for inspection. The uniform for inspection is__." Officers' call. "All hands to quarters for inspections."
0915		Knock off work. Shift into clean uniform of the day."
0930		Church call. "Maintain quiet about the decks during divine service." Hoist the church pennant. At the end of divine service, haul down the church pennant. Commence holiday routine. Inspection of mess cooks.
1115	1115	Early mess for mess deck MAA and cooks.
1115	1115	Pipe sweepers. "Sweepers, man your brooms. Make a clean sweep down fore and aft. Empty all trashcans."
1145		"Knock off all ship's work."
1150	1150	Quartermaster reports "Chronometers wound and compared" to the OOD.
1155	1155	Mess call. "Clear the mess decks." OOD reports "12 o'clock" to the admiral if embarked, and "123 o'clock, chronometers wound and compared; request permission to strike 8 bells" to captain.
1200	1200	Pipe to dinner. Hoist meal pennant.
1220	1220	"Relieve the watch." "Commence holiday routine." Extra-duty persons muster at MAA office.
1300		"Turn to."
1300	1300	Pipe sweepers. "Sweepers, man your brooms. Make a clean sweep down fore and aft. Empty all trashcans."
1300		Haul down meal pennant. Friday, or when ordered, inspection call. "Stand by for inspection of lower decks."
1545	1545	"Relieve the watch."
1600	1600	Pipe sweepers. "Sweepers man your brooms. Make a clean sweep fore and aft."
1600		"Knock off all ship's work." All hands shift into the uniform of the day. Extra-duty persons muster at MAA office.
1600	1600	Test anchor and boom lights. Rig and test gangway lights and floodlights. Report their readiness to OOD (or at least one hour before sunset). Liberty call.
Half hour before sunset		Guard of the day.
5 minutes before sunset		First call. Hoist PREP.
Sunset		Evening colors. Turn on anchor, boom, accommodation ladder (brow), and flood lights. If ship is to be darkened, "Darken ship. The smoking lamp is out on the top side."
1645	1645	Early mess for mess-deck MAA and mess cooks.
1700	1700	Closure of watertight doors, etc. "Set material condition__ throughout the ship."
1730	1730	Pipe to supper. Hoist meal pennant, if before sunset.
1750	1750	"Relieve the watch."
1800	1800	Haul down meal pennant at sunset or completion of supper. Rig for movies.
1815	1815	Pipe sweepers. "Sweepers man your brooms. Make a clean sweep down all lower decks and ladders. Empty all trashcans. Lay below to the MAA office for muster, all restricted persons." Movie call. Time and place as designated in POD.
1930	1930	"On deck, all the 8 o'clock reports."
1950	1950	"Relieve the watch."
1955	1955	OOD reports " "8 o'clock" to the admiral if embarked, and "8 o'clock, lights out and galley ranges secured. Request permission to strike 8 bells" to the captain. XO (or command duty officer) reports to captain (if on board), "All departments secured for the night (or as appropriate)."
2000	2000	Hammocks. "Out all lights and silence in all berthing spaces."

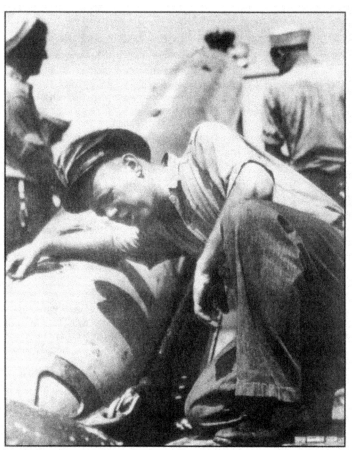

A chief petty officer directs the loading of a torpedo on board a submarine at New London, Conn., in 1943. (Courtesy of U.S. Navy National Archives, Edward Steichen)

2100	2100	MAA reports to OOD "9 o'clock, lights out." Tattoo. "Turn in, keep silence about the decks." Taps (5 minutes after tattoo).
2200	2200	MAA reports to OOD "10 o'clock, lights out."
2350	2350	"Relieve the watch."

Standard Organization and Regulations of the US Navy

If you had to learn a new set of regulations and an entirely different organization every time you moved from one division, department, or ship to another, you would waste time. The Navy has standardized everything—routine, regulations, and organizations—as much as possible on all ships.

This information is contained in the current edition of the Standard Organization and Regulations of the US Navy (OPNAVINST 3120.32). You will be loaned a copy. Your daily and weekly routine aboard ship will be governed by it, as will the organization of your division and department. You must first know and become familiar with the general regulations, a list of which follows. You must read all of the regulations, and sign a statement to the effect that you have done so and that you understand them. Whether you are aboard a minesweeper or an aircraft carrier, the titles and numbers of the general regulations will be the same.

Communications

When you mention communications in the Navy, most sailors think of a radioman copying a radio message or a signalman handling flag hoists during fleet operations. But communications is much more than that—it involves everyone. The bow lookout using a sound-powered telephone to report to the bridge is communicating; so is the officer of the deck (OOD) using the talk-between-ships (TBS) link to advise another ship of a course change. The lookout is using internal communications; the OOD is using external communications. All naval communications can be classified as one or the other.

Internal Communications

Internal communications are relayed aboard a single ship using both sound and visual methods. Communication by messenger, probably the oldest of all means, is still the most reliable. Other means include everything from "passing the word" over the intercommunications voice (MC) circuits to using "squawk boxes," sound-powered telephones, dial phones, bell and buzzer systems, or boat gongs.

Internal communications also means printed or written material such as the plan of the day (POD), visual-display systems such as the rudder-angle indicator and engine-order telegraph on the bridge, the combat information center (CIC) plot, and even the "onboard/ashore board" for officers on the quarterdeck. Everyone aboard the ship must be aware of the internal communications system at all times.

Passing the Word

In the old Navy, before the days of loudspeaker systems, the boatswain's mate (BM) passed any orders for the crew by word of mouth. The BM of the watch (BMOW) sounded "Call mates" on his pipe to get the BMs together, and they answered repeatedly with the same call while converging on the bridge or quarterdeck. Upon hearing the word, they dispersed fore and aft to sing it out at every hatch.

While this procedure was colorful, it took a lot of time. Today a single BM can quickly pass the word over the MC network while the others stay where they are. The basic MC circuit is the IMC, the general announcing system, over which the word can be passed to every space in the ship. The general alarm system is also tied into it. Transmitters are located on the bridge and quarterdeck; additional transmitters may be installed at other points.

An announcement is preceded by a boatswain's call or pipe. "All hands" is piped before any word concerning drills and emergencies. "Attention" is piped before the passing of routine messages.

Common shipboard events are listed below, with the appropriate orders following each one. The orders may differ slightly from ship to ship.

Air bedding: "All divisions, air bedding."

Arrivals and departures: Title of officer, preceded by proper number of boat gongs, for example (chief of naval operations) arriving (departing)."

Boats: "Away, the motor whaleboat (gig, barge), away."

Church call: "Divine services are now being held in (space). Maintain silence about the decks during divine services."

Eight o'clock reports: "On deck all 8 o'clock report."

CSP-Q Marion F. Hutchison was a chief near the end of World War II.

Extra-duty personnel: "Lay up to the quarterdeck for muster, all extra-duty personnel (or other special groups)."

Fire: "Fire! Fire! Class (A,B, etc.), compartment A205L (or other location, including deck, frame or side). This is not a drill. Away the duty fire party."

Flight quarters: "Flight quarters. Flight quarters. Man all flight quarters stations to launch (recover) aircraft (helicopters)."

Hoist boats: "First division stand by to hoist in (out) no. (1, 2, etc.) motor launch (gig)."

Inspection (personnel): "All hands to quarters for captains' personnel inspection."

Inspection (material): "Stand by all lower deck and topside spaces for inspection."

Knock off work (before evening meal): "Knock off all ship's work." (First pipe "All hands.")

Late bunks: "Up all late bunks."

Liberty: "Liberty to commence for the (first) and (third sections at 1600; to expire on board at (hour, date, month)."

Mail: "Mail call."

Meals: "All hands, pipe to breakfast (noon meal or dinner, evening meal or supper)." (First pipe "Mess call.")

Mess gear (call): "Mess gear (call). Clear the (all) mess decks." (First pipe "Mess call.")

Mistake or error: "Belay my last."

Muster on stations: "All divisions muster on stations."

Pay: "The crew is now being paid (space is given)."

Preparations for getting under way: "Make all preparations for getting under way."

Quarters for muster: "All hands to quarters for muster, instruction and inspection."

Quarters for muster (inclement weather): "All hands to quarters for muster. Foul-weather parade." (First pipe "All hands.")

Rain squall: "Haul over all hatch hoods and gun covers."

Readiness for getting underway reports: "All departments, make readiness for getting underway reports to the OOD on the bridge."

Relieving the watch: "Relieve the watch. On deck the (no.) section. Lifeboat crew on deck to muster. Relieve the wheel and lookouts." (First pipe "Attention.")

Rescue and assistance: "Away rescue and assistance party, (no.) section."

Reveille: "Reveille. Reveille, all hands heave out and trice (lash) up." Or, "Reveille. Up all hands, trice up all bunks." (First pipe "All hands.")

Shifting the watch: "The OOD is shifting his watch to the bridge (quarterdeck)."

Side boys: "Lay up on the quarterdeck, the side boys."

Smoking: "The smoking lamp is lighted (out)." (Unless the word applies to the whole ship, the space should be specified.)

Special sea detail: "Go to (man) your stations, all the special sea and anchor detail." Or, "Station the special sea and anchor detail."

Sweepers: "Sweepers, start (man) your brooms. Make a clean sweep down fore and aft." (First pipe "Sweepers.")

Taps: "Taps, lights out. All hands turn in to your bunks and keep silence about the decks. Smoking lamp is out in all living spaces." (First pipe "Pipe down.")

Turn to: "Turn to (scrub down all weather decks, scrub all canvas, sweep down compartments, dump trash)."

The OOD is in charge of the IMC. No call can be passed over it unless it is authorized by him, the executive officer, or the captain, except for a possible emergency call by the damage-control officer.

Normally, the IMC is equipped with switches that make it possible for certain spaces to be cut off from announcements of no concern to them. The captain, for instance, does not want his cabin blasted with calls for individuals to lay down to the spud locker. If the BMOW is absent, and you are required to pass the word yourself, be sure you know which circuits should be left open. Some parts of the ship have independent MC circuits, such as the engineers' announcing system (2MC) and the hangar-deck announcing system (3MC).

The bull horn (6MC) is the intership announcing system, but it is seldom used for communication between vessels. It is, however, a convenient means of passing orders to boats and tugs alongside or to line-handling parties beyond the range of the speaking trumpet. If the transmitter switch is located on the 1MC control panel, do not cut the bull horn when you are passing a routine word.

Navy History

Why History?

History can be dry and dull. But a glance at the history of an organization can give us a good idea of what that organization is, and what it has done. Knowing that, we can more easily figure out our place in it. History can show us where mistakes were made before. If enough of us are aware of those mistakes, we can avoid making them again.

What follows is a bare-bones record of the accomplishments and failures of the US Navy, to help you find out more about this organization to which you now belong. Remember, today is tomorrow's history. You are helping to make it.

The Earliest Years

America was born of the sea. The people who made this nation came from over the sea, and they were sustained by goods exchanged by the shipload. Trade went on for 150 years before the desire to be master of their own destiny led the colonists to strike for independence. The first efforts at sea power were often feeble and fruitless, and yet they had their impact on the course of events. And at the critical juncture, it was the timely actions of the French navy that resulted in the isolation of British General Cornwallis and his subsequent surrender.

12 Jun 1775	First engagement at sea during the Revolution. Citizens of Machias, Maine, under the command of Jeremiah O'Brien, seized a cargo sloop and with her captured the cutter HMS *Margaretta* (TB 30 and DDs 51, 415 and 725 were named *O'Brien*).
6 Sep 1775	The schooner *Hannah* sailed as the first unit of a number of armed fishing vessels sent to sea by the Continental Army to intercept British supply ships during the siege of Boston.
13 Oct 1775	The Continental Congress authorized the outfitting of a 10-gun warship "for intercepting such transports as may be laden with stores for the enemy" - the start of the Continental Navy. The same act established the Marine Committee, though Congress directed its naval efforts.
3 Dec 1775	The first man-of-war of the Continental Navy, the *Alfred*, was commissioned at Philadelphia. Her "first lieutenant" (XO) was Lieutenant John Paul Jones.
3-4 Mar 1776	A Continental squadron under the command of Commodore Esek Hopkins, composed of the *Alfred* (24 guns), *Columbus* (20), *Andrea Doria* (14), *Cabot* (14), *Providence* (12), *Hornet* (10), *Wasp* (8), and *Fly* (8), successfully attacked the British at Nassau in the Bahamas. (Sailing men-of-war were rated by the number of broadside guns they mounted.) Captured were 71 cannon and 15 mortars. This was also the first amphibious assault by American Marines, under the command of Captain Samuel Nicholas. (DDs 311 and 449 were named for him.)
4 Apr 1776	The brig *Lexington* (16) under John Barry, defeated HMS *Edward* (8) in lower Delaware Bay. This was the earliest of Barry's successes. (DDs 2, 248, and 933 were named for him.)
7 Sep 1776	Sergeant Ezra Lee of the Continental Army made the first "submarine" attack on a warship, an unsuccessful attempt to attach a powder charge to the hull of an anchored British ship from the submersible *Turtle*, designed by David Bushnell. Submarine Tender AS-2 and AS-15 were named for Bushnell. (The deep-submergence craft DSV-3 is named *Turtle*.)
11 Oct 1776	A Continental Army squadron of gunboats under Colonel Benedict Arnold fought a British force on Lake Champlain in the Battle of Valcour Island. This caused the British to delay the invasion of the Hudson River Valley for a year, by which time the Continental Army defeated it.
15 Nov 1776	Continental Congress set pay rates for officers and men. Petty officer rates were prescribed, though these were not divided into classes until 1885.
16 Nov 1776	The US flag was saluted for the first time by the Dutch governor of St. Eustatius Island in the West Indies.
24 Apr 1778	John Paul Jones, in command of the sloop *Ranger*, defeated the sloop HMS *Drake* off Belfast, Ireland. The Drake became the first major British warship to be taken by the new Navy.
4 May 1780	An insignia, adopted by the Board of Admiralty, set up by the Continental Congress to direct naval operations, became the Navy's first official seal.
23 Sep 1780	John Paul Jones, now commanding the converted merchantman *Bon Homme Richard* (42), defeated the frigate HMS *Serapis* (50) in a night fight off Flamborough Head, England. His ship badly battered (she would sink after the fight), Jones rejected the British surrender question with his defiant," I have not yet begun to fight!" (DDs 10 and 230 and DDG 32 were named in honor of Jones and DDs 24, 290 and 353, and CG 19 in honor of his gallant first lieutenant, Richard Dale.)
5 Sep 1781	The French fleet, under Admiral Comte de Grasse, blockaded Hampton Roads to keep reinforcements from General Cornwallis' British Army at Yorktown, Virginia, under siege by General George Washington's Continental troops and by French forces under General Rochambeau. The *Comte de Grasse* (DD-974) honors this ally.
17 Oct 1781	General Cornwallis surrendered at Yorktown.

Bullion Eagle Rating Badges

In an attempt to provide some means of dating rating badges with the eagles embroidered in silver bullion, the writer has identified and classified six distinct patterns of bullion eagles from samples. While many variations exist, they generally appear to match to a large degree one of the patterns illustrated here. It should be noted that the time periods applied to the six patterns are estimated and there seems to be a considerable amount of overlap, as well as numerous minor changes in patterns from one time period to the next. In the Introduction, this section was referred to as a survey and is not intended as absolute, simply an aid to the collector's attempts to establish the age for rating badges of this type.

The silver bullion eagle on the rating badge has been used by all petty officers when the gold chevrons have been earned. The silver eagle is also worn on the scarlet chevron blue rating badge of chief petty officers who are not entitled to the gold chevrons. Silver bullion eagles are found on green, gray and khaki rating badges as well, although only authorized for use on the khaki rating badge, and that since only 1949.

As described in Part 1, prior to 1941, all rating badges had the eagle facing to its left. During the period 1941 to 1948, only rating badges of the seaman branch had left facing eagles; all others faced to the eagle's right. After 1948 all rating badges were to be worn on the left sleeve with an eagle that faced to its right. These configurations also apply to the silver eagles.

TYPE I

From early samples known to be of the World War I period: A coil of bullion forms the upper edge of the wings, with a twist of two threads forming the division between the wing shoulder and the feathers. The shoulder portion of the wing is of diagonal stitches and the feathers of single horizontal stitches. The body is formed by twists of threads in a vertical pattern, with a knot of red thread for the mouth and a black knot for the eye. The perch is of single diagonal stitches and the tail is formed by five straight vertical stitches of the same length.

TYPE II

(A) The wings made of short diagonal stitches forming the shoulder and each feather, with long diagonal stitches overlapping the bases of the four topmost feathers. The body is of diagonal stitches with three twists of thread vertically down the center. The legs are formed of horizontal stitches with a vertical stitch and three short diagonals for the claws. The perch is of short diagonal stitches. The tail is made of 13 vertical stitches, alternating short and long.

(B) This variation has each wing feather separate, without the overlapping stitches on the top four feathers. Five twists of thread form the body; nine stitches for the tail, and a separate stitch for the beak, brow, and the upper part of the peak. The upper edge of the wing is outlined by a bullion coil.

TYPE III

World War II and post-war period: Embroidered in lightweight silver thread rather than bullion. The pattern of the stitches is basically the same as that of eagles on cloth rating badges. There is no row of short inner wing feathers as on the cloth rating badges and later types of bullion eagles. The eye is formed by a knot of black thread. There are five pointed tail feathers below the perch. There is a current variation of this type which is embroidered of heavy silver-colored metallic thread.

TYPE IV

Post World War II period: A coil of bullion outlines the upper edge and point of the wing tip. The wing shoulder is formed by short diagonal stitches. The outer feathers on the wing are made of long curved stitches, and the inner feathers are of extra bright bullion. The body is formed of five vertical twists of thread over a series of diagonal stitches. The legs are of horizontal stitches with one vertical and three short diagonal stitches for each claw. The body and legs are thickly padded to stand out.

TYPE V

Post-Korean War period: A coil of bullion forms the upper edge of the wing and first feather. The shoulder and the center stitch of each inner wing feather is of extra bright bullion. The outline of the inner feathers and the diagonal stitches of the outer feathers are of plain silver bullion. The body is formed of diagonal stitches, with a single stitch for the lower part of the beak. The tongue is a stitch of red thread. The brow is a stitch of extra bright bullion, with the lower part and center of the eye of black thread. The legs are of horizontal stitches of extra bright bullion. A vertical stitch and three short diagonals form the claws. The tail feathers are outlined in plain bullion with a center stitch of extra bright bullion. The perch is made of short diagonal stitches of plain bullion.

Current bullion rating badges also include patterns similar to Type II-B and Type IV. Current variations of these types incorporate a greater use of extra bright bullion than samples of earlier rating badges. The increased cost of gold and silver in recent years has brought about the use of rating badges made of gold-tone and silver-tone metallic thread, embroidered in much the same style as Type III and the eagles on the cloth embroidered rating badges.

Specialty Marks

From 1886 to the present day, a mark indicating the occupational specialty of a petty officer has been the central element of the rating badge. There had been specialty marks in use by the Navy since 1 December 1866, but not as part of a standard rating badge which was the same basic design for all petty officers.

Specialty marks which have been used since 1886 are illustrated and dated in this section. The marks are presented in alphabetical order, using the present title for marks still in use, and the last title used for marks that have been discontinued. In addition, other titles and other ratings for which each mark has been used are listed, along with the dates of use.

Two notes of clarification are in order at this point. First, many of the ratings were in existence before their association with a given specialty mark. The dates included in this study are the dates for which the illustrated mark was used to represent a rating in the period since 1886. Second, there are US Navy rating badges in existence with specialty marks other than those illustrated here - marks for which the writer has found no authorization or approval by the Naval Department, and which have, therefore, been excluded from this study.

For each mark illustrated and title listed, the grades authorized for the rating, along with applicable dates, are listed. Grades above chief petty officer are not listed for any of the marks, however. With the creation of the grades of senior and master chief petty officer effective 1 June 1958, subsequent advancement was made to all of these grades in all specialties except Teleman which was a transitional rating at the time. The precise date when such grades became active in any specialty group is not included here since that would depend on the promotion of individual chief petty officers within their specialties, and that individual data is outside the scope of this study.

Some of these specialty marks were also used as distinguished marks. This information is included with each mark so used. The manner of wearing each mark when used as a distinguishing mark is listed as of the date instituted.

History of the Aviation Boatswain's Mate

Roy L. Warman, senior chief aviation boatswain's mate/handler, works at NAS Memphis and is currently the vice-president of the ABMA Mid-South Chapter. As a sponsor of the aviation boatswain's mate chiefs of past and present, he sent this little tidbit in hopes to help and have their unique history mentioned.

The rating of aviation boatswain's mate (ABM) was established 6 September 1944 and promulgated to fleet personnel on 14 September 1944 by Bupers Circular Letter 268-44.

Approval for the aviation boatswain's mate rating badge was granted 30 November 1944 by Bupers Circular Letter 363-44. Also approved by the same letter was the badge for the rating of mailman.

The Bureau of Naval Personnel training bulletins were published monthly throughout the war years and beyond. The 15 December 1944 issue (Issue #14925) listed the following training schools:

Naval Aircraft Factory, Navy Yard, Philadelphia, PA. Catapult and arresting gear training for officer and enlisted. Length - four weeks.

Manchester, Washington, NAVSTA, Middle & Orchard Point. Carrier gasoline systems training for officer and enlisted. Length - two weeks. Started 20 November 1944.

Rating qualifications for aviation boatswain's mate were first printed and promulgated in change #4 to the Bupers Manual in February 1945.

Naval Training Bulletin 14939 of 15 February 1946 lists the following: Philadelphia, Pennsylvania, Naval Base, Naval Air Material Center.

School	Length	Eligible
MK-4 Arresting Gear	5 weeks	Off/Enl
MK-5 Arresting Gear	5 weeks	Off/Enl
H2-1 Catapult	5 weeks	Off/Enl
H-4 Catapult	5 weeks	Off/Enl
P-6 Powder Catapult	10 days	Off/Enl

Naval Training Bulletin 14942 of 15 May 1946 shows the carrier gasoline system course moving to the Fleet Training Center at Newport, Rhode Island. The same bulletin establishes a CV firefighting course at the same FTC in addition to their DC courses.

From September 1944 to April 1948 ABMs were split into four groups as follows: ABM (AG) - Arresting gear and barriers; ABM (CP) - Catapults; ABM (PH) - Plane handlers; ABM (GA) - Gasoline.

In *Naval Training Bulletin 14945* of 15 August 1946, the following is published for the first time: Philadelphia, Pennsylvania, Naval Base, Naval Air Material Center, NAVSCH (Aviation Boatswain's Mate) Class "A." Length, 16 weeks. Open to all enlisted. Starts first Monday of each month. The MK-4 and MK-5 arresting gear as well as the H2-a, H-4, and P-6 Catapult courses were listed as Class "C."

Bupers Circular Letter 189-17 of 30 September 1947 grouped the rating together making everyone straight "AB." Also in 1947 Lo-Cap Gas School moved to Philadelphia, and Hi-Cap gas was taught aboard the USS *Block Island* in the Severn River Command near Annapolis, and aboard the USS *Mission Bay* at Bayonne, New Jersey. Fuels courses were taught at Bayonne until 1967.

Naval Training Bulletin 14962 of January 1948 changes the Aviation Boatswain's Mate Class "A" School to 14 weeks vice 16. The same issue lists "criteria for selection of candidates for Class "A" and "P" service schools." The following is an excerpt from that list: Aviation Boatswain's Mate, GCT+ARI-100, Normal Color Perception, 20/20 vision uncorrected.

AllNAV - 44 of 12 March 1948 established the Naval Reserve emergency service ratings of ABU (utility) and ABG (gas). Also established was the "exclusive emergency service rating" of airship rigger - ESA, which existed until 1955. (Airship rigger was not grouped officially with ABs at this time.) Naval Reserve Notice 12-48 realigned the reserve rating structure as follows: Reserve ABUs encompassed the Pre-1948 ratings of ABM (AG), ABM(CP), and ABM(PH). ABM(GA) became ABG.

Training Bulletin 14979 of July 1949 provides a list of Navy training publications. Among them was, *Enlisted Men's Guide to Aviation Ratings*. NAVPERS 10301 and *Aircraft Fuel Systems*, NAVPERS 10335.

Training Bulletin 14994 of October 1950 lists the new training courses aviation boatswain's mate, Vol. I NAVPERS 10382 and aviation boatswain's mate, Vol. II NAVPERS 10383.

The first steam catapult training was in Philadelphia and began in 1953.

On 3 August 1955 Bupers notice created the emergency service rating of aviation boatswain's mate (airship) - ABA. The old "exclusive emergency service rating" of airship rigger was eliminated. Personnel performing airship rigger duties were encouraged to submit requests for a change in rating to ABA.

On 1 September 1956 the two emergency service ratings of aviation boatswain's mate (utility) third class (ABU-3), and aviation boatswain's mate (gasoline) - third class (ABG-3) were activated to selected emer-

Rating Badges

The rating badges of all petty officers, male and female, are worn on the left sleeve, half way between elbow and shoulder. The specialty marks are shown between eagle and the inner angle of the upper chevron. This type of petty officer rating badge was instituted in 1886 and brought into the present form in 1894.

Enlisted specialty marks are an integral part of a petty officer's rating badge. From the small beginning of eight distinctive marks in 1866, the number has grown to sixty-three. The large number of marks now employed indicates the complexities of the modern Navy–in the air, on the land, and on and under the sea. Many of the specialty marks are used to identify men of the construction battalions who build the Navy's shore facilities overseas and in combat areas in time of war.

Chief
Petty Officer

Petty Officer
First Class

Petty Officer
Second Class

Petty Officer
Third Class

Enlisted Specialty Marks
General Ratings

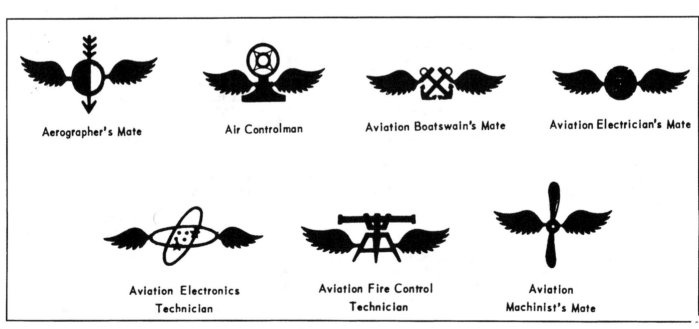

Aerographer's Mate

Air Controlman

Aviation Boatswain's Mate

Aviation Electrician's Mate

Aviation Electronics
Technician

Aviation Fire Control
Technician

Aviation
Machinist's Mate

Enlisted Specialty Marks
General Ratings (Continued)

Aviation Ordnanceman

Aviation Storekeeper

Aviation Structural Mechanic

Boatswain's Mate

Boilermaker

Boilerman

Builder

Commissaryman

Communications Technician

Construction Electrician

Construction Mechanic

Damage Controlman

Dental Technician

Disbursing Clerk

Electrician's Mate

Electronics Technician

Engineering Aid

Engineman

Equipment Operator

Fire Control Technician

Gunner's Mate

Hospital Corpsman

Illustrator Draftsman

Interior Communications Electrician

Instrumentman

Journalist

Lithographer

Machine Accountant

Machinery Repairman

83

Enlisted Distinguishing Marks

The Navy uses a system of embroidered devices to indicate a man's special qualifications, in addition to those which are required for the general ratings indicated as part of the petty officer's device. The distinguishing marks are worn on the right sleeve between the elbow and shoulder as indicated in the cut.

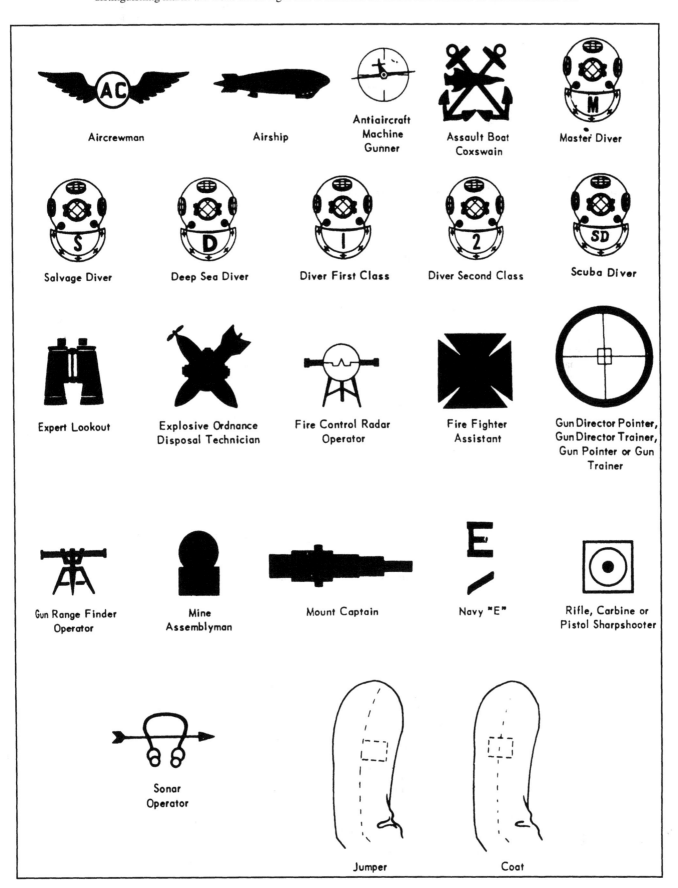

Aircrewman

Airship

Antiaircraft Machine Gunner

Assault Boat Coxswain

Master Diver

Salvage Diver

Deep Sea Diver

Diver First Class

Diver Second Class

Scuba Diver

Expert Lookout

Explosive Ordnance Disposal Technician

Fire Control Radar Operator

Fire Fighter Assistant

Gun Director Pointer, Gun Director Trainer, Gun Pointer or Gun Trainer

Gun Range Finder Operator

Mine Assemblyman

Mount Captain

Navy "E"

Rifle, Carbine or Pistol Sharpshooter

Sonar Operator

Jumper

Coat

gency service ratings. Selective emergency service ratings were applicable to USN as well as USNR personnel. Regular Navy AB-3s were not changed to selective emergency service rates. However, ABU-3 and ABG-3 were established as the normal path of advancement for USN non-rated strikers at the R-3 level. For USN personnel, AB-2 served as the normal path of advancement from either ABU-3 or ABG-3.

On 1 June 1958 the Navy established the grades of E-8 and E-9. A selection board was convened and on 17 October 1958 Bupers Notice 1430 listed the one AB to be advanced to master chief and the nine to be advanced to senior chief. All were advanced on 16 November 1958. All were E-7 at the time and the list is as follows:

To ABCM: Joe M. Harrison Jr.

To ABCS: Rural R. Boyd, Robert O. Coffelt, Lee Me. Davis, Troy W. Davis, James C. Holley, Jesse S. Horton, James C. Huckaby, John W. Hurel and Herbert J. Reece.

In 1958 the naval training bulletins changed to quarterly publications and became NAVPERS 14900. In the fall 1960 issue the new course ABH 3&2 was introduced to replace AB 3&2 Vol. I, "Necessitated by the division of the AB rating into service ratings." It further reads: "This training course will be followed by courses for the aviation boatswain's mate "F" (Fuels) 3&2, and the aviation boatswain's mate "E" (equipment) 3&2 which are now being prepared.

Bupers Notice 1440 of 22 September 1960 considerably revised the aviation boatswain's mate family of ratings, both USN and USNR. This notice established the service ratings of ABE, ABF and ABH which became effective on 1 January 1961. Regular Navy personnel serving with the aviation boatswain's mate rating at third through chief petty officers were shifted to one of the service ratings. Those USN members holding the selective emergency service rate of ABU-3 were changed to either ABE-3 or ABH-3 whereas persons holding the ABG-3 rate were re-rated to ABF-3.

Reservists designated senior or master chief aviation boatswain's mate (utility), (gasoline) and (airships) were changed to the general ratings of aviation boatswain's mates (ABCS) and (ABCM).

Reserve ABGs at third class through chief petty officer were changed to ABF at equal paygrades. Reserve ABUs were shifted rate for rate to either ABE or ABH. Reserves ABAs at third class through chief were changed to ABH only.

In 1966 the Aviation Boatwain's Mate School moved to Lakehurst, New Jersey after more than 20 years in Philadelphia.

In 1967 the fuel school was moved in its entirety from Bayonne, New Jersey to Lakehurst, marking the first time that all phases of AB training had been centrally located. Also in 1967 the NATTC detachments were established for Norfolk, Virginia and San Diego, California.

Endnotes:

[1] *The Recruits Handy Book, USN was prepared under the direction of the Bureau of Navigation, 1910, ninth edition, Published by the Naval Institute, Annapolis, MD*

[2] *The ratings so marked are not available to the incoming recruit.*

[3] *The ratings so marked are not available to women.*

Chapter 10
History of the Boatswain's Pipe
History of the Boatswain's Call

The boatswain's call, or whistle, was once the only method other than the human voice of passing orders to men on board ship. Today more sophisticated communications systems exist but the Royal Navy, always believers in tradition, still use the boatswain's call as a mark of respect to pipe the captain or special visitors on board, or for emphsizing important orders.

The boatswain was the officer in charge of rigging, sails and sailing equipment. He therefore needed to issue orders more often than other officers and so the whistle was named after him.

In the old days men were rigidly trained, almost like sheepdogs, to respond immediately to the piping of the call. At sea, in moments of danger, particularly in storms, they could be counted on to hear the high-pitched tones of the call, and react without delay. A shouted order may not have been heard above the sound of howling winds and lashing waves.

Instructions to hoist sails, haul or let go ropes were conveyed by different notes and pitches.

It is known that the galley slaves of Rome and Greece kept stroke to the sound of a flute or whistle similar to the Boatswain's call. It was first used on English ships in the 13th century, during the Crusades, and became known as "The Call" about 1670 when the Lord High Admiral of the Navy wore a gold whistle as a badge of rank. This was known as the "Whistle of Honour." The ordinary whistle of command was issued in silver and often each officer had his own call decorated with rope designs and ship's anchors.

Each section of the boatswain's call has a nautical name. The ball is the buoy; the mouthpiece is the gun; the ring is called the shackle and the leaf is called the keel.

Using the Boatswain's Call

The call should be held between the index finger and thumb, with the thumb on or near the shackle. The side of the buoy rests against the palm of the hand and the fingers close over the gun and buoy hole in position to throttle the exit of air from the buoy to the desired amount. Care should be taken not to touch the hole of the gun, or the sound will be choked.

There are two main notes, the "low" and the "high" and there are three tones, the "plain, warble" and the "trill." The plain note is made by blowing steadily into the mouth of the gun with the hole of the buoy unobstructed by the fingers. The plain high note is produced by partly throttling the exit of air from the buoy. This is done by wrapping the fingers round the buoy, taking care not to touch the edges of the hole. Intermediate notes can be obtained by throttling to a greater or lesser degree. The warble is obtained by blowing in a series of jerks, with a result like the call of a canary. The trill is produced by vibrating the tongue while blowing, as if rolling the letter "R."

Piping the Side as a Mark of Respect

The best known use of the boatswain's call is for "Piping the Side," the signal of respect which, in the Royal Navy, is reserved for the Sovereign, Senior Royal Navy officers and for all foreign naval officers. A corpse when being brought on board or sent out of a ship is also piped, but the side is never piped at any shore establishment.

This mark of respect owes its origins to the days when captains used to visit other ships when at sea.

Chapter 11
Leadership - Interview with McPon John Hagan
An Interview With Master Chief Petty Officer of the Navy - John Hagan

1. What is your perception of the role of the chief petty officer (CPO) in the USN?

A: I equate the job of the CPO to that of a middle manager or supervisor in the private sector; however, it's really not a fair comparison. A CPO has to do much more than manage and lead eight hours a day, especially at sea. The CPO must also be a counselor, a father, a teacher, a disciplinarian, a motivator. While underway, a CPO is responsible for his people and their jobs 24 hours a day. He or she is the technical expert who is responsible for the well-being of the entire crew and he or she must take part in preparing the ship to fight and help keep it afloat when its hit so it can fight again. Furthermore, the CPO must be dedicated to the tradition of the Navy and care deeply about the Navy as an institution. Ultimately, it is up to the PO to train his or her relief and to pass along the Navy's tradition.

2. Which of your accomplishments in the Navy are you most proud?

A. I break my accomplishments into two categories, singular deeds and collective deeds. It always gives me great satisfaction to help individual sailors. Each time I helped a sailor, I felt a great deal of satisfaction. That's the best part of being a CPO. Collectively, many people helped me qualify as OOD underway while I was assigned to the USS *Philippine Sea* as the command master chief. I experienced a great amount of pride when the wardroom officers presented me with the watch flag for the last watch I stood aboard the ship.

McPon John Hagen - 1993

3. Where do you see the best opportunities for enlisted men and women in service in today's and tomorrow's Navy?

A: I see the best opportunities in the hardest, toughest jobs; the jobs other people do not want - especially at sea. I encourage people to seek these jobs, then to learn as much as possible about them, the sea and the maritime service. I encourage people to get assigned away from the hometowns and, when stationed overseas, to learn as much as possible about the host country's culture, history and language. Even during the down-sizing, the Navy has excellent opportunities for people who get qualified and work hard. As a side note, one of the best opportunities in the Navy is that if a person feels stifled or becomes dissatisfied with their jobs, they can cross train into another profession.

4. During your career, what has been the most unpleasant task you have been required to do as a CPO?

A: That's an easy question to answer. Serving on the Early Selective Retirement Board where we had to select good sailors to retire early was by far the most unpleasant task I've done. There have been other unpleasant tasks, too. For example, serving as a CACO and notifying parents or spouses that their service member has been killed, or presenting an American flag to a next of kin at a funeral is always unpleasant.

Master Chief Petty Officer of the Navy ETCM(SW) John Hagan

Master Chief Hagan was born in Luton, England, on May 20, 1946. He was reared and attended schools in Asheville, North Carolina. After high school, he enlisted in the Navy in December 1964 and attended basic training at Recruit Training Center, San Diego, California. He then attended Electronics Technical "A" School at Naval Training Center, Treasure Island, California and completed a short assignment at Naval Air Test Center Patuxent River, Maryland.

After he completed Ground Control Approach Radar Technician School at Naval Air Technical Training Center, Glynco, Georgia, he reported to Naval Air Station Whidbey Island, Washington as the Leading Petty Officer for the Maintenance Division. During his tour there, he earned an Associate of Arts degree.

Hagan's next assignment was aboard the USS *Lester* (DE-1022), homeported in Naples, Italy. During a subsequent tour of sea duty as a maintenance technician at Underwater Demolition Team 21 in Little Creek, Virginia, he was advanced to chief petty officer and qualified as a naval parachutist. While assigned to a shore tour at Naval and Marine Corps Reserve Center in Louisville, Kentucky, he was advanced to senior chief petty officer. While there, he earned his bachelor of business administration degree from McKeendree College.

In September 1980, he reported aboard USS *Richmond K. Turner* (CG-20), homeported in Charleston, South Carolina. While there, he qualified as an enlisted surface warfare specialist and was advanced to master chief petty officer. Shortly after reporting to his next assignment at the Naval Air Technical Training Center at Memphis, Tennessee, he was selected as the force master chief for the chief of naval technical training.

In April 1988 Master Chief Hagan reported to pre-commissioning unit *Philippine Sea* (CG-58) in Norfolk, Virginia, as the command master chief. After commissioning, the USS *Philippine Sea* reported to her homeport in Mayport, Florida, and subsequently deployed to the Red

and Mediterranean seas in support of Operation *Desert Shield* and *Storm*. During this tour of duty, he qualified as officer of the deck (Underway).

Soon after reporting to Helicopter Anti-Submarine Squadron (Light) 48 at Mayport, Florida, as the command master chief, Hagan was selected as the eighth master chief petty officer of the Navy. He assumed the job on August 28, 1992.

Hagan's personal awards include the Meritorious Service Medal, Navy Commendation Medal, Navy Achievement Medal with Gold Star, as well as unit and campaign awards. He is married to the former Catherine Mosher. They have three children: Robert (a first lieutenant in the Marine Corps), Melissa and Melody.

Chapter 12

National Chief Petty Officers' Association
Hail To The Chiefs

The National Chief Petty Officers Association was formed in 1988 in order to provide a platform for airing issues which affect us as active duty, retired and veteran chiefs. Its specific purpose is to generate world wide awareness of the importance of regular and reserve chief petty officers in the USN and Coast Guard, of the past, present and future, to encourage young sailors to appreciate the importance of study and advancement within the various ratings of these services; to promote reunions of members for remembering, camaraderie and good fellowship; to maintain true allegiance to the Government of the United States of America and to foster true patriotism.

Chapter 13

Naval Terms and Terminology
Examples of Slang Terminology and Expressions

Ge Dunk
Slop Chute
Pogey Bait
"Ask the Chief"
If it moves, salute it; if it doesn't move, paint it!
RHIP
Doc
Liberty Cal
Ship Calls
Heave Out and Trice Up
Rope Yarn Sunday
Field Day
Ole Man
Belly Robber
Whiz Kids

Naval Terms and Terminology

Abandon ship: To leave the ship in an emergency such as sinking. Traditionally, ordered in extremis by the senior surviving officer.

Aboard: In a ship, on or at any activity such as a naval station. If one ship comes aboard another, they have come into actual contact, although perhaps with little damage. See close aboard. Distinguished from "on board," which refers only to a temporary condition. "He is regularly aboard the naval station, but came on board our ship to observe target practice."

Absence indicator: Pennant flown by a ship to indicate absence of commanding officer or embarked flag or staff officer. Also called absentee pennant. See absentee.

Absence without leave (AWOL): See unauthorized absentee.

Accommodation ladder: Portable steps from a ship's gangway down to the waterline or a pier alongside, rigged from a davit or crane on board. Sometimes incorrectly referred to as the gangway. See bail.

Acey-deucy: Nautical version of backgammon.

Action report: Detailed report of combat with the enemy.

Active duty: Full-time service as distinct from inactive, retired, or reserve duty.

Addressee: Activity or individual to whom a naval message is directed for action or information.

Administration: The management of all phases of naval operations not directly concerned with strategy or tactics.

Admiral: Highest rank in the Navy, equivalent to general (numerical designation: O-10). An officer of four-star rank. Rear admirals (upper half) and major generals (0-8) wear two stars; vice admirals, lieutenants and generals (0-9) wear three. Fleet admirals and generals of the Army had five stars; this was a special rank created by Congress for WWII leaders only. The four-star admiral is sometimes called full admiral to distinguish him or her from other admirals, inasmuch as the title of admiral is loosely used for all. Commodores wore one star and were equivalent to brigadier generals, but recent changes designate them as rear admirals of the lower half (0-7), and commodore is, by consequence, again only a courtesy title. Admirals of all grades are flag officers, authorized to fly flags with stars to denote their ranks. The Army, Air Force and Marine Corps corresponding term is general officer.

Admiral's March: Ceremonial music for flag officers and officials of equivalent ranks. (Unofficial lyrics go: "He's a bum, he's a bum, he's a rotten lousy bum," etc.) The number of ruffles and flourishes receding the ceremonial tune denotes the number of stars authorized for the individual honored.

Admiralty law: Laws that deal with maritime cases: ships, collisions, etc.

Adrift: A boat or ship is said to be adrift when not made fast to the dock or bottom. A sailor who leaves personal possessions scattered about is said to have gear adrift. Gear that is adrift may find its way into the lucky bag.

Advance force: Task force preceding the attack force in an amphibious assault, conducting preparatory minesweeping, bombardment and reconnaissance.

Affirmative: Communications term meaning: Yes, permission granted, authorized, approved, approval recommended, etc. Often used in conversation.

Afloat: Supported by the water. Also, at sea, as in forces afloat.

Aft, after: According to nautical and naval usage, aft is an adverb (let us go aft) and after is an adjective (the after cabin). US usage permits the use of aft as an adjective, and some non-nautical people use it that way. The Oxford English Dictionary lists aft as an adverb.

Aground: Fast to the bottom and in contact therewith. Immobile by consequence and probably damaged as well. Distinguished from deliberate grounding, as by a landing ship in amphibious warfare, or being fast to the bottom via anchor and chain. A ship runs aground or goes aground.

Ahoy: A distinctly nautical hail. Supposedly once the dreaded war cry of the Vikings.

Aide: Officer assigned as administrative or personal assistant to a flag officer or senior civilian official. The aid wears aiguillettes.

Air bedding: An order, aboard ship to bring bedding topside for exposure to the sun and fresh air. The word "air" is used as a verb.

Air boss: Slang: the air officer aboard a carrier who directs aircraft launching and recovery. His assistant is known as the mini boss.

Aircraft carrier: Major offensive ship of the fleet. Its chief weapon is its aircraft.

Airdale: Jocular term for naval aviation personnel. See brown shoe.

Air group: The aircraft of an escort or antisubmarine warfare carrier, made up of squadrons. The more numerous aircraft of an attack (first line) carrier are organized into an air wing, and may consist of two or more air groups.

Air lock: A double door giving access to and preserving air pressure in a fireroom or similar space under pressure. See light lock.

Airman (AN): An enlisted person in paygrade E-3 who performs aviation duties.

Air officer: Officer responsible for aviation matters in an aircraft carrier. Heads the air department.

Airspeed: Speed of aircraft through and relative to the air and distinct from groundspeed. Indicated airspeed is an uncorrected reading of the airspeed indicator. Calibrated airspeed is the indicated airspeed corrected for instrument errors. True airspeed is corrected for altitude and temperature.

Air Wing: The aircraft of an attack aircraft carrier, made up of squadrons. See air group.

Alert, dawn or dusk: Special precautions, normally all hands to battle stations, at time when attack is most likely - prior to first light and at sunset.

Alive: Alert, as in "Look alive!"

All hands: All those aboard ship (except, under certain circumstances, those on watch). Name of a call on boatswain's pipe. *All Hands* magazine is a monthly publication produced for free distribution to naval personnel.

All hands parade: A designated assembly place for all hands on board ship. Used for such events as change-of-command ceremony.

Allotment: Portion of an individual's pay, or of an appropriation or fund, regularly assigned to a specific account.

Allowance: Authorized personnel on a peacetime level, reduced from the wartime complement. Based on peacetime operations, habitability, budgetary considerations, and upkeep requirements. See complement, manning level.

Almanac, Air and Nautical: Naval Observatory publications providing astronomical data needed for navigation.

ALNAV: Message intended for general distribution to all naval personnel.

Aloft: Up high, as on a mast, or strong winds aloft.

Alongside: Near the side of the ship, pier, dock, etc.

Altimeter: An aneroid barometer that measures in feet, yards, or meters an aircraft's elevation above a given reference plane, such as sea level. Must be corrected constantly for barometric pressure at ground level. Radio, electronic, and radar altimeters are now common.

Altitude: The height of an aircraft above a reference point. True altitude is height above sea level, corrected for temperature. Absolute altitude is height above ground.

Amidships: In or toward the middle of a ship.

Amphibious: Capable of operating on land and sea.

Amphibious force: Naval force and landing force, together with supporting forces, who are trained, organized and equipped for amphibious operations.

Amphibious transport dock (LDP): Ship designed to transport and land troops, equipment, and supplies by means of embarked landing craft, amphibious vehicles and helicopters.

Amphibious troops: Troop components, ground and airborne, assigned to land in an amphibious operation. Synonymous with landing forces as defined in the National Security Act of 1947.

Anchor (n): A device used to hold a ship or boat fast to the bottom. Old-fashioned anchors were in the form of the traditional hook. They are so difficult to rig for sea that, as ships became bigger and required heavier anchors, the currently familiar stockless type was developed. Such anchors can be hoisted snugly into their hawsepipes when the ship gets under way, in contrast to the old-fashioned ones that required fishing, catting and stowing and then had to be strongly secured—all very hazardous in a seaway.

Anchors may be: bower, stream, stern, kedge, boat, and sheet. Naval stockless anchors may be Dunn, Baldt or Norfolk, but such distinctions are of little interest to operating personnel. See patent anchor.

Anchor detail: Those on forecastle assigned to handle the ground tackle.

Anchor pool: A sort of shipboard lottery in which people buy tickets that show a specific time of anchoring. Winner takes all.

Anchor's aweigh: Expression used to report that an anchor has just been lifted clear of bottom. The ship now bears the weight of her anchor and is considered to be under way (not "underweigh"), although not necessarily with way on.

Anymouse: Naval aviation slang for anonymous. Anymouse reports of aircraft accidents are sent in to Approach Magazine where Grandpa Pettibone turns them into dramatic safety lessons. See Dilbert.

Apprehension: According to the Uniform Code of Military Justice, means clearly informing a person that he/she is being taken into custody. Same as arrest in civil life. See restraint.

Appropriation: Government funds provided by Congress for specific purposes. May be continuing or annual. Not the same as authorization. Most defense legislation requires both authorization and appropriation.

Armed forces of the United States: Collective term for all the components of the Army, Navy, Marine Corps and Air Force. Includes Coast Guard, when part of the US Navy during time of war. See United States Armed Forces.

Arrest: Restraint of a person by competent authority. Involves relief from military duties. See apprehension, restriction.

Articles for the Government of the Navy: No longer effective, replaced by the Uniform Code of Military Justice. See Rocks and Shoals.

ASAP: As soon as possible.

Ashore: On the beach or shore. A sailor may go ashore on liberty, but if a ship goes ashore, she is aground.

Astern: Toward the back or after end of a ship or formation. Generally used in the sense of behind, or out of the ship. Thus, a fast ship leaves a slower one astern. Someone who falls overboard would be left astern. One would go aft in one's own ship, not astern. See aftermost.

As you were: Command that means resume condition of at ease or former activity or formation.

At ease, stand at ease: In a strict military sense, as command to those at attention to assume a more relaxed but carefully defined posture, still standing in ranks. In another and informal sense it is a command to an assembled group to relax.

Athwart, athwartships: At right angles to the fore and aft centerline of a ship or boat. Sometimes pronounced "thwart-ships." Thwarts are always althwartships.

Atoll: A ring-shaped coral reef usually found in the Pacific and Indian oceans, often with low sand islands. The body of water enclosed by the reef is a lagoon.

Attend the side: To be on the quarterdeck to meet important persons. Same as tend the side.

Attention to port, starboard: Command given to topside personnel when ship is rendering passing honors. Personnel in view of the honored ship are required to come to attention facing the designated side, and salute if ordered.

Authentication: Communication security measure designed to prevent fraudulent transmissions.

Authorization: Congressional permission to carry out a program (generally procurement of ships, aircraft, or weapons) involving expenditure of funds. Actual appropriation of the funds is necessary before they can be spent (or obligated).

Awash: So low that water washes over.

Away: Term used in passing the word aboard ship, e.g.: "Call away the gig." Refers to prospective departure from the ship on a mission or errand, e.g.: "Away rescue and assistance party, away!"

Aweigh: Said of an anchor when clear of the bottom. See anchor's aweigh.

Aye aye: A seamanlike response to an order or instruction signifying that the order is heard, is understood, and will be carried out. Aye is Old English for yes. Pronounced "eye."

Bad conduct discharge (BCD): A punitive discharge awarded to an enlisted person for severe infractions of regulations. The only type of discharge that carries greater prejudice is the dishonorable discharge (DD).

Baiting: Tactic designed to lull an enemy, especially a submarine, into a false sense of security and induce it to take action, making it liable to detection or attack.

Barge: Boat for official use of a flag officer. A non-self-propelled cargo carrier in harbors or rivers. Also called a lighter or scow. As a verb, "The coal was barged alongside." See scow.

Barn burner: An achiever, one who gets things done.

Barrier reef: Name given to offshore reefs separated from land by channels or lagoons, as distinct from fringing reefs.

Base, Naval: A shore command providing administrative logistic support to the operating forces.

Battalion landing team (BLT): Battalion of troops specially organized for an amphibious landing.

Battle bill: List of battle assignments based on ship's armament and ship's complement. See watch, quarter and station bill.

Battle dress: Flash and splinter protective clothing worn in battle by crews of surface ships.

Battle group: An aircraft carrier, a battleship if available, one or more Aegis cruisers, a unit of destroyers, and a logistic support vessel make up this typical offensive unit of the fleet.

Battle lantern: Electric lantern, battery powered, for emergency use.

Battle lights: Dim red lights below decks for necessary illumination during darken ship periods. The red spectrum has less of the temporary blinding effect on the retina of the eye than any other color, hence, red lighting permits the quickest possible dark adaptation and has the least deleterious effect on night vision. The blue battle lanterns previously used had a devastating effect on the ability to see in the dark.

Battle port: Hinged metal cover for an air port, or porthole.

Battleship (BB): Derived from "line of battle ship" or "ship of the

battle line." The battleship, whether of wood or steel, was originally the largest and most powerful man-of-war that could be built. Development of the airplane produced the aircraft carrier, which in World War II decisively replaced the battleship as the primary capital ship of navies. The biggest battleships ever built were the 80,000-ton Yamatoes of Japan. The only ones now in existence, recently recommissioned from the reserve fleet, are the four 65,000-ton Iowa class (nine 16-inch guns, 30 knots speed) of the United States. USS *Missouri* was the site of the surrender of Japan in World War II; the others are the USS *Iowa, New Jersey* and *Wisconsin.* Note: A number of World War II battleships, and other ships as well, have been preserved in state memorial parks. See armored cruiser, battle cruiser, cruise, deck(er), warship.

Beach: As used in amphibious operations, portion of shoreline required for landing one battalion landing team. To run a ship or boat ashore is to beach it. Slang for shore. In oceanography, area extending from shoreline inland to a marked change in physiographic form, or to line of permanent vegetation.

Beachhead: The initial objective of an assault landing. A section of enemy coast used for continuous landing of men and equipment in an amphibious operation. After consolidation of the beachhead, the next move is to break out of same, and at this point the operation takes on the characteristics of regular land warfare, except that until capture of a suitable harbor or port, the beachhead remains the support base, under the charge of a beachmaster.

Beachmaster: In amphibious operations, the officer designated to take charge of logistic activities on the beach after the assault phase of the landing has been concluded.

Beach party (amphibious): Naval shore party to control boats, survey channels, salvage landing craft, etc.

Beacon: A navigational aid for establishing position of ship or aircraft. May be lighted, aerial, radar, radio, radio-marker, radio-range two-marker, or infrared.

Bear a hand: Hurry up; expedite.

Beer muster: Slang: beer party ashore.

Belay: To make fast or secure, as in "Belay the line." To cancel, as in "Belay the last word."

Bell bottoms: A sailor's uniform trousers that widen at the bottom.

Bell, ship's: Used for sounding fog or distress signals, as fire signal, and to denote time. See **ship's bell.**

Berth: Anchorage or mooring space assigned a vessel. Sleeping place assigned on board ship. A margin in passing something, as a wide berth. To inhabit, as "He is berthing in the forward compartment."

Bilge(s): The inside bottom of a ship or boat. The turn of the bilge refers to the curved plating where a ship's side joins the bottom. A ship is said to be bilged if her bottom has been damaged sufficiently to take on water, as when running aground, but the expression is almost never used to refer to battle damage, even though torpedo damage could technically be so described. To bilge an examination is slang for receiving an unsatisfactory grade. To bilge someone is to fail that person, if one is an instructor or superior, or to get a higher grade if a peer. To refer to something (e.g., the opinion of another) as bilge or bilge-water is to be contemptuous.

Billet: Duties, tasks, and responsibilities performed by one person. Also, a specific assignment in a ship or station organization.

Binnacle list: A list of personnel excused from duty because of illness or injury, customarily placed in the binnacle for the information of the officer of the watch. Although the binnacle list survives with the same meaning, it is no longer placed in the binnacle.

Black Cats: The Navy and Marine Corps seaplanes that flew and fought so effectively during World War II. They were usually painted black because they operated at night.

Black gang: Slang: personnel of the engineering department of a ship. Now obsolete because the reference was to coal and the coal dust with which all old-time engineers had to contend.

Black shoe: Slang: an officer who is not in the aviation or submarine communities. See brown shoe.

Blockade: Naval operation barring ships from certain ports or ocean areas. Although many different types of blockages have been proclaimed, including "paper" blockages, international law has generally held that to be meaningful, a blockade must be physically enforced at the shop (i.e., a neutral station blockade runner is not subject to capture merely because she had, sometime previously, run a blockade).

Blow tubes: To inject steam into fireside of boilers to remove soot from tubes.

Blue Angels: A team of Navy aviators that performs precision-formation aerobatics.

Bluejacket: Navy enlisted person below the rank of CPO (E-7). Slang: white hat, tar, swabbie.

Boarders: Also called boarding party. Personnel detailed to go aboard an enemy ship to capture or destroy it. The term boarders is more generally applied to those who spontaneously board an enemy vessel when the opportunity presents itself, as when antagonists foul each other. This occurred in the battles between the *Constitution* and the *Guerriere* and between the *Chesapeake* and the *Shannon*, in 1812-13. A boarding party, having been detailed and organized, may board the enemy ship directly but is more frequently sent by boat or other means. One of the most famous boarding parties was Stephen Decatur's in the *Intrepid* when he boarded and burned the *Philadelphia* in 1804. See boarding party for use in a social sense.

Boarding call: Official visit by a naval boarding officer to another ship of war or public vessel of importance. Generally made to foreign ships when they call in US ports, but may be made in foreign ports also, and in such cases may be made to US ships as well. The visit is initiated by the senior officer in port to ships arriving, with the purpose of exchanging courtesies and information leading to mutual beneficial cooperation. The boarding officer should be as senior as possible, but always junior in rank to the senior officer of the ship on which the call is made. The call is always returned in kind as soon as possible, again by an officer junior to both seniors. Such calls are preliminaries to official calls between the seniors themselves.

Board of investigation: An investigatory body of one or more persons. No power of subpoena. General term for all such bodies below court of inquiry.

Boat box: First-aid kit for use in a boat.

Boatswain: Pronounced "bo-sn." A warrant officer whose major duties are related to deck and boat seamanship.

Boatswain's call: A rudimentary tune played on a boatswain's pipe announcing or calling for some standard evolution such as meals for the crew, piping the side, lower away, etc. It can refer to the boatswain's pipe. See pipe down.

Boatswain's chair: Sea sent aloft or over the side on a line to facilitate repairs or painting.

Boatswain's locker: Compartment where deck gear is stowed.

Boatswain's pipe: A small, highly stylized instrument held in the palm of the hand, actually a specially shaped whistle (though never admitted to be such), capable of several different high-pitched and piercing notes at the hands of a skillful operator, used to call attention before passing the word; to render honors, piping the side; and to give orders to winchmen, crane operators, etc. A call is the notes played on the pipe, but sometimes the pipe itself is referred to as a boatswain's call.

Bolo line: A nylon line with a padded lead weight or a weighted monkey fist, thrown from ship to ship or from ship to pier in underway replenishments and mooring. Used to bring aboard a larger line.

Boomer: Slang: a submarine that carries large ballistic missiles.

Boot: Slang: recruit. A newly enlisted marine or sailor.

Bow: The front or forward part of a ship. Sometimes referred to as the bows, because every ship has a starboard bow and a port bow.

Bow book: Member of a boat's crew who mans the boat hook forward and who handles lines. See also stern hook.

Brassard: Arm band, e.g., shore patrol brassard.

Bread and water: Reduced rations authorized with confinement as punishment. Slang: cake and wine.

Breakwater: Structure that shelters a port or anchorage from the sea. Also, a low bulkhead forward that prevents solid water from sweeping the deck of a ship.

Bridge: Ship's structure, topside and usually forward, that contains control and visual communication stations. Sailing ships were conned from the main deck aft, abaft the aftermost mast. The huge paddle wheel boxes of early steamers so interfered with vision that underway OODs stood their watches on the cross-over bridge built between them. The bridge has been the underway conning station ever since, except in submerged submarines.

Bridle: A span or rope, chain, or wire with ends secured, and the strain on the mid-part, as in towing a ship or pulling an aircraft on a catapult.

Brief: To instruct people for a specific mission or operation. Debriefing means a verbal report after the operation has been completed.

Brig: A place of confinement. A prison. A two-masted, square-rigged sailing ship.

Broadcast: Originally a naval term meaning to transmit radio messages to the fleet.

Brown bagger: Slang for a person who carries his or her lunch to work.

Brown shoe: Slang: Aviation or submarine officers. The brown shoe officer is entitled to extra hazardous duty pay for flight or submarine duty. The term originally referred to uniforms; before World War II, only aviators and submariners wore khakis and greens and the brown shoes that went with them. See black shoe.

Bucket of steam: Nonexistent term, requested of new personnel aboard ship. See relative bearing.

Bugle: A horn with limited notes, all controlled by the player's lips, used for military purposes to broadcast a general order to all hands within hearing range, such as taps, reveille, retreat, liberty call, torpedo defense, general quarters. The first bugle calls were supposedly written by Joseph Hayden, the celebrated musician, in about 1793, but of course the bugle has been used for military purposes since antiquity. The first bugles were made from the horns of wild oxen. The Horn of Roland was one of these.

Builder's trials: Trials conducted at sea or at a dock by the builder to prove the readiness of a ship for preliminary acceptance trials.

Bulkhead: Walls or partitions within a ship, generally referring to those with structural functions such as strength or water-tightness. Light partitions are sometimes called partition bulkheads.

Bull horn: High-powered, directional, electric megaphone.

Bumboat: A civilian boat selling supplies, provision, and other articles to the crews of ships. Supposedly derived from "boomboat," signifying a boat permitted to lie at the ships' booms. Bumboats and bumboatsmen had a bad reputation because they frequently were the source of trouble among a ship's crew. Now a possible security risk.

Bunting: Cloth from which signal flags are made. Also, the flags themselves, as in the order "Air bunting."

Buoy: Floating object, anchored to the bottom, that indicates a position on the water, an obstruction, or a shallow area, or that provides a mooring for a ship. Buoys may be of various shapes and types, with special markings. Many are lighted, with their characteristics shown on harbor charts, and some have a whistle or bell actuated by wave action. See can, nun, spar, mooring, dan buoys.

Bust: Slang: to reduce in rate. Also, to fail or make a mistake.

Butts: That part of a rifle range where targets are tended.

By your leave: A courteous expression or greeting voiced by a junior who overtakes a senior while walking. Traditional and recommended. Under virtually all circumstances, the expression would be, "By your leave, sir/ma'am," signifying appreciation of the relative ranks, and would almost always be accompanied by a salute in passing. A friendly senior might use such an announcement of intention to pass, or he (or someone accompanying him) might just sing out "gangway!"

Cabin: Quarters aboard ship for the captain or an admiral.

CAG: Carrier Air Group. Also a nickname for the commander of a Carrier Air Group.

Caisson: Any temporary structure of wood or metal built to hold back water for repairs or construction. Also, the floating gate of a drydock.

Call: Formal social visit by an officer and spouse to the home of another. It involves leaving calling cards and was once rigidly prescribed and carefully followed, including return calls. Now a call is largely passe' as a custom, although large stations may have an annual party with the understanding that it also constitutes "all calls made and returned." An informal visit of courtesy to another ship just arrived, made by an officer junior to the commander of the arriving unit and the visiting unit. The purpose is to exchange necessary information. It requires no special ceremonies, other than piping the side. See official visit, boarding call. If made by a principal, the call is termed an official call or official visit. Also, a tune played on a boatswain's pipe, calling for certain prescribed evolutions.

Camel: Float used as a fender between two ships or a ship and a pier. Also called breasting float.

Can-do: Slang: efficient, capable, and willing; e.g., a repair ship might be praised as a can-do-ship.

Cannibalize: To remove serviceable parts from one item for use in another.

Capsize: To turn over; to upset.

Captain: A military rank; 0-6 in the sea services and some government services, equivalent to Army colonel. In Army, Marine Corps and Air Force, however, a captain is an 0-3, equivalent to a Navy lieutenant. A title by which a ship's master or anyone who has master's papers should be addressed. The form of address for any commanding officer of a Navy ship, regardless of rank. The senior of a yacht's crew is the captain; the owner or any amateur in charge is the skipper.

Captain of the head: Person responsible for cleaning washrooms and toilets. Known to Marines as the head orderly.

Captor: Deep water mine involving an encapsulated torpedo. The torpedo homes in on the source that activates the release mechanism.

Cardinal point: One of the four principal points of the compass—north, east, south and west.

Cargo: Material carried in ships or aircraft. Classified as dry, bulk, general, heavy lift, deck, dangerous, liquid, refrigerated, etc. May be palletized, breakbulk or containerized.

Carriage: The part of a gun mount that supports the gun itself.

Carrier: Aircraft carrier. Also any organization that operates ships for ocean transport. General use of the term has now extended it to include transporters of freight on land as well.

Carry on: An order to resume or continue previous activity, usually after personnel have come to attention.

Casualty report (CASREP): Report required when casualties of a specified nature occur either to personnel or material. See CASREP.

Catapult: A device for launching aircraft from a ship's deck at flying speed.

Cat-o'-nine-tails: A short piece of rope fashioned into an instrument for flogging. Traditionally the victim to be punished was required to make his own "cat" by unlaying a portion of a three-strand line, separating each of the strands into three parts, then tarring and braiding the parts into nine "tails." Tradition and pride dictated that the victim made the cat as fearsome as possible by knotting the tails and even including small nails or other metal objects in the knots. A poorly marked or soft cat was considered a mark of the craven, and in any case, if it did not pass muster, it would be replaced by one that did. See room to swing a cat.

Catwalk: A walkway constructed over or around obstructions on a whip for convenience of the crew.

Chaff: General name for radar confusion reflectors. Includes rope, a long roll of magnetic foil or wire for broad, low-frequency response and rope-chaff, which contains one or more rope elements.

Chain locker: Compartment where anchor chain is stowed.

Challenge: A demand for identification or authentication. Can be as simple as the flashing light AA signal that internationally requires the recipient to respond with name and destination (warships need only identify themselves). May be a coded signal, transmitted by any of a number of means which must be properly replied to, such as in IFF (identification, friend or foe).

Chaplain: Minister, rabbi, or priest of a recognized religious order, commissioned in the Navy. Slang: sky pilot.

Charlie Noble: Sailors' nautical name for the galley smokepipe. Derived from the British merchant service Captain Charlie Noble, who required a high polish on the galley funnel of his ship. His funnel was of copper and its brightness became known in all ports his ship visited. To "shoot Charlie Noble" meant to fire a blank inside the pipe to shake down the soot.

Cherry picker: Slang: wheeled crane to handle crashed planes on aircraft carrier flight deck. See **crash dolly, karry krane**.

Chevron: V-shaped mark that denotes rate, located beneath the specialty mark or a rating badge.

Chief of the boat or chief of the ship: Senior chief petty officer in a submarine; the executive officer's right-hand man in the administration of the crew.

Chief petty officer (CPO): An enlisted person in paygrade E-7. Until 1958 this was the highest rank attainable while still in the enlisted category. For further promotion one had to look to the warrant grade. Now, however, two superior grades have been established, senior chief petty officer, paygrade E-8 and master chief petty officer, paygrade E-9.

Chipping hammer: small hammer with a sharp peen and face set at right angles to each other; used for chipping and scaling metal surfaces. Also called scaling hammer or boiler pick.

Chow: Slang: food. Chow line is the mess line. Chow down means mess call or dinner is served.

Church pennant: A blue and white pennant flown during church service. By tradition, the only flag or pennant flown on the same hoist above the national colors, and then only during services aboard ship. Now being flown ashore at naval facilities as well—reflecting ignorance of naval tradition.

Classified information: Any information the revelation of which must be controlled in the interest of national security.

Clearance: Determination that an individual is eligible to have access to classified information of a specific category. Also, an aspect of mine countermeasures. Permission for a ship to enter or leave harbor; clearance through quarantine, etc.

Clothes stop: Small cotton lanyard used for fastening clothes to a line after washing them, or for securing clothes that are rolled up.

Clothing and small stores: A government-operated shop on a base or large ship that stocks standard articles of uniforms for officers and enlisted personnel, with such related articles as buttons, brushes, etc.

CO: Commanding officer.

Cockpit: The pilot's compartment in an aircraft; a well, or sunken place in the deck of a boat, almost always protected with a coaming, for use of the crew or passengers. In the days of fighting sail, the cockpit was a compartment below the waterline where the ship's surgeons would try to cope with battle injuries.

Collision mat: A mat of canvas and fiber designed to be hauled down over the hole in a ship's hull caused by a collision, grounding, or battle damage. See thrums, thrumming, hogging line.

Colors: The national flag. The ceremony of raising the flag at 0800 and lowering it at sunset aboard a ship not under way, or at a shore station.

Combat distinguished device: A small metal V worn by the Legion of Merit, Bronze Star, and Commendation Ribbon to indicate the medal was awarded for actual combat operations.

Command master chief: A large ship's ombudsman. The command master chief's office is usually next to that of the chaplain.

Commission: (v) To put a ship in active service under a commissioned commanding officer, who breaks the commission pennant and sets the watch. (n) The document from which a commissioned officer gets rank, status, pay and emoluments.

Commissioned officer: One who derives authority from a commission under authority of the President confirmed by the Congress.

Commission pennant: Narrow, red, white and blue pennant with seven stars, flown day and night at the main truck of ship in commission, under command of a commissioned officer. See distinctive mark.

Company, ship's: Everyone assigned to a ship or station; all hands.

Condition watch: Watch stood under a particular condition of readiness.

Continental shelf: The sea bottom from shore to a depth of 200 meters. Width varies from nearly zero to 800 miles. Generally speaking, the depth increases very gradually to about 100 fathoms. (200 meters), at which point it increases more rapidly at a steeper slope.

Convoy: A number of merchant ships or naval auxiliaries, or both, usually escorted by warships and aircraft. A single merchant ship or naval auxiliary under surface escort.

Corpsman: Short for hospital corpsman, an enlisted medic.

Court of injury: Three or more officers convened by any person authorized to convene a general court-martial to investigate something. Has subpoena power. Designated witnesses may have counsel. See board of investigation.

Courts-martial: Courts of law for military personnel.

Covert: In intelligence work: secret, clandestine. Opposite of overt.

Coxswain: Person in charge of a small boat, pronounced "cox-un."

Crossing the line: Crossing the equator.

Crow's nest: Lookout station aloft, generally on the foremast.

Cruise: Tour of sea duty. A period of enlistment. Also, a voyage to several ports, such as a shakedown cruise, or to a specific ocean area. See deployment.

Cumshaw: Something procured without official payment. Free; a gift. Comes from the beggars of Amoy, China, who said "kam sia," meaning grateful thanks. The historic reference is to graft for personal gain. In the US Navy the term now relates to unauthorized work done for or equipment given to a ship or station, and usually no connotation of personal gain exists. A "cumshaw artist" is adept at getting cumshaw work done, usually by liberal handouts of food and coffee to shipyard workers.

Customs of the service: Unwritten naval practice having the force of usage and tradition. An example is the removal of caps or hats by officers entering a compartment where the crew is eating.

Damage control: Measures to preserve and re-establish shipboard watertight integrity, stability, maneuverability and offensive power; to control list and trim; to make rapid repairs of materiel; to limit the spread of and provide adequate protection from fire; to limit the spread of, remove the contamination by, and provide adequate protection from toxic agents; and to care for wounded personnel.

Damage control central: Compartment located in as protected a position as practicable from which measures for damage control and preservation of the ship's fighting capability are directed.

Darken ship: Blackening out all lights visible from outside the ship.

Davy Jones: The traditional mythological spirit inhabiting the sea, usually considered evil. Sea bottom is sometimes referred to as Davy Jones' locker.

Day's duty: A tour of duty or a watch lasting 24 hours.

D-day: Term used to designated unnamed day on which an operation commences. D+7 means 7 days after D-day.

Deck: The uppermost complete deck is the main deck. Complete decks below it are numbered from the top down: second deck, third deck, etc. Partial decks between complete decks are called half decks; those below the lowest complete deck are platform decks, or flats. Partial decks above the main deck, if they extend to the sides of the ship, are called, according to location, the forecastle deck, middle deck, or poop deck. Those that do not extend to the side are superstructure decks. Weather decks are those exposed to the weather. In aircraft carriers, however, the topmost deck is the flight deck, and the next one below is the hangar deck. The main deck is the one below the hangar deck, after which the numbering system proceeds normally, and partial decks above the hangar deck are called gallery decks. Decks often get their names from construction, as armored flight deck, protective deck, splinter deck (all three of which are fitted with armor), or from employment, as boat deck, gundeck, berth deck. As an added use of the word deck, the officer of the deck's watch is called the deck, as in the expression, "...has the deck," meaning he has charge of all deck functions and, if under way, is supervising all maneuvers of the ship. See flat(s).

Deck gang: All persons attached to the ship's deck departments, as opposed to those attached to engineering, radio or electronics departments. In the old days the deck gang wore rating insignia on the right arm, hence the expression, right-arm rate. All others wore their rates on the left arm and, of course, were known as left-arm rates.

Decoration: A medal or ribbon awarded for exceptional courage, skill or performance.

Decrypt: To convert a cryptogram into plain text by a reversal of the encryption process (not by cryptanalysis). Same as decipher. See decode.

Department of the Navy: The executive part of the naval establishment; the headquarters, US Marine Corps; the entire operating forces of the US Navy and US Marine Corps, including reserve components; all field activities, headquarters, forces, bases, installations, activities and functions under the control of the Secretary of the Navy. See Navy Department.

Deploy: Specifically, to change from a cruising or approach formation to a formation of ships for battle or amphibious assault. Generally, to send ships or squadrons aboard for duty.

Deserter: Person absent from his or her command without authority whose apparent purpose is to stay away. Absence over 30 days is presumptive, but not full legal proof, of desertion, and the word, therefore (at least in the "old Navy"), was, when in danger of apprehension, run in the direction of the ship so that you can claim you were "heading back."

Detail officer, detailer: An officer in the Bureau of Naval Personnel who assigns personnel.

Dinghy: Small (less than 20 feet), handy pulling boat with a transom stern. May be rigged for oars or sail.

Disbursing officer: Officer who keeps pay records and pays salary allowances and claims.

Dispensary: A medical or dental facility or both offering services less elaborate than those of a hospital.

Dirty bag (box): Small canvas bag or a box used by sailors and marines to stow odds and ends of gear.

Division officer: Junior officer assigned by the commanding officer to command a division.

Dock trials: Test of ship's operating equipment, including engines, while alongside the pier or other docking facility prior to sea trials and after construction or overhaul. Part of a builder's trials for new construction. See fast cruise.

Dog down: Tighten dogs or clamps on a port, hatch or door.

Dog tag: Slang: identification disk.

Dowse (douse): To put out, to lower a sail quickly, or to wet down or immerse in water.

Dressing ship: Displaying national colors at all mastheads and the flagstaff; full dressing requires a rainbow of flags from bow to stern over the mastheads.

Drill: Training exercise in which actual operation is simulated, such as a general quarters drill. In the Naval Reserve a person in drill status is paid for a four hour drill.

Dutch courage: The courage obtained from drink. Comes from the custom initiated by the famous Dutch admirals Tromp and deRuyter of giving their crews a liberal libation before battle with the English. the practice was naturally belittled by the English, who nevertheless were forced to admit to the effectiveness of the Dutch navy.

Eagle screams, the: Slang: payday.

Ear banger: Slang: one who is over-anxious to please.

Eight-o'clock reports: Reports received by the executive officer shortly before 2000 (8:00 p.m.) from the heads of departments. The executive officer, in turn, reports eight o'clock to the commanding officer. Sometimes mistakenly called the twenty-hundred reports.

El Nino: A complex and, as yet, little understood oceanic phenomenon mainly in the Pacific but with an effect on the frequency of hurricanes in the Atlantic. It involves an interruption of Pacific trade winds with marked changes in ocean currents. This includes a temporary shift of the Peru current, which interrupts upwellings vital to coastal fishing off South America. Called El Nino (the Christ child) because it usually occurs at Christmas.

Embarkation officers: Landing force officers who advise naval unit commanders on combat loading and act in liaison with troop officers. Formerly called transport quartermasters.

Enlisted man or woman: Navy personnel below the grade of warrant officer.

Ensight: The most junior commissioned officer. Also the national flag flown by a man-of-war from the gaff underway, flagstaff in port.

Escape hatch: In general any hatch, usually small, installed to permit escape from a compartment when ordinary means of egress are blocked. Developed to a high degree in modern submarines, which have escape trunks fitted to receive a rescue chamber or a deep-submergence rescue vehicle, an additional hatch for unassisted escape, and numerous specialized operating mechanisms and devices.

Executive officer: The second in command of a ship, station, aircraft squadron, etc. Slang: exec, XO.

Extra duty: Additional work assigned as punishment under the Uniform Code of Military Justice (UCMJ).

Eyes of the ship: The forward-most portion of the weather deck, as far forward as a person can stand, where fog lookouts are customarily stationed, because a fog often is less dense, even clear, near the surface of the sea. (Other lookouts are of course in masts and anywhere else the ship's officers believe might be useful.) Most old ships had figureheads on their bows, and the term is supposed to refer either to their eyes or to those of the fog lookouts. Interestingly, the Chinese, with an entirely different cultural history, have carved or painted eyes on the bows of their junks for centuries.

Fall in at quarters: Command to form ranks at quarters. The command to disband or fall out is: "Leave your quarters."

Fantail: The aftermost deck area topside in a ship.

Fathom: Measure of length, 6 feet. See mariner's measurements.

Feather merchant: An uncomplimentary term of mild scorn, applied to those new to the service (especially reservists).

Field day: Cleaning day, traditionally Friday, the day before inspection.

First-aid kit: Emergency medical equipment in a repair party locker or on a life raft. At a gun, material is in a first-aid pouch or gun bag. In a boat, the first-aid gear is in the boat box.

First call: A bugle call sounded five minutes before quarters, colors, or tattoo.

Fitness report: Periodic evaluation of an officer's performance of duty and worth to the service, by his commanding officer.

Flagship: The ship from which an admiral or other unit commander exercises command.

Flat hat: Blue cap formerly worn by enlisted men. Similar in shape to the officer's cap, but without the visor.

Fleet: An organization of ships, aircraft, marine forces, and shore-based fleet activities, all under a commander or commander-in-chief who may exercise operational as well as administrative control. Also, all naval operating forces. See task organization.

Fleet Marine Force (FMF): A balanced force of combined arms comprising land, air, and service elements of the US Marine Corps. A Fleet Marine Force is an integral part of a US Fleet and has the status of a type command.

Flight quarters: Manning of all stations for flight operations aboard ship.

Flight surgeon: Medical officer specially qualified for duty with an aviation unit.

Fog lookouts: Special lookouts because of fog. Generally sent to the eyes of the ship, because in the thick fog they might well be the first to see another ship or object when danger of collision might exist. The Rules of the Road specify that in a fog a ship must proceed sufficiently slowly that she can come to a complete stop in half the distance of visibility. Lookouts may also be sent aloft when the fog is dense but close to the surface of the water.

Foul weather: Rainy or stormy weather.

Frigate: In days of sail, a full-rigged ship mounting guns on a single gundeck and on forecastle and poop, or spar deck, depending on construction of weather decks, i.e., two decks of guns. Fast and maneuverable, comparable to the cruisers of a later day. The British navy used frigates as adjuncts to their battle fleet, for scouting, signaling, etc. The US Navy designed and used them for ocean raiders, able to defeat anything fast enough to catch them. In modern navies, the frigate is smaller and slower than a destroyer, an ocean escort for anything less important than an aircraft carrier. For a time, the US Navy was alone in using the designation frigate for its large destroyers (so-called destroyer leaders) in recognition of the place the early frigates held in our history, but it has now reverted to the common practice of all other navies. The large destroyers have, at the same time, been redesignated as cruisers. See deck(er), double-banked frigate, razee.

Frogmen: Slang: underwater demolition team (UDT) personnel.

Full speed: A prescribed speed that is greater than standard speed but less than flank speed. Highest sustainable speed. See speed.

Galley: A shipboard kitchen.

Gangway: An order to stand aside or stand clear. An opening in the rail or bulwarks of a ship to permit access on board; not a synonym for accommodation ladder. In the days of sail, ships with raised forecastles and poop decks frequently had walkways connecting the two, one on either side. Originally called gangways, these walkways gradually were constructed of much heavier timbers, until they were indistinguishable from the decks they connected, and the once open area between forecastle and poop evolved into a large hatch. The combination of forecastle, poop deck, and gangways became known as the spar deck. In some particular heavy warships, guns were mounted along its entire length. See double-banked frigate.

General alarm: The signal for manning battle stations. Nowadays given by musical notes over a ship's general announcing system. In days past, various other means, such as bugle calls, fife and drum, drum alone, a loud rattle, or the boatswain's pipe were used.

General quarters: Stations for battle. To sound general quarters is to give the signal, or the general alarm, that will bring all hands to their battle stations as quickly as they can be there.

Gibson girl: Portable radio for sending distress signals, carried on life rafts.

Gob: Slang: an enlisted man (not good usage).

Goldbrick: Slang: a loafer or to loaf.

Go-to-hell hat: Slang: overseas or garrison cap.

Grandpa Pettibone: See anymouse, Dilbert.

Greens: A pre-WWII uniform worn by aviators and submariners. Heavy woolen material, green in color, with black buttons, black stripes on sleeves, khaki shirt, black tie, and no shoulder boards. Warm weather version was similarly cut cotton khaki (same as now except for black buttons and black stripes). Today, members of the Seabees use the term for their green utility uniform and aviation personnel for their aviation green working uniform.

Greenwich mean time: The mean (average) solar time as measured from the meridian of Greenwich. The navigational time reference point.

Grog: Slang: any alcoholic drink, particularly rum. In old British navy, it was a mixture of rum and water, invented by Admiral Vernon, whose affectionate nickname was Old Grog. Mount Vernon, Washington's home was named for him.

Guard mail: Mail delivered by guard mail petty officers between naval activities.

Guidon: Company identification pennant for naval units ashore.

Gulf Stream: That part of the Gulf Stream system from off Cape Hatteras to the Grand Banks. See Gulf Stream System.

Gun salute: Blank shots fired to honor a dignitary or in celebration. The national salute is 21 guns, fired for a chief of state. Lesser dignitaries rate progressively fewer guns, according to rank. The number of guns is always odd.

Half deck: A partial deck between complete decks. See deck, gallery deck.

Half hitch: Usually seen as two half hitches; a knot used for securing a line to a post.

Half mast (half staff): To fly a flag halfway up the mast, as a sign of mourning.

Hand salute: Gesture of respect exchanged between military men and women. See salute, hand.

Happy hour: Period of entertainment aboard ship, including refreshments. Same as smoker. The name has been plagiarized to signify a period at a bar or club ashore when prices are reduced.

Hash mark: Slang: service stripe.

Haven, submarine: Sea area in which no attacks on submarines are permitted, allowing a safe passage for allied submarines in wartime. Submarine havens may be fixed (stationary, defined by navigational or geographic coordinates) or moving at a specified course and speed. See moving havens, submarine haven.

Head: Toilet and washroom. Derived from days of sail when the comfort station for the crew was forward on either side of the bowsprit. Also, the upper corner of a triangular sail.

Headers: Reservoir into which or from which the tubes of a boiler or heat exchanger terminate. Also, another name for tube sheets. See nipples.

Heave in: To haul in.

Heave to: To stop-in an affirmative sense, to bring the ship to a halt, dead in the water. In case of heavy weather, a ship may heave to in order to take the most comfortable and safest heading. She is in this case considered to be hove to even though making considerable way through the water from the action of wind and sea.

Helm: The helm proper is the tiller, and thus the order to put the helm to port, for example, is the same as an order to put the rudder right. The term has now developed to mean the rudder and gear for turning it, and the helmsman is, of course, the person who steers. Because of possible confusion as to intended direction, however, orders are today given with respect to the rudder, and never, except in small pleasure craft, with respect to the helm, using "right" and "left," never starboard and port.

Highline: A single line rigged between two ships under way transferring stores. The simplest transfer rig. Stores and personnel are transferred on a wheeled trolley riding on the highline and hauled back and forth between the ships.

High water: The maximum height of a tide because of tidal and weather conditions.

Hitch: A knot whose loops jam together in use, particularly under strain, yet remain easily separable when the strain is removed. Method of securing a line to a hook, ring, or spar, e.g., clove hitch. Slang for a term of enlistment.

Hit the deck: Get up; same as rise and shine.

Hit the sack: Slang: to turn in, go to bed. See rack, sack.

Hoist: Display of signal flags at a yardarm. To lift. Also, the vertical portion of a flag alongside its staff.

Holiday: Unscrubbed or unpainted section of a deck or bulkhead. Any space left blank or unfinished through inadvertence.

Holiday routine: Schedule aboard ship involving no work or drills; normal for Saturday (sometimes Wednesday) afternoons and Sunday. See rope yarn Sunday.

Home port: Port or air station on which a ship or aircraft unit normally bases.

Homeward bound pennant: Pennant flown by ships returning to US after absence of a year or more. Traditionally very long (1 foot of length for each day away). After arrival it is cut up and passed out to the crew for souvenirs, a rare procedure now.

Honey barge: Garbage scow.

Honors and ceremonies: Collective term: official guards, bands, salutes and other activities that honor the colors, celebrate a holiday, or greet a distinguished guest or officer.

Hospital ship (AH): An unarmed ship, marked in accordance with the Geneva Convention, staffed and equipped to provide complete medical and surgical facilities.

Hurricane: Destructive cyclonic storm with winds more than 65 knots. In the eastern hemisphere it is called a typhoon or cyclone. See breeze, gale, storm.

Ice fields: Mass of drifting ice, offshore. A form of pack ice.

In: Navy personnel serve "in" a ship, not "on" her. A carrier's aircraft and aircrews are however based "on" the ship and fly "off" her.

Inland Rules of the Road: The Inland Navigational Rules (in accordance with the Inland Navigational Rules Act of 1982) govern the conduct of vessels in the inland waters of the US.

Inspection: There can be many types, ranging from a careful and critical examination of personnel, material, and record keeping to the ceremonial inspection of an honor guard at a spit-and-polish formation.

Intercom: Ship's voice intercommunication system. Also called a squawk box.

International Rules of the Road: Officially known as *The International Regulations for Preventing Collisions at Sea.* Rules of the nautical road made effective by agreement of the major maritime powers for use on the high seas and most inland waterways of the world. They are binding on public and most inland waterways of the world. They are binding on public and private ships of all signatory powers when on the high seas (international waters), and they apply on the inland waters of all such countries unless (international waters) and they apply on the inland waters of all such countries unless superseded by duly enacted special rules. Masters of ships operating anywhere in the world are required by law as well as custom to be aware of and comply with all applicable rules. See *Inland Rules of the Road, Sailing Directions, Pilot Rules.*

Irish Pennant: Loose, untidy end of line left adrift. Also called deadman or cow's tail.

Jack: Short for union jack, a blue, white-starred flag flown at the bow jackstaff of a vessel at anchor. To jack over the engines is to turn them over. Short for cablejack.

Jackass: Cover over hawsepipe to keep water out. Also, similar cover over entrance to piple leading to chain locker, after anchor chain comes off wildcat. See buckler.

Jack-of-the-dust: Person in charge of the provision issue room.

Jacob's ladder: Portable ladder, with rope or wire sides and wooden rungs, slung over the side for temporary use. See sea ladder.

Jew's harp: Ring or shackle at upper end of shank of an anchor to which anchor chain is secured. Any harp-shaped shackle.

Job order: Order issued by a repair activity to its own people to perform a repair job in response to a work request from the unit to be repaired.

Joe pot: Slang: coffee pot; short for jamoke, which was once the naval term for coffee.

Judge advocate general (JAG): The senior legal officer in the Navy.

Jump ship: Slang: leaving ship without authority or permission. See over the hill.

Junior officer: Technically, lieutenant commanders and below (i.e., without brass hat). In practice, usually ensigns and lieutenants are considered junior officers. Lieutenant commanders are often addressed as commander, like the traditional Army practice of addressing lieutenant colonels as colonel. The navy is said to be composed of "young, studs, old fuds, and lieutenant commanders."

Kamikaze (divine wind): Japanese suicide aircraft in WWII that crashed into ships. Also the pilots of such craft. Term is now used to refer to any deliberate suicidal tactic in war.

Kapok: Natural, light, waterproof fiber used in stuffing life jackets. From seeds of tropical kapok or silk tree.

Keeping ship: Observing the routine of a ship when not engaged in exercises or operations.

Keep the sea: Refers to staying at sea and remaining effective under all conditions. A term of praise, admiration, professional regard.

Knife and fork school: Slang: the Officer Indoctrination School to which officers commissioned directly from enlisted ranks are sent.

Knock off: To stop; cease.

Knot: Unit of speed equivalent to 1 nautical mile (6,076 feet) per hour. A collective term for hitches and bends (any combination of loops, mostly interlocking) used to attach two or more lines to each other, or to a fixed object such as a post or bollard, or to form an eye or loop. A knob (knot) in a line or rope, used for climbing or any other purpose, including the measurement of a ship's speed through water (the origin of knot as a unit of speed) by how many knots were pulled across the taffrail in a given unit of time. See nautical mile.

Landing force: The troops organized for an amphibious assault. Also, a portion of a ship's crew detailed to go ashore in an organized unit for any military operation, a term obsolete since WWII.

Landing party: An organized force of infantry from the ship's company detailed for emergency or parae duty ashore. Formerly called landing force.

Landing signal officer (LSO): Officer who directs pilots in landing on an aircraft carrier.

Landlubber: Seaman's term of derision for one who has never been to sea, or who has never learned seamanlike ways; hence a lubber, or lubberly.

Lash: To secure with line or wire by wrapping and tying with seamanlike knots in the case of line, or with an approved hitch in the case of wire. Lashing or wire lashing are the materials used. LASH (lighter aboard ship) is a method of shipping in which loaded lighters or barges are transported in a specially designed merchant ship.

Lay: To go, as in "Lay aft on the fantail." The direction of the twist of strands of a rope.

Lay before the mast: To assemble or fall in, usually to make reports. Reference is to ancient custom when crew members stood forward of the mizzenmast facing officers who stood on the quarterdeck, aft of it.

Leadership: The art of accomplishing the Navy's mission through people. It is the sum of those qualities of intellect, human understanding, and moral character that enables someone to inspire and manage a group successfully. From General Order 21 as originally promulgated.

Leatherneck: Slang: a marine.

Leave: Authorized vacation or absence from duty other and longer than liberty. The term shore leave is no longer used. Leave carries with it permission to travel beyond the allowed radius of liberty. See liberty.

Letters of censure: Nonjudicial punishment imposed on an officer by the commanding officer or reporting senior and entered in officer's record. Most severe is a letter of reprimand; of lesser severity—letter of admonition; least severe—letter of caution.

Liberty: authorized absence of an individual from place of duty, normally not more than 48 hours. Sometimes incorrectly called shore leave. Liberty sometimes carries geographical limits to provide possibility of quick recall. See leave.

Lifeboat: In merchant ships, a boat required by international law that can be quickly launched for safety of passengers and crew. The number required is determined by the number of persons permitted to voyage in a ship (as a result of the *Titanic* disaster, when there were not enough). In a warship, the lifeboat is a ready boat that can quickly be launched with a crew trained and equipped for rescue day or night. The motor whaleboat is traditionally the most seaworthy of all ship's boats and is much the preferred type for this use.

Lifeline: Line secured along the deck to lay hold of in heavy weather; any line used to assist personnel; knotted line secured to the span of lifeboat davits for the use of the crew when hoisting and lowering. The lines between stanchions along the outboard edges of a ship's weather decks are all loosely referred to as lifelines, but specifically the top line is the lifeline, middle is the housing line, and bottom is the footline. Any line attached to a lifeboat or life raft to assist people in the water. Also called a grab rope.

Lighter: Barge-like vessel used to load or unload ships.

Lighthouse: A building in which an aid-to-navigation light is located.

Limited duty officer (LDO): An ex-enlisted person who has won a commission because of his or her high quality. Because of the branch and experience from which risen, the LDO is limited to duties of his or her specialty.

Line: General term for rope; the equator; a formation in which ships or personnel are formed in any direction from the guide. See rope.

Log, deck: Official record of a ship in commission submitted to the Chief of Naval Personnel in accordance with *Navy Regulations and the Bureau of Naval Personnel Manual.*

Longevity pay: The increase over base pay that is computed on years of service. See fogy.

Loofa sponge: A type of sponge formerly used in the hot well of a feedwater heater aboard ship.

Look alive: To be alert, move quickly.

Lookout: Crew member stationed as a visual watch; air, horizon, surface, fog, etc. See reports, lookout.

Lubberly: Unseamanlike; clumsy. A lubber is an unseamanlike person. See landlubber.

Lucky bag: Container or stowage for articles found adrift. Unclaimed articles aboard ship are periodically sold at auction. A sailor losing something might be lucky enough to find it in the lucky bag and reclaim it.

Machinery index: Comprehensive listing of all machinery and related equipment, other than electronic, installed on board.

Mae West: Pneumatic life jacket. So named for the obvious reasons.

Mailbuoy: An ancient sailor's joke: the mid-ocean buoy in which mail is kept for delivery to passing ships. See bucket of stream, hammock ladder, relative bearing grease.

Man: to assume station, as in "Man your planes."

Manned and ready: Report made by a gun or station when all hands are present and ready for action. See ready.

Manning the rail: All-hands evolution in which the crew lines up along the ship's rail to honor some personage or occasion.

Man-of-war: Fighting ship; armed naval ship. A warship. Most commissioned vessels of the Navy are men-of-war, but not al, e.g., a hospital ship.

Mast, captain's: A sort of court held in a designated place at which the commanding officer awards punishment, listens to requests (request mast), or commends men for special services (meritorious or commendatory mast). Derived from the old practice of holding this court at the mizzenmast, which, in a three-masted ship, separated officers' quarters from those of the crew. The term "before the mast" also relates to the significance of the mizzenmast. See office hours.

Master-at-arms (MAA): Ship's police, headed by a chief MAA. There may be special MAAs, such as the one in charge of the mess decks. In the Old Navy, the chief MAA had custody of the ship's hand weapons and trained the crew in their use.

Mayday: International distress signal, voice radio. Derived from the French "Maidez" (help me).

Meatball: The battle efficiency pennant. In naval aviation, the meatball is an amber light that appears at the center of a "mirror," which in reality is a stack of five lenses. If the aircraft is properly positioned in the glide path, the meatball will be aligned with the horizontal line of green reflected light on either side of the center lens. To be on the ball means that the pilot has the meatball in the right place and is on the glide path.

Mess: To eat. A group of people eating together. Crew's mess is called the general mess.

Mess decks: Area where the crew dines (messes). In older times the entire crew generally messed on a single deck that, for the period of the meal, was known as the mess deck. Now the term more accurately refers only to the compartment(s) in which the meals are served.

Mess Kit: Ingenious assemblage of cooking and eating utensils, favored by troops planning a landing. Loosely used to denote table setting on board ship.

Messman: An enlisted person detailed to serve food to the crew. Slang: bean jockey. Now called a mess management specialist.

Mess treasurer: Person who administers finances of any mess.

Midshipman: A student officer, enrolled at the US Naval Academy or at a civilian university under the Naval Reserve Officers Training Corps (NROTC) program, who is commissioned as a Navy or Marine Corps officer upon graduation. ROTC refers to the comparable Army program. Unlike an ROTC cadet, he or she wears gold braid and has a legal status in the Navy between chief petty officer and warrant officer. Acceptable slang: mid. Middie is a term of derision.

Midwatch: The watch from midnight to 4:00 a.m. (0000-0400).

Mission: The general objective or purpose of a military operation.

Morning colors: The act of hoisting the national flag aft, and the union

jack forward, at exactly 0800 local time. Done with maximum precision by all ships in port, following exactly (and with pride) the motions of the senior ship present. Usually preceded for five minutes by the Blue Peter (P:papa) at the force yardarm.

Mustang: Slang: an officer who was formerly enlisted and was not commissioned directly from college or from the Naval Academy.

Muster: Roll call.

Muster on stations: Roll call taken aboard ship while personnel are at work or drill.

Nadir: That point on the celestial sphere vertically below the observer, or 180 degrees from the zenith. See zenith.

Napalm: Jellied gasoline for use in flame throwers and incendiary bombs.

Naval air station (NAS): Provides operating testing overhaul training, and personnel facilities for naval aviation.

Naval attaché: A naval officer on duty at an embassy abroad whose major tasks are advising and representing the ambassador on naval matters and collecting intelligence. See Alusna.

Naval base: A shore command in a given locality that includes and integrates all naval shore activities in the assigned area.

Naval district: A geographically defined area in which one naval officer, designated commandant, is the direct representative of the Secretary of the Navy and the Chief of Naval Operations. The commandant has the responsibility for local naval defense and security and for the coordination of naval activities in the area.

Naval Reserve: A force of qualified civilians who are trained and available, in an emergency, to meet the needs of an expanding Navy while an adequate flow of new personnel is being established. Component parts are fleet, organized volunteer and merchant marine reserves.

Navigator: Officer who is head of the navigation department, responsible for the safe navigation of the ship.

Navy Department: The executive part of the Navy located in Washington, D.C. Differs from Department of the Navy in that by long usage it refers only to the administrative offices at the seat of government. See naval establishment.

Navy enlisted classification (NEC) code: A system for identifying and designating special skills and knowledge for enlisted personnel. Every sailor has at least two of the four digit NEC codes: a primary code and a secondary code. Details are contained in the *Manual of Navy Enlisted Classifications.*

Navy exchange: A store for naval personnel and their dependents that sells at a small profit for the benefit of the welfare and recreation fund.

Navy League of the United States: A national organization that believes in and works for a strong Navy; headquarters, Washington, D.C. Completely independent of the Navy, active duty officer many not belong.

Navy Relief Society: The quasi-official relief agency operated for the benefit of naval personnel and their dependents.

Navy Unit Commendation (NUC): An honor accorded a naval unit for distinguishing itself in combat or other operations; of lesser degree than Presidential Unit Citation.

Navy yard: The Washington Navy Yard in Washington, D.C. headquarters of the Washington Naval District is now the only "yard" in the USN. The CNO's officially assigned quarters are in this yard (the historic Tingey House), as well as quarters for the District Commandant and other officers. There are also numerous other resident activities, such as the Navy Department Library, headquarters of the Navy Blue Band, the Navy Museum, etc.

Negative:" In any naval message, means no, not granted, do not concur, not approved etc.

Neptune: mythical God of the sea.

Night stick: A short wooden club carried by personnel on shore patrol duty.

Night vision: Faculty of seeing well at night. See dark adaptation.

Nonjudicial punishment: Punishment by a commanding officer imposed without trial by court-martial. Specified and limited by the Uniform Code of Military Justice. See mast, captain's; office hours.

Nonrated man/nonrated woman: Enlisted person in the first three paygrades. Not a petty officer.

Notice: A specially numbered announcement or one-time directive. Not permanent, as an instruction is. Also, notice for getting under way;

required material and personnel condition of readiness (e.g., four hours' notice). See instruction.

November: Phonetic word for the letter N.

Now do you hear there: Traditional preface to the word when passed by the boatswain's mate throughout the ship. Usually shortened to "Now hear there (this)."

Oars: Command given to a crew of a pulling boat, directing them to stop rowing and stand by with oars in rowlocks, extended horizontally, ready for the next order.

Officer candidate: Person under instruction at the Officer Candidate School. Has status of enlisted.

Officer of the deck (OOD): An officer on duty in charge of the ship representing the commanding officer. There may also be a junior officer of the watch (JOOW) or junior officer of the deck (JOOD). The officer on watch under instruction is the assistant officer of the watch.

Officer's call: A bugle call (or word passed) for officers to take their stations. Precedes general call for all hands.

Old Man: Slang: the commanding officer of an activity (almost always capitalized). The corresponding expression for an admiral is old gentleman (usually not capitalized).

Ombudsman: An officer in the Navy Personnel Command who receives and acts on complaints by naval personnel who do not receive satisfaction through normal channels.

On the double: Quickly; with speed; as: "Man your stations on the double."

Operating forces: Fleet, seagoing, sea frontier, and naval district forces, and such other activities and forces as may be assigned by the President of the United States or the Secretary of the Navy.

Operational: Pertaining to operations; capable of operating; in contrast to administrative.

OPNAV: Overall title for the entire Pentagon staff that is under control of the CNO. The CNO's own personal staff is only a small part of OPNAV.

Order: An order directs that a job be done but does not specify how. See command.

Orderly: Messenger or personal attendant, usually for a senior officer.

Overhaul: Repair, clean, inspect, adjust. To overtake. Also to separate the blocks of a tackle. Opposite of round in.

Over the hill: Slang: deserting. To desert is to go over the hill.

Padre: Slang: chaplain

Painter: A line in the bow of a boat for making fast. The corresponding line in the stern is the stern fast. See bow painter, sea painter.

Parade: An area aboard ship where the divisions fall in for muster or inspection. Fair weather parades are topside; foul-weather parades below. See all hands parade.

Passageway: Corridor or hall aboard ship.

Passing honors: Those honors, except gun salutes, rendered by a ship when ships or embarked officials or officers pass close aboard.

Pass the word: Broadcast of information.

Pay: Slang: short for paymaster. Disbursing, or supply, officer of a ship. Also, to fill the seams of a wooden vessel with pitch or other substance. "The devil to pay and no pitch hot." (Devil was the longest and toughest seam to pay, alongside the keel.)

Paygrade: Level of military pay, from E-1 (recruit) to E-(master chief petty officer); from W-1 (warrant officer) to W-4 (commissioned warrant officer); and from O-1 (ensign) to O-10 (admiral).

Paymaster: General term for any disbursing officer.

Peacoat: The heavy topcoat worn by seafarers in cold weather. Cut short, well above the knees. The coat was originally made of material, called pilot cloth, so it is probable that the name was successively pilot-cloth coat, P-coat, and finally peacoat. Also called reefer in the Old Navy.

Pennant: A flag that is longer in the fly than in the hoist, and usually tapers to a point. Some pennants terminate in a swallow-tail (two points). Examples are commission and broad command pennants, etc.

Per diem: Additional expense money for a person on temporary additional duty or in a travel status.

Periscope: An optical device of mirrors and prisms used to project one's vision over an obstacle. Most usually a submarine periscope.

Petty officer: Noncommissioned officer in the grades of master chief, senior chief, chief, and first, second and third class.

Pilot: An expert on local harbor and channel conditions who advises the commanding officer in moving a ship in or out of port; one who operates an airplane; a book of sailing directions.

Pipe, boatswain's: Distinctive silver whistle used to sound calls when passing the word aboard ship or during quarterdeck honors. Also used to direct operations when hoisting in a boat or handling cargo. Sometimes referred to as a boatswain's call, although this term is also used to designate a tune played on the pipe.

Pipe down: An order to be silent or reduce noise, when used alone. In combination with other instructions, means to take down (pipe down aired bedding), announce meals (pipe down chow for the crew), or initiate an activity (pipe down sweepers).

Plankowner: Person who has been on board since ship was commissioned.

Plan of the day: Schedule of the ship's activities for the day including work, training, meals, recreation, etc. Also called morning orders.

Pogey bait: Slang: candy, also soda fountain items.

Police: to inspect and to clean up.

Polliwog: One who has not crossed the equator. See shellback.

Poop deck: A partial deck, aft above the main deck, in wooden men-of-war. See spar deck.

Port: Seagoing term for left as opposed to starboard, which means right. A coastal city accessible to sea commerce. Short for air port, cargo port, etc.

Porthole: Round opening in the side of a ship. Also an air port. British term is sidelight.

Presidential Unit Citation (PUC): An honor accorded a naval unit for distinguishing itself in combat or other operations. Of higher degree than Navy Unit Commendation.

Prisoner at large (PAL): Person under arrest whose restraint to certain specified limits is morally enforced.

Public reprimand: A rare but legal form of punishment given an officer pursuant to the sentence of a general court-martial.

Quarterdeck: Ceremonial area of the main deck, kept especially neat and clean. The specific domain of the officer of the deck while in port. Always located near the accommodation ladder or brow—or the principal one if more than one is rigged. In days of sail, the ship would be canned from the quarterdeck.

Quarterdeck: Ceremonial area of the main deck, kept especially neat and clean. The specific domain of the officer of the deck while in port. Always located near the accommodation ladder or brow—or the principal one if more than one is rigged. In days of sail, the ship would be canned from the quarterdeck.

Quarters: An assembly, as quarters for inspection; or a gathering on stations, as fire quarters. Government-owned houses or apartments assigned to naval personnel. Living spaces aboard ship.

Rail: An open fence aboard ship, made of pipe or other rigid material as opposed to lifelines, usually along the edge of the weather deck. Also used below in large spaces such as engine rooms. Also a plank, timber, or piece of metal forming the top of a bulwark. A hand rail, ladder rail, or safety rail associated with ship's ladders. Rail screens of canvas are sometimes rigged along a ship's rail.

Rank: Relative position of officer or petty officer within a particular grade. See grade for usage.

Rate: Relative position of officer or petty officer within a particular grade. See grade for usage.

Rat guard: A conical metal shield secured around mooring lines to prevent rats from coming aboard.

Rating badge: Insignia of rating and rate worn by petty officers.

Rations: Food. May be abandon ship, flight, aircraft, emergency, landing party, travel, or leave rations.

Readiness, conditions of: Comprehensive term that includes: material—what closures have been made, engineering—maximum speed immediately available, armament—what weapons are manned. A ship's battle and damage control bills will indicate the strictures and nomenclature for each of the conditions. In addition, condition 1 is general quarters, the maximum readiness attainable; condition 2, half the crew on stations and ready; condition 3, normal cruising. See condition watch.

Recall: Signal flown from a ship summarily directing all her boats to return immediately. Sometimes used in emergency, this is one of the preliminaries to getting under way.

Red Cross flag: The distinctive mark of a hospital ship and medical services.

Red tide: A marine phenomenon that turns the sea reddish and kills marine life. Caused by plant-like protozoans called "dinoflagellates." Red tide occurs world-wide and may be toxic to humans, forcing resorts to close.

Reef: Rock, coral, sand, or any bottom material extending so near the surface of the water that boats or ships cannot pass over safely. To reef a sail is to reduce its effective area, either by tying its reef points and rolling or folding it up at the head if a square sail, or at the foot if a fore-and-aft sail. In modern yachts it may be rolled around the boom or even taken into a split metal mast with a roller built into the mast's hollow interior. The part of the sail that can be reduced is a reef, as in take in a reef or shake out a reef. In days of sail it also meant to lower or take in a portion of a spar such as top gallant masts, top masts and the outer end of a bowsprit.

Repair party: Group of specialists organized to control damage and make repairs throughout the ship.

Replenishment at sea: The process or procedure of supplying fuel, food, stores, ammunition, and personnel to fleet combatant units while under way. If helicopters are used it is called vertical replenishment.

Request mast: Process by which crew members can submit requests to the commanding officer or executive officer.

Rescue and assistance party: Group of specially qualified personnel sent off the ship with special equipment to assist in rescue, firefighting and salvage operations. Formerly called fire and rescue party.

Restraint: According to the *Uniform Code of Military Justice,* involves confinement, arrest, and restriction in lieu of arrest. Arrest is the moral restraint of a person by an oral or written order to certain specified limits. It is not punishment, but it does relieve the person of all military duties. Thus restriction is often used instead of arrest to permit working and watchstanding. Confinement is actual physical restraint for good cause.

Retreat: Bugle call or word passed that means fall out (disband) from a formation, e.g., retreat from inspection. Also sounded at evening colors.

Reveille: Arousing the ship's company in port for work and breakfast. At sea, idlers are called, and the expression reveille is not properly used.

Rig ship for visitors: Word passed to all hands to have ship prepared for expected visitors. This involves closing off restricted areas, stationing sentries, providing guides, etc.

Rise and shine: To get up, go to work, get going. See show a leg.

Rocks and Shoals: Extracts from the Uniform Code of Military Justice periodically read aloud to all hands.

Roger: Used in voice radio, meaning "I have received your transmission." Not to be confused with wilco ("I understand and will comply").

Roger doger: Slang: affirmative, yes, okay, will do.

Rope: Term used in the Navy for special items such as manrope and bellrope, wheel rope, and wire rope. Line is used in general sense for all cordage and fiber rope, and for special ones such as mooring line, heaving line, shot line, etc. Fiber rope is made of natural fibers (manila, hemp, sisal, cotton) as well as synthetic fibers (nylon, polyester, etc.) Strips of metal foil used in radar countermeasures also are called rope.

Rope yarn Sunday: Any afternoon, except a weekend, that is free of work and drills. Usually Wednesday afternoon is rope yarn Sunday if no work is scheduled at sea or if liberty is granted in port. British equivalent is make and mend, or mend clothes.

Ruffles and flourishes: The roll of the drum (ruffles) and short burst of music (flourishes) that make up one of the honors rendered to high-ranking military and civil officials.

Rules of the Road: Regulations designed to prevent collisions of ships at sea and in inland waters. Proper full term is International Regulations for Preventing Collisions at Sea. Short term is Colregs. See *Inland Rules and International Rules.*

Running lights: Required lights carried by a vessel or aircraft under way between sunset and sunrise. See riding lights.

Sack: Slang: bunk or bed. To sack out is to take a nap. Slightly earthier derivation than the synonym rack.

Sailor: A person who has spent time at sea and is accustomed to the ways of the sea and ships. Applied to officers as well as enlisted personnel as a term of approbation.

Sea legs: Adaptation to the motion of a vessel in a seaway. To find one's sea legs is to have recovered from seasickness.

Seaman guard: Enlisted person who performs guard duty in the absence of marines.

Sea painter: Line used for towing a boat alongside a ship under way. Led from well forward in the ship to the near bow of the boat, where it is secured with a loop and a toggle, for quick release. Not to be confused with bow painter, which is secured to the stem of the boat and cannot be used for towing alongside.

Sea stores: Cigarettes and other luxuries sold at sea and abroad free of federal tax.

SECNAV: The secretary of the Navy. The term comparable to OPNAV is properly the SECNAV Secretariat, but in Pentagonese is often rendered simply SECNAV, as "He is in SECNAV."

Secret: Information or material whose disclosure would endanger national security or cause serious injury to the interests or prestige of the United States. See classified matter.

Secure: To make fast in a permanent sense, as to "Secure the forward hatch for sea." Well-fastened or safe. To cease or stop, e.g., "Secure from fire drill." To quit, give up, or knock off.

Selection board: Panel of officers who review records and recommend for promotion. Used for all grades above that of lieutenant junior grade. Also called promotion board.

Semaphore: Rapid method of short-range visual communications between ships, using hand flags.

Service medal: Medal for service in specific campaign or theater of operations.

Service number: Record identification number for Navy and Marine Corps enlisted and for Marine Corps officers. Has been replaced by social security numbers. See file number.

Service schools: School offering advanced technical training for enlisted personnel.

Set the watch: To establish the regular routine of watches on a ship or station.

Shakedown: Period of adjustment, clean-up, and training for a ship after commissioning or a major overhaul. After commissioning, a ship also makes a shakedown cruise, usually including a visit to several foreign ports.

Shellback: One who has crossed the equator and was accepted by Davy Jones.

Sheriff: Slang: master-at-arms.

Shift colors: To shift the ensign and jack between the steaming and in-port positions and vice versa.

Ship: Any large, sea-going vessel. Specifically, in days of sail, a vessel with bowsprit and three masts, entirely square-rigged, except for the jib(s) and the lowest sail on the aftermost mast, or mizzenmast, which was fore-and-aft rigged and called the spanker. Also called a full-rigged ship. In days of sail each type of rig had its own name, such as brig, schooner, sloop, snow, brigantine, hermaphrodite brig, bark or barque, barkentine, topsail schooner. Sloop-of-war was a full-rigged ship mounting guns on only a single deck. Also, to set up, to secure in place; e.g., ship the rudder. To re-enlist, as to ship over. To take something aboard, as to ship a sea, or to send freight. See sloop, vessel.

Shipmate: Person with whom one is serving or has served, particularly at sea.

Ship over: To re-enlist. See ship aboard.

Shipping-over chow: Any particular good meal, supposedly served to encourage re-enlistment.

Ship's bell: Struck every 30 minutes. At seven bells the next watch is called. At eight bells the new watch should be entirely in place. Also used to sound fog signals and as a fire alarm. See watch.

Ship's company: All hands, everyone on board who is attached to the ship. Does not include passengers.

Shipshape: Neat, orderly, as a ship should be.

Ship's secretary: Officer who assists the executive officer with ship's correspondence.

Ship's service store: The ship's retail store. Carries supplies for health, comfort and personal cleanliness of the crew.

Shooter: Slang: an alcoholic drink.

Shore leave. Obsolete term; now considered poor usage in the Navy. See leave, liberty.

Shore patrol (SP): Personnel ashore on police duty.

Short timer: Person nearing the end of enlistment, service or tour of duty.

Shoulder mark: Device indicating rank worn on an officer's overcoat and the jacket or coat of summer uniforms. Also called shoulder board.

Shove off: Slang: depart, leave, go. Proper naval usage as in "Shove off on the bell, coxswain, make the liberty landing and return."

Sick bay: Infirmary or first-aid station aboard ship.

Sick call: A call (and word) passed daily aboard ship for those who require medical attention to report to sick bay. A scheduled time each day at all medical facilities when patients may be seen without appointment.

Side boys: Nonrated crew members stationed in two ranks at the gangway on the arrival or departure of officers or officials for whom side honors are being rendered. Number varies from two to eight, depending on rank of visitor.

Sight the anchor: To heave an anchor up far enough to see that it is clear, then again let go. Necessary when holding ground is very soft; anchor may sink beyond easy recovery.

Signal: Short message using one or more letters, characters, flags, visual displays, or special sounds. Any transmitted electrical impulse.

Skipper: Slang: commanding officer, or any captain of a ship or boat, composed of saddles making up a cradle and gripes to hold the boat down. See captain.

Skivvy: Slang: underwear.

Skylark: A distinctly nautical expression meaning to play or to have fun, i.e., to have a lark. Derived from the practice of young sailors laying aloft and sliding down the backstays. Thus, to engage in horseplay, noisy banter, or friendly scuffling.

Sky pilot: Slang: Chaplain on board ship, sometimes also called padre.

Sloop-of-war: A fully rigged ship mounting her main battery on only a single deck, usually the spar deck, as distinguished from a frigate, which mounted guns on two decks. Smaller than a frigate and faster in light airs or moderate breeze. Same as corvette. Spar-decked sloops-of-war, developed after the War of 1812, had covered gundecks and sometimes were more powerful than earlier frigates.

Slop chute: A chute hung over the side for the discharge of garbage. Slang: anyone who is dirty and disorderly in appearance or habits.

Small arms: Rifles, shotguns, pistols and carbines.

Smoker: Shipboard entertainment, including food, boxing, humorous skits and movies. Same as happy hour when used at sea.

Smoking lamp: A lamp aboard old ships, used by men to light their pipes; now used in phrase "the smoking lamp is lit" to indicate when smoking is allowed. Today it is entirely a figurative term.

Soldier: Slang: to loaf on the job. One who loafs, as in "He is soldiering on the job." Note, however, that to address an Army person or a marine as soldier, or to use the word in reference to him or her, is a term of approval.

Sound: To measure depth of water at sea or the depth of a liquid in a ship's tanks. Result is a sounding. A whale sounds when it dives. Also to blow as a bugle. A long wide body of water connected to the sea, larger than a strait or channel. The air bladder of a fish with which it controls its buoyancy.

Special duty officer: An officer who specializes, for example in public information, but usually does not command.

Sponsor: The woman who christens a ship at its launching.

Spud locker: Slang: topside shipboard potato storage.

Squadron: Administrative or tactical organization consisting of two or more divisions of ships. Also, administrative unit of aircraft, for tactical purposes divided into divisions and sections.

Squall: Short but intense windstorm.

Square away: To straighten, make shipshape, or to get settled in a new job or home. Also to inform or admonish someone in an abrupt or curt manner.

Squilgee: Wooden, rubber-shod deck dryer. Pronounced "Squee-gee. Also squillagee.

Staff: Personnel without command function who assist a commander in administration and operation.

Staff officers: Those who perform staff functions, such as doctors, chaplains, dentists, civil engineers, supply officers, and medical service officers, as distinct from line officers.

Stand by: To wait. To substitute for someone who has the day's duty. A substitute. A preparatory expression, e.g. "Stand by, commence firing."

Standing lights: Dim red lights throughout interior of ship.

Standing order: Semipermanent order or directive.

Starboard: Directional term for right, as opposed to port, which means left.

Star shell: Projectile that detonates in the air and releases an illuminating parachute flare.

Stateroom: An officer's living space aboard ship. The captain and an embarked admiral have cabins.

Station: To assign. A post of duty, as a battle station. A position in formation of ships. A naval activity. In ship's plans, a section perpendicular to the keel is sometimes called a station.

Steady as you go or steady as she goes: An order to the helmsman to steer the course upon which the ship's head lay at the moment the order was given, even though she might be swinging through that particular heading. The helmsman should respond with, e.g., "Mark! Two five seven!" to indicate the reading of the lubbers line at the instant designated by the order, and the understanding that this is the ordered course to which the ship's head will be returned.

Stern hook: Member of a boat's crew who stands aft and makes the stern of the boat secure. See bow hook.

Stevedore: A person who loads and unloads ships' cargo. Generally, same as longshoreman.

Storekeeper (SK): Petty officer who performs clerical and manual duties in the supply department.

Storm: Meteorological disturbance; literally, a wind of 56-65 knots. See breeze, gale, hurricane.

Stow: To put away; to store.

Striker: An apprentice or learner.

Submarine (SS): Warship designed for under-the-surface operations. Attack submarines have the primary mission of locating and destroying ships, including other submarines. Missile submarines have primarily the mission of attacking land targets.

Submariner: Officer or enlisted person assigned to duty in submarines. US Navy pronunciation is "submarine-er." Royal Navy says "sub-mariner," as though "mariner" was a separate word.

Subsistence allowance: Money paid in lieu of food furnished. For officers only, unless there is no general mess. See commuted rations.

Sundowner: An extremely strict officer. The term nowadays carries a connotation of sadism in the application of the rules. See martinet for the same meaning without the sadistic implication. Derived from the ancient regulation that officers and men of a ship in commission must spend the night on board and must in fact be back aboard by sundown. A captain who insisted on observance of this regulation after it had outlived its purpose was called a sundowner.

Swab: Mop. Also a first-year cadet at the US Coast Guard Academy.

Swell: Wind-generated waves that have advanced into a calmer area and are decreasing in height and gaining a more rounded form. The heave of the seas. See roller.

Tactics: The employment of units in battle. Distinguish from strategy.

Tailhook: The hook lowered from the after part of the carrier aircraft that engages the arresting gear upon landing. Slang: a tailhooker is a naval aviator qualified in carrier operations.

Tanker: A ship that transports fuel to a base or service squadron. An oiler fuels other ships at sea or at an anchorage. An aircraft that refuels planes in the air.

Taps: Bugle call sounded as last call at night for all hands to turn in.

TAR: Designates a naval reserve officer or enlisted person on active duty in the training and administration of the naval reserve.

Tarpaulin muster: Collection of funds aboard ship for some common purpose, such as to assist a shipmate's widow. A tarpaulin is spread and contributions tossed into it. Now rare.

Tattoo: Bugle call sounded just before taps as signal to prepare to turn in.

Taut: Tight, without slack. Well-disciplined, as a taut ship.

Temporary additional duty (TAD): A short assignment in addition to regular duties. Pronounced "tee-ay-dee." Navy and Marine Corps only. Other services use TDY—temporary duty.

Tender: Logistic support and repair ship such as a destroyer tender (AD).

Texaco: The tanker plane that refuels carrier aircraft when airborne. Flies off the carrier.

Tidal Wave: A slow-moving bulge of water on the earth's surface caused by the gravitational attraction of the sun and the moon. Also called tide wave. As this bulge or wave advances and then retracts from the shore, high and low tides result. Also an inaccurate term for a seismic sea wave. See tsunami.

Tide: The vertical rise and fall of the ocean level caused by the gravitational forces of the moon and the sun. Rising tide is a flood tide; falling tide is an ebb tide.

Tie-Tie: Cloth straps or strings that tie together, as on a kapok life jacket.

Tilly: A large crane on the flight deck of an aircraft carrier.

Time, Greenwich mean: Mean or civil time at the meridian of Greenwich and universally used in almanacs as well as in worldwide communications.

Time, Navy: Expressed in four digits, 0000-2400, based on 24-hour day. 1430 ("fourteen-thirty") is 2:30 p.m. 0230 ("oh-two-thirty") is 2:30 a.m.

Tincan (can): Slang: a destroyer.

Tongue of the Ocean: Deep natural basin in the Bahamas running more than 100 miles along the eastern shore of Andros Island. Site of the Atlantic Undersea Test and Evaluation Center (AUTEC).

Topgun: A naval aviation advanced school for strike and fighter pilots.

Top off: To fill up, as a ship tops off in fuel oil before leaving port.

Top secret: Refers to national security information or material that requires the highest degree of protection. The test for assigning such a classification is whether unauthorized disclosure of the information could reasonably be expected to cause exceptionally grave damage to the national security. See classified matter, confidential, secret.

Topside: Above, in a ship, referring to the deck above, as distinguished from overhead, which refers to the ceiling of a compartment. The topside (or topsides) means the upper deck (or decks); any deck or area that is exposed to the weather is considered topside.

Torpedo: Self-propelled underwater explosive weapon designed to be aimed or to seek a target, and detonated by contact, sound, or magnetic force. Various types identified by mark and modification numbers. Slang: fish or tin fish.

Touch and go: A near thing. Reference is to a ship barely touching ground, but coming loose immediately and not damaged. In such a case, reporting the incident becomes a matter of ethics and honor. In any situation, to be touch and go is to have a close call. Aviators practice touch and go landings.

Tow: The vessel being towed. To pull along through the water.

Tracer: A message sent to ascertain reason for nondelivery of a prior message. Also, a projectile trailing smoke or showing a light for correction in aim; most usually used in machine-gun ammunition.

Tractor: Term referring to landing ships and craft; as the tractor group of a task force moving to the objective area. Also an aircraft that tows target for antiaircraft practice.

Trice: To haul up, as to trice up all bunks, which means to push up all bunks and secure them.

Trouble Central: The chaplain's office in a large ship.

Turn in: Go to bed. To turn in all standing is to do so fully clothed.

Twilight: The period before sunrise and after sunset during which light is reflected from the sun. The four kinds, depending on angular distance of the sun below the horizon, are civil at 6 degrees, observational at 10 degrees, nautical at 12 degrees and astronomical at 18 degrees.

Under way: Said of a vessel when she is not made fast to the ground in any manner. She may or may not have way on (i.e., she may be hove to), but she is free floating in the sea, subject to wind, currents and of course her own propulsion system. Correct use is two words, except when modifying and immediately preceding a noun, as in underway replenishment.

Underway replenishment (UNREP): See replenishment at sea.

Uniform Code of Military Justice (UCMJ): Enacted by congress for all armed services. For the Navy it replaces the Articles for the Government of the Navy. See Rocks and Shoals.

Uniform of the day: Prescribed by the commanding officer or by the senior officer present afloat (SOPA). To be worn at all times except when working uniform is authorized.

United States Armed Forces: A collective term for the regular components of the Army, Navy, Air Force, and Coast Guard in time of war. See Armed Forces.

United States Naval Institute: The nonprofit, nongovernmental, self-supporting professional society that publishes the *Proceedings* as a forum for the sea services, as well as *Naval History* magazine and a variety of books.

United States Naval Ship (USNS): A ship owned by the US Navy, but not commissioned as part of the Navy. Normally manned with civilian crews and operated by the Military Sealift Command.

United States Navy Regulations: Principles for guiding the naval establishment, particularly the duties, responsibilities and authority of all offices and individuals, issued by the Secretary of the Navy and approved by the President.

Up anchor: The order to weight anchor and get under way.

Very good; very well: Response by a senior to a report by a junior.

Vessel: By US statutes, includes every description of craft, ship or other contrivance used as a means of transportation on water. "Any vehicle in which man or goods are carried on water."—Dr. Samuel Johnson. At one time vessel differed from ship in that a ship was defined as a square rigged vessel with three masts, distinguished from a brig, bark, schooner, snow, etc. This distinction for ship no longer holds, although those for the others still do. Thus, the traditional three-masted square rigger is the only one that does not possess its own specific, exclusive semantic identification. See ship.

Vice admiral: The rank between admiral and rear admiral. See admiral.

Visit and search: A visit to a private vessel to determine its nationality, character of cargo, nature of employment, etc.

Voice tube: Tube for voice communication within the ship. Now generally replaced by telephones.

Wake: The disturbed water astern of a moving ship.

Walk back the cat: Expression meaning to start all over again or to retire to a previously held position and start a process or procedure again.

Walkways: Space adjacent to a flight deck aboard an aircraft carrier.

Wardroom: The compartment in which officers gather to eat and lounge aboard ship.

Warrant officer: An officer, senior to all chief petty officers and junior to all commissioned officers, who derives his authority from a warrant issued by the Secretary of the Navy. A commissioned warrant officer is the highest grade of warrant officer who holds a commission under authority of the President and confirmed by Congress.

Watch: Duty period, normally four hours long. A day's watches are: evening or first watch, 2000-2400; midwatch, 0000-0400; morning watch, 0400-0800; forenoon watch, 0800-1200; afternoon watch, 1200-1600; first dog watch, 1600-1800; second dog watch, 1800-2000. A buoy is said to watch when it is floating in its proper position and attitude.

Watch, quarter and station bill: List showing the duties and billet assignments of all crew members. Opposite the names are listed battle, cleaning and emergency stations, etc.

Water sky: Dark streak on sky caused by reflection of leads, polynyas and open water.

Waterspout: A tornado-like phenomenon occurring over water in which very low pressure in center sucks up water.

Water taxi: Shore boat, available for hire like a taxi.

Way: A ship's movement through the water, as the ship has way on. See under way.

Weather deck: Topmost deck of a ship, or any exposed deck. See deck.

Weekend warrior: Slang: member of the Naval Reserve.

Weigh: To lift the anchor off the bottom in getting under way.

Well deck: Part of the weather deck that has some sort of superstructure both forward and aft of it.

Wharf: Structure parallel to the shore line to which ships moor for loading, unloading, or repairs. Sometimes called a quay, which is usually a solid masonry structure.

White hat: Slang: enlisted personnel.

Windward: Toward the wind. Similar to weather but not used interchangeably.

Wolf pack: Coordinated submarine attack group of two or more submarines.

Wooly-pully: Marine slang: green uniform sweater.

Word, the: News; information. Slang: the dope.

Working party: Group of persons assigned to a specific job.

XO: Slang: executive officer, exec.

Yacht ensign: A modified ensign, flown by yachts, whose dip is answered by men-of-war. May be saluted upon arrival aboard or upon departure from the yacht.

Yardarm: Either end of a yard, referring to the spar. In the modern navy, a light yard, high on the foremast, for signal flags.

Yeoman (YN): Petty officer who performs clerical and secretarial duties.

Zenith: That point of the celestial sphere vertically overhead. See nadir.

Zigzag: Series of relatively short, straight-line variations from the base course. See evasive steering.

Zone inspection: In times past, active commanding officers prided themselves on personally inspecting all spaces of their ships at regular intervals, followed by the inspecting party that made notes of discrepancies found. In large, modern ships this is no longer practicable, and it has been replaced with the zone inspection, in which a designated officer is responsible for each zone. with an inspection party. Frequently, the captain will join one of the zone inspection parties.

Chapter 14

Traditions and Customers: Initiation Rites of Passage

Customs and Traditions

Even though change is inevitable, these references to our naval customs and traditions will provide a frame of reference for not only the crusty older Navy persons, but also the new generation and casual readers. I want to emphasize that this book is written for everyone. First, some required terminology beginning with customs:

The origins of some of the naval customs are hidden behind a veil of time. We are taught that the observance of naval laws and regulations all have become identified with navies and related entities to such an exacting extent that ancient customs are automatically adopted. The chief in that role of leadership and management plays a vital part in seeing these accepted customs are constantly before our minds. The most frequently observed naval custom and its true meaning will be discussed or defined in the glossary of naval terms in a separate section of the book. The present form of military salute is perceived to have emanated from the Middle Ages when knights were said to have raised their visors in greeting when passing each other. Roman seamen of all ranks came over the side; they faced aft and saluted the authority of the emperor. This place has evolved to be known as the quarterdeck - revered by the emperor as a place for ceremonies, quarterivs or quarterdeck. The chief spends a lot of time on the quarterdeck, both aboard ship or at shore commands, and, in many instances, is the official greeter for the command. Today, on boarding a United States naval vessel, we face aft, salute our National Ensign, then the official representing the ship's commanding officer, state our name, rank, business, and "request permission to come aboard."[1]

Benefits

See The World And Get Paid For It
A Man in the Navy is Somebody
Free Medical and Dental Attention
First Clothing Outfit Free, Board and Room Furnished
Good Pay
Travel Combined With Training
Investigate This Opportunity
List of Anniversaries

February 6, 1802	War with Tripoli recognized
February 15, 1898	Sinking of the USS Maine
March 3, 1883	Birth of new Navy
March 8, 1862	Battle of Hampton Roads
March 9, 1862	The Monitor and Merrimac
March 31, 1854	First Treaty with Japan
April 1, 1899	Samoan Incident
April 6, 1909	Peary's discovery of the North Pole
April 6, 1917	War with Germany declared
April 12, 1861	Civil War began
April 19, 1775	Beginning of the Revolutionary War
April 25, 1898	War with Spain declared
April 29, 1862	Surrender of New Orleans
April 30, 1798	Navy Department founded
May 1, 1898	Battle of Manila Bay
May 4, 1917	US Navy forces in war zone
May 9, 1926	Byrd's flight over North Pole
May 10, 1776	John Paul Jones given first command
May 13, 1846	War with Mexico
May 27, 1919	First trans-Atlantic flight by Read
June 1, 1813	Chesapeake captured by the Shannon
June 3, 1898	Hobson and the Merrimac
June 18, 1812	War of 1812 declared
June 19, 1864	The Alabama and the Kearsarge
June 25, 1859	Fighting on the Peiho River
July 3, 1898	Battle of Santiago
July 4, 1776	Declaration of Independence
August 5, 1864	Battle of Mobile Bay
August 14, 1900	Legations in Peking rescued
August 19, 1812	The Constitution and the Guerriere
September 10, 1813	Battle of Lake Erie
September 11, 1814	Battle of Lake Champlain
September 23, 1779	The Bonhomme Richard and Serapis
October 25, 1812	The United States and the Macedonian
October 27, 1864	The Albemarle sunk by Cushing
November 21, 1918	Surrender of German Fleet
November 29, 1929	Byrd's flight over South Pole
December 22, 1775	First American fleet organized

Dining In

A celebration - a feast - a special ceremony for chief petty officers. The "Dining In" features a formal dinner. The term derives from an old Viking tradition celebrating great battles and feats of heroism by formal ceremony, which spread to monasteries, early-day universities and to the military when the officers' mess was established. A dinner dress uniform is prescribed for each member's service, medals are given by all numbers of the mess and other military guests and retired personnel. The essential elements of a "Dining In" include: a formal sitting, a fine meal, the camaraderie of members of the mess and guests, the traditional toasts, limericks, "two-bell" ceremony and dedication, perceptions of chief petty officers - past, present and future.

Taps - The Final Farewell
by Tom Joyce (AFIS)

Memorial Day: A holiday originally observed as an occasion for decorating the graves of soldiers killed in the Civil War. But since World War I, Memorial Day has been a day to commemorate those who died in all wars.

This year, as in the past, those attending Memorial Day observances around the country will undoubtedly hear a lonely bugle sound *Taps*. And even though no other bugle call is more recognizable to service people and other Americans, little is known about its origin.

Its composer, Civil War Union General David Butterfield, said of *Taps*, "It brings down the curtain on the soldier's day and upon the soldier's life." It's not generally known, but there are words to *Taps*:

> Fades the light;
> And afar
> Goeth day,
> Cometh night;
> And a star
> Leadeth all,
> Speedeth all
> To their rest.
>
> Another version:
> When your last
> Day is past
> From afar
> Some bright star
> O'er your grave
> Watch will keep
> While you sleep
> With the brave

But *Taps* wasn't always taps. The French "L'Extinction des Feux" ("Lights Out") was originally used by the American Army to signal the end of the day. A particular favorite of Napoleon, it didn't suit Butterfield.

Butterfield, a Medal of Honor awardee, composed *Taps* one night in the hopes it would comfort the men under his command. Their morale had sunk to a new low during heavy fighting near Richmond, Virginia. Their hopes for an early end to the war and reunions with their wives and families had been dashed when they were informed that help would not be coming. Dug in for the night, Butterfield, without knowing a note of music, composed what we know today as *Taps*.

The next morning, Butterfield summoned the brigade bugler and hummed the melody he had composed. After a few tries, Oliver W. Norton had mastered the call. At nightfall, he played Taps officially for the first time.

Norton would later write about the reaction to *Taps*: "The music was beautiful on that still summer night, and was heard beyond the limit of the Butterfield brigade as it echoed through the valleys. The next morning, buglers from other brigades came to visit and inquire about the new *Taps* and how to sound it."

Later in the same campaign, *Taps* was used for the first time in connection with a military funeral.

In this case, a soldier from Captain John D. Tidball's Battery A of the 2nd Artillery had been mortally wounded. Since the unit was so close to the enemy in a forward position and hidden in the woods, Butterfield decided it would be unwise and dangerous to fire the three traditional volleys over the soldier's grave. He decided instead to have *Taps* played as a custom at military funerals.

The Army officially adopted *Taps* in 1874; by 1900, the other services followed suit.

It was, and still is, the custom to fire three volleys over the grave of a deceased soldier to signify the end of the funeral ceremony. This custom originated in the early funeral rites of the Romans. The Romans would throw dirt over the coffin three times and call the dead three times by name to mark the end to the funeral ceremony. Today, when three volleys are fired at military funerals, it is a way of bidding a soldier farewell three times.

National Anthem

The Star Spangled Banner

Oh, say, can you see, by the dawn's early light,
What so proudly we hailed at the twilight's last gleaming?
Whose broad stripes and bright stars, thro' the perilous fight,
O'er the ramparts we watched, were so gallantly streaming.
And the rockets' red glare, the bombs bursting in air,
Gave proof through the night that our flag was still there.
Oh, say, does that star-spangled banner yet wave
O'er the land of the free and the home of the brave?
On the shore dimly seen, thro' the mists of the deep,
Where the foe's haughty host in dread silence reposes,
What is that which the breeze, o'er the towering steep,
As it fitfully blows, half conceals, half discloses?
Now it catches the gleam of the morning's first beam,
In full glory reflected, now shines on the stream;
Tis the star-spangled banner; oh, long may it wave
O'er the land of the free and the home of the brave.

Oh, thus be it ever when free men shall stand
Between their loved home and the war's desolation;
Blest with victory and peace, may the heaven-rescued land
Praise the Power that has made and preserved us a nation.
Then conquer we must, when our cause it is just,
And this be our motto: "In God is our trust;"
And the star-spangled banner in triumph shall wave
O'er the land of the free and the home of the brave.

Imagine a clear morning, unusually beautiful with the red glow of the son slowly permeating in the east. You are ready to face a new day and to experience the touching custom of hoisting "Old Glory." Some call it the National Ensign, yet it is morning colors to a lot of us. Raising the flag of the United States represents freedom, and all people respond in a different way as the event heralds a new day. It happens every place in the world where the naval establishment lies; at naval stations, aboard ships plowing through a sea of foam, and on ships lying at harborage or anchorage. There is always a sudden hush, deepened, rather than interrupted by the typical sprinklings of noise. At some places one witnesses a sharp, precise group of men and women whose honor it is to ceremoniously execute the drill for colors. At the flag staff, the halyards are broken free and clear. One person attaches the top hoist buckle of the American Flag, then passes it, neatly folded, to another, and another short wait ensues. The mind is assaulted with a myriad of thoughts when the first vibrant hum sounds from the amplifying systems, and those chills that usually run up and down the viewer's spine attest to a slumbering patriotism deeply rooted in your being. The ceremony continues - what a feeling it is to be a free person watching the flag being lifted. As it shakes its long folds, climbing and waving in the early morning light, the days seem full of promise, challenge; even foreboding, but full of purpose.

The Union Jack - A Navy Tradition

Whenever a Navy ship is in port or moored, the union jack is flown off her bow. In fact, the instant the final mooring line is set, or the anchor is dropped at sea, the boatswain's mate of the watch signals "shifts colors," and the star-studded, blue union jack is hoisted. But what is the significance of the union jack, and how did this tradition evolve? John Reilly, naval historian in the ship's histories department, Navy Historical Center (Washington Naval Yard), and an authority on Navy traditions, explained: "First of all, a union jack is a sort of abbreviation of a nation's ensign or flag. In other words, when you speak of a union jack, you may be referring to any country's ensign.

"The display of the union jack off the bow of a ship is an age-old tradition, the origin of which we can only guess about. Displays of flags, ensigns, pennants and jacks have traditionally been communication signals for ships at sea. They provided decoration, identification and sent specific messages," Reilly explained.

As far back as the Middle Ages, sailing ships displayed flags from their bows as evidenced in paintings and narratives from that era, but flag and pennants among navies began towards the end of the 17th century.

Just before the 18th century, according to Reilly, the maritime nations: Britain, Spain, France and Holland, to name a few, developed regular, full-time navies. "Tacking a few weapons on a passenger or cargo ship to meet immediate defense requirements began to give way to construction of larger, heavier ships with built-in weapons systems that were deployed and manned by specially trained crews even in peacetime," he explained. These warships, however, didn't look remarkably different from cargo ships, so a long narrow "commissioning pennant" was flown from the mast to identify the ship and its purpose. Today's naval ships carry the same type of pennant, only much smaller, and it is still referred to as the commissioning pennant.

"Moving these heavy ships through the water required many, many sails. The sails frequently obscured the view of the ensign or commissioning pennant, so a smaller flag - a sort of condensed version of the ensign - was flown from the bow of the ship.

Today the union jack flies from the bows of naval ships from 8:00 a.m. until sunset, while moored, anchored or pierside. "The practical significance of the jack these days," said Reilly, "is to allow immediate determination whether a naval ship is underway, moored or anchored. If you are out on the water in a sailboat and in close proximity to an aircraft carrier, you had better know whether or not it's anchored," Reilly emphasized. "A ship barely making way may not seem to be moving, and you might receive an unpleasant surprise."

Traditions associated with ships and the sea are often as timeless as the sea itself. The art and science of communications has found surer methods for ship-to-ship communication and identification, but flags and pennants still color Navy ships with a sense of pride and history. The union jack is proudly hoisted among them.

The Origin of Khaki

Traditionally, we know that khaki is the name for the olive drab cloth used in the construction of military uniforms. The term khaki originated from the Urdu word of the same spelling in English and means "dusty or dust colored." The British army first utilized the khaki uniform while in India in 1857.

The khaki uniform surfaced in the United States Military in the late 1800s to early 1900s. The Marine Corps Aviators were the first to wear it as an official authorized uniform in 1912.

The naval aviators were wearing khaki as early as 1914 but in an unauthorized state. The first naval khaki uniform consisted of a khaki coat copied after a white service coat and khaki Marine Corps breeches.

The khaki uniform was adopted by the CPO Community in 1941 where it is still proudly worn.

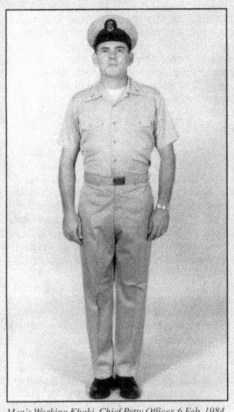

Men's Working Khaki, Chief Petty Officer, 6 Feb. 1984. (Courtesy of U.S. Navy)

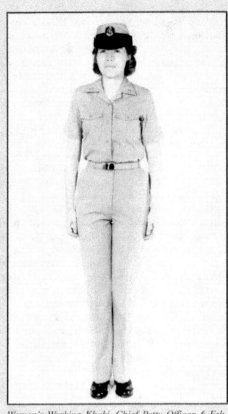

Women's Working Khaki, Chief Petty Officer, 6 Feb. 1984. (Courtesy of U.S. Navy)

Men's Winter Working Blue, Chief Petty Officer, 6 Feb. 1984. (Courtesy of U.S. Navy)

Women's Winter Working Blue, Chief Petty Officer, 6 Feb. 1984. (Courtesy of U.S. Navy)

Third Class Petty Officer's Blue Dress Uniform, 1886.

Chief Petty Officer in Blue Dress Uniform, 1905.

Second Class Petty Officer, White Dress Uniform, 1886.

Men's Dinner Dress White Jacket, Chief Petty Officer, 6 Feb. 1984. (Courtesy of U.S. Navy)

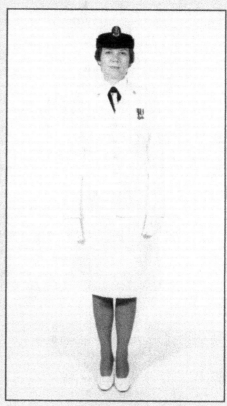

Women's Dinner Dress White, Chief Petty Officer, 6 Feb. 1984. (Courtesy of U.S. Navy)

Lieutenant, White Service Dress Uniform, 1886.

Petty Officer in Dress Whites, 1905.

Navy Hymn (Eternal Father)

Eternal Father, strong to save,
Whose arm hath bound the restless wave,
Who bidd'st the mighty ocean deep
Its own appointed limits keep;
Oh, hear us when we cry to Thee,
For those in peril on the sea!

O Christ! Whose voice the waters heard
And hushed their raging at Thy word,
Who walked'st on the foaming deep,
And calm amidst its rage didst sleep;
Oh, Hear us when we cry to Thee
For those in peril on the sea!

Most Holy Spirit! Who didst brood
Upon the chaos dark and rude,
And bid its angry tumult cease,
And give, for wild confusion, pease;
Oh, hear us when we cry to Thee
For those in peril on the sea!

O Trinity of love and power!
Our brethren shield in danger's hour;
From rock and tempest fire and foe,
Protect them wheresoe'er they go;
Thus evermore shall rise to Thee
Glad hymns of praise from land and sea.
Amen

It was written in 1860 by the Reverend William Whiting, clergyman of the Church of England, after he had come through a terrible storm in the Mediterranean Sea.

Anchors Aweigh

Stand, Navy, out to sea,
Fight our battle cry;
We'll never change our course,
So vicious foe steer shy-y-y-y.
Roll out the TNT
Anchors A-weigh
Sail on to victory
And sink their bones to Davy Jones hooray!

Anchors A-weigh my boys,
Anchors A-weigh
Farewell to college joys,
We sail at break of day-day-day-day!
Through our last night on shore,
Drink to the foam,
Until we meet once more
Here's wishing you a happy voyage home.

Words by Captain Alfred H. Miles, USN, 1907. Revised lyrics by George D. Lottman.

Foreword

This pamphlet was originally compiled in 1929 by officers attached to the Ninth Naval District under the direction of Captain D.W. Bagley, US Navy.

It is now issued by the Navy Recruiting Service for the benefit of recruits, to familiarize them with the traditions and achievements of the United States Navy and to enable them to converse intelligently with civilians upon the part played by the Navy in our National development.

Adolphus Andrews
Rear Admiral, USN
Chief of Bureau of Navigation
Navy Department

Subpoena and Summons Extraordinary

The Royal High Court of the Raging Main

County of Roosevelt
Vale of Pacificus,
Domain of Neptunus Rex
To Whom May Come These Presents:

Greetings and Beware

WHEREAS, The good ship *Kermit Roosevelt*, bound southward for Nauru, Gilbert Islands, has never before entered our domain; and

WHERAS, The aforesaid ship carries a large and slimy cargo of land-lubbers, swabs, hay-tossers, park-bench warmers, chicken-chasers, soda-inhalers, dance-hall engineers, strap hangers, tea slingers and sand crabs, falsely masquerading as seamen, of which low scum you are a member, having never appeared before us; and

WHEREAS, The Royal High Court of the Raging Main has been conveyed by us on board the good ship *Kermit Roosevelt* on the 4th day of May 1950, in Latitude 00° 00'' 00,' Longitude 167 East together with such surgeons, dentists, barbers, police and executioners as may be necessary to execute its judgments; and

BE IT KNOWN, That we hereby summon and command you Joe B. Havens now a SN, USN, to appear before this Royal High Curt and Our August Presence on the aforesaid date at such time as may best suit our pleasure to be examined as to fitness to become one of our trusty shellbacks and to answer to the following charges:

Charge I - In that Joe B. Havens, now a SN, USN, has hitherto willfully and maliciously failed to show reverence and allegiance to our Royal Person, and is therein and thereby a vile landlubber and pollywog.

Charge II - Leaving the State of Arkansas in a state of confusion.

Charge III - Attempting but failing at - "Bringing in the Sheaves."

Charge IV - Acquiring the appetite of a "McJoyner."

Charge V - Painting without any bristles in the paint brush.

Disobey this summons under pain of our swift and terrible displeasure. Our vigilance is ever wakeful, our vengeance is just and sure.

Given under our hand and seal
Neptunus Rex
Attest for the King:
Davy Jones

Operation "Miki" Finn

Pilot's License No. 606

"Drink and be merry for tomorrow ye may die!"
Down the Hatch!
Brown Derby
Honolulu, Hawaii
"He who does not love wine, women and song, remains a fool, his whole life long."

This license, is issued to Joe Havens, SN, USN, by the Brown Derby, 1166 Nuuanu Avenue, Honolulu, T.H., and certifies that he has fully qualified as a first class pilot on the troubled seas of beverages, has a complete knowledge of bars, knows all Harbors where the biggest schooners can be unloaded, and is willing to do his share of emptying such schooners. Can steer a straight course for any bar, can sail on an even keel when fully loaded AND FURTHERMORE, is entitled to rank as chief pilot on any vessel using beer, whiskey, gin, cognac, vodka, wines, or any mixture thereof. He is also recommended as a helmsman for young, beautiful, or old and battered hulks looking for a snug harbor to lay up in while waiting for the storm to blow over.

Witnesseth this 16 day of May 1950
Master Drink Pilots

With the 1959 amendment to the Career Compensation Act of 1949, the pay grades of E-8 and E-9 were created, effective June 1, 1958. Eligibility for promotion to E-8, the senior chief level, was restricted to chiefs (permanent appointment) with four years in grade and a total of 10 years of service. For elevation from E-7 to master chief, E-9, a minimum of six years service as a chief petty officer with total of 13 years service was required. The E-8 and E-9 levels included all ratings except teleman and printer which were being phased out of the naval rating structure. Personnel holding those ratings were in the process of being absorbed or converted to yeoman or radioman from teleman, and to lithographer from printer. Service-wide examinations for outstanding chiefs were held on August 8, 1958, with the first promotions becoming effective on November 16, 1958, with the first promotions becoming effective on November 16, 1958. A few months later, from the February 1959 examination, a second group of chiefs were elevated to E-8 and E-9, effective on May 16, 1959. For information purposes the names of the first two groups of selectees are listed in Bureau of Naval Personnel Notices (BPN) 1430 of October 17, 1958, and May 20, 1959. It is noted that after the May 1959 elevations that promotions to E-9 was through senior chief only.

On July 1, 1965, compression of several ratings at the two top grades was enforced. Created were six new rating titles: master chief steam propulsionman, master chief aircraft maintenanceman, master chief avionics technician, master chief precision instrumentman, master chief constructionman and master chief equipmentman.

Conversely, about four years later, on February 15, 1969, some decompression at the senior and master chief grades transpired which eliminated master chief steam propulsionman. Decompressed were the rates of master and senior chief torpedoman's mate quartermaster and storekeeper. Seven ratings were re-established at the E-8 and E-9 grades presenting the opportunity for chiefs to again advance within their specialty to E-9. The seven affected ratings were signalman, mineman, aircrew survival equipmentman, aviation storekeeper, aviation maintenance administrationman and boiler technician.

The only substantial rating changes of recent implementation that has had an effect on the chief petty officer community occurred on January 1, 1991. That was the merger of the anti-submarine warfare technician, aviation fire control technician and the aviation electronics technician ratings at the E-3 (apprenticeship) and E-4 through the E-8 petty officer grades into the single rating of aviation electronics technician. Secondly, the rating of avionics maintenance technician (E-9 only) remains as the normal path of advancement from the rates of senior chief aviation electrician's

Endnotes:
[1] Blue Jackets Manual, US Naval Institute, 11th edition, Annapolis, MD. 1943

Chapter 15

Epilogue

Last night - very close to embarking on a long deployment away from home and family - a myriad of thoughts taunt him as he holds his wife fondly - What's ahead? Will he be prepared? What more can he do to ensure the children's education and future? Could he had done better if given the chance again? Will all these unknowns enhance a creative and visionary person to tie all those loose ends together? He drifts back into a fitless slumber and awakens to the smell of breakfast wafting through the house. Too late to ponder further, he responds to the challenges of the present day.

The Last Cruise

The boatswains pipe heralds an uneasy silence as the chief is piped over the side. Staring straight ahead, he musters all the pride and energies available as he proudly holds his salute as he passes through his fellow chiefs manning each side of the aisle as another chapter in his life, his Navy career ends.

The chief petty officer creed, quoted below, brings a sense of encouragement.

Chief Petty Officer Retirement Creed

You have on this day experienced that which comes to all of us who serve on active duty in "Our Navy" because your departure from active duty in no way terminates your relationship. By law and by tradition, US Navy Retirees are always on the roles ever ready to lend their service when the need arises. The respect that you earned as "The Chief" was based on the same attributes that you will now carry into retirement. You

should have no regrets. Do not view your retirement as an end of an era but rather as orders to a new and challenging assignment in the form of independent duty. Remember well that you have been, and always will be, an accepted member of the Petty Officers. The active duty chiefs salute you and the retired chiefs welcome you. I wish you the traditional "Fair Winds and Following Seas."

What will the next 100 years be like for the chief petty officer, he muses. Are chiefs more sophisticated now? No, but he does feel they are more technical. The crystal ball is unclear about the future of the Navy and the elitists, the chief petty officers. One thing is for sure, whether ashore, aboard or with the fleet, whether they are serving as volunteers in their communities, they have one thing in common...they

A most senior chief: Retired senior chief Rhea Rohn appears at the April 1 ceremony of the "Chiefs Bell" presentation in Annapolis. Rohn, who is 82 years old, joined the Navy in 1929 and first made chief in 1949. (Times photo by Steve Elfers)

are well-trained, dedicated to the mission of the US Navy and prepared to go in harms way.

A sequel" Perhaps!

The Future of the Navy

The Navy will always employ new weapons, new techniques, and greater power to protect and defend the United States on the sea, under the sea, and in the air.

Now and in the future, control of the sea gives the United States her greatest advantage for the maintenance of peace and for victory in war.

Mobility, surprise, dispersal, and offensive power are the keynotes of the new Navy. The roots of the Navy lie in a strong belief in the future, in continued dedication to our tasks, and in reflection of our heritage from the past.

Never have our opportunities and our responsibilities been greater.

The clique used where every man in the ??? until he has written a book" cannot atten??? to my heartfelt appreciation to all who have made this creative happening less difficult. I particularly applaud the support and efforts of all those who unknowingly encouraged me from a humble beginning.

About the Author

Joe B. Havens retired from the US Navy as a senior chief hospital corpsman. During his career that spanned from 1949-70, he served on board naval hospitals, naval air stations, national naval medical center, basic and advanced hospital corps schools, USS *FDR*, (CVA-42), *Kermit Roosevelt* (ARG-12), USS *Estes* (AGC-12), USS *Cleveland* (LPD-7) and the 1st Marine Division in Korea. His numerous decorations include the Bronze Star, Purple Heart and Navy Achievement Medal. He has earned a BS from University of the state of New York and has completed graduate studies at Memphis State University. Mr. Havens was instrumental in the development, marketing and management of the Trezevant Episcopal Home (CCRC) in Memphis, wherein he serves as its president. He is an avid proponent of naval affairs and supports local units wholeheartedly.

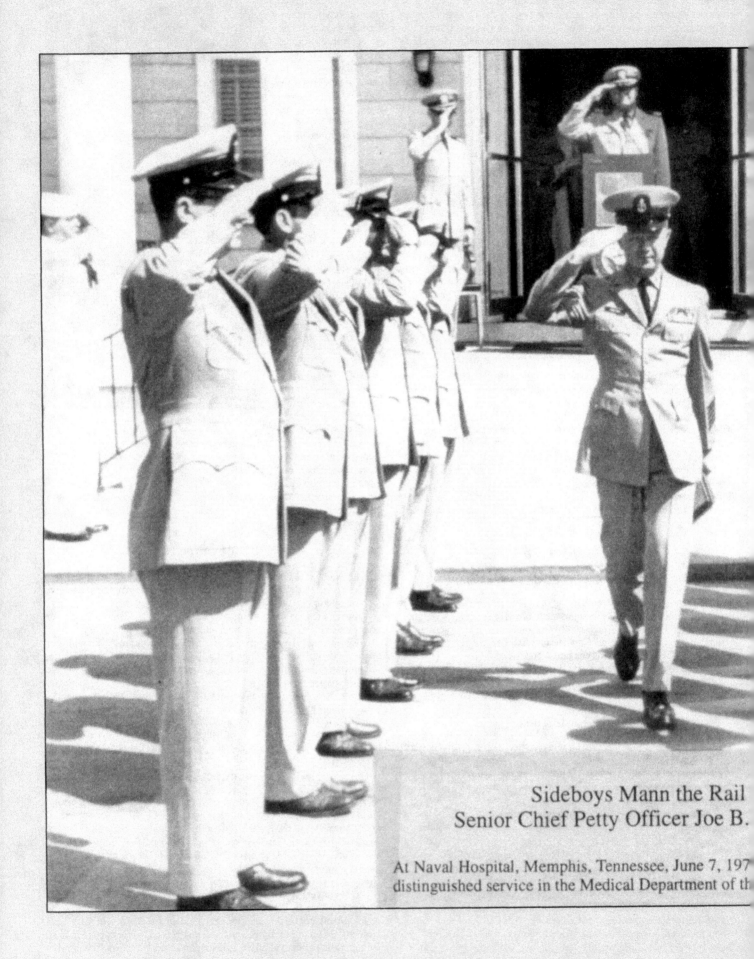

Sideboys Mann the Rail
Senior Chief Petty Officer Joe B.

At Naval Hospital, Memphis, Tennessee, June 7, 197
distinguished service in the Medical Department of th

Ashore

pleting 22 years of
ates Navy.

Chief Petty Officer
Veterans' Biographies

Editor's Note: All members of the Chief Petty Officer's Association were invited to submit their biography for inclusion in this publication. The following biographies are from those members who chose to participate. Many of the biographies were typeset exactly as received, with minimum of editing; therefore, neither the Publisher nor the Chief Petty Officer's Association is responsible for errors or omissions.

HARVEY J. ADAMS, OSCM, USN (Ret), was born at Clarkesville, GA, Sept. 8, 1927, and joined the USN July 2, 1942, at Macon, GA. Attended boot camp at Norfolk, VA, as part of Co. 229, July 1942 before being assigned to the AMPH TR, Solomons Island, MD, September—October 1942. Later assigned to LCT-381 from October 1942—January 1944 and LST-457 from January 1944—August 1945 when he was discharged Dec. 6, 1945.

Rejoined the USN, Feb. 10, 1947. Finished Radarman's Class A School, Norfolk, VA, Class of September 1948. Assigned to the USS *Albany* (CA-123), February 1947—February 1950; USS *Carpenter* (DDK-827), March 1950—September 1951; USS *Bremerton* (CA-130), Sept. 1, 1951—Oct. 7, 1953; USS *Stickell* (DDR-888), October 1953—September 1954.

Assigned to FADTC, Dam Neck, VA, September 1954—January 1957; assigned again to the USS *Stickell* (DDR-888), January 1957—February 1958; assigned to FLTGR, Guantanamo Bay, Cuba, March 1958—March 1961; assigned to ASW Tact School, Norfolk, VA, May 1961—April 1964.

Retired from USN April 20, 1964. Advanced from GM2 to GM1/c, May 30, 1944, LST-457. Believed to be youngest GM1/c in modern Navy. Advanced to RDC Aug. 8, 1954, on the USS *Stickell*. Advanced to E8 May 1959 at FLTGR, Guantanamo Bay, Cuba and to E9 May 1962 at the ASW TACT School, Norfolk, VA. LCT-381 as a 20mm pointer assisted in shooting down one JAP O at Mariobe, New Guinea, September 1943; one JAP O at Arawa, New Brittan, December 1943; LST-457 pointer 40mm, one JAP Q, Leyte Gulf, October 1944. Pictured in *Life* magazine, January 1945.

Married Frances Cranford and has three children and two step-children. Owner of Adams' Rib Restaurant at Clarkesville, GA and he is the owner of Adams Black Angus Farm, Clarkesville, GA.

DALE JAY ANDERSEN, CWO4, was born Sept. 26, 1939, Omaha, NE. Graduated HS and the University of Nebraska where he attained his BA degree from the Arts & Sciences College. On April 29, 1962, he began duty with the USN.

Duty Stations include MCB-3 (Seabees) on Guam, Vietnam and Port Hueneme, CA; N&MCRC, Omaha, NE, and made YNC with Fleet Hospital Unit 1984. Moved through the ranks becoming a CPO and command chief. In 1986 he received his commission to CWO2 and now serves as CWO4, NR PHIB CB-1 DET 218 Administrative Officer and drills at the Naval and Marine Corps Reserve Center, Omaha, NE.

Personal awards include the Naval Achievement Medal, Combat Service Ribbon, Naval Meritorious Medal (4), Armed Forces Reserve Medal (2), Navy Unit Commendation Medal, Fleet Marine Force Combat Operations Insignia, National Defense (2), Vietnam Service (3), Vietnam Campaign, RVN Meritorious Unit Citation, Gallantry Cross, RVN Meritorious Unit Civil Actions, Navy Sea Service, Batt. E, Reserve Overseas Service and Rifle and Pistol Expert.

In civilian life Anderson has been in sales, sales management, small business ownership and manage-

ment. He is currently self-employed in the environmental products industry. He is active in numerous civilian/professional affiliations.

Service oriented activities include The Navy League (pres.), ROA (pres.), Naval Enlisted Reserve Assoc. (past-pres.), Naval Reserve Assoc., Fleet Reserve Assoc., TROA, The Naval Institute, American Legion and VFW (past adjutant).

Married Patricia E. Funk, March 19, 1994. He has two children, Kye and Travis.

ANNA MARIE (WILBER) ANDERSON, YNCS, born in Woodson, IL, and entered the USN in Los Angeles, CA, Jan. 5, 1944. Boot camp, Hunter College, NY, Yeoman School, Cedar Falls, IA. First duty station at Camp Elliott, CA, was followed by NTC San Diego. Was honorably discharged in August 1945.

After a 20 year break, she re-enlisted Jan. 23, 1967, and retired as YNCS, Oct. 16, 1967. During the interim Anna Marie was married for 23 years to Leo Patrick Murtha, YNC; five years to Robert Powell, ADR1; and seven years to John A. Anderson, AQC, who died in 1967. She has one son, Paul E. Murtha, who retired from the Army as LTC in January 1993.

During her Navy time she had 10 years Reserve and 10 years active duty. Her medals include regular ones from WWII plus the Navy Achievement and the Navy Commendation which she earned during active duty, second enlistment.

As a result of the GI Bill, she earned a BS degree in business administration and spent most of her life as a business education teacher. She also worked 10 years in federal civil service retiring in 1986 from Point Mugu, CA. Her membership affiliations include life memberships in WAVES National (Nat. Pres. 1984-88), American Legion, Fleet Reserve Assoc. and AMVETS; also in Naval Enlisted Reserve Assoc.

ROBERT M. "CHESTY" ARNOLD , CY, born Sept. 19, 1918, Ewan, WA, and has lived in Orlando, FL, since 1920 except during naval duty starting July 7, 1941, in Intelligence Dept.; nine months in Jacksonville; two years in Miami; and nine months in Nassau. Went to Bainbridge "boot camp" as 1st class yeoman—my, my, what a disgrace.

Went to Pearl Harbor aboard the USS *Wisconsin* FFT to Ulithi. During cruise he threatened to change name Arnold to Karnold to avoid "letter 'A' dirty work assigners" and might have stayed plush at Pearl with letter "k."

At Ulithi, aboard: USS *Prairie*, USS ARD-15 and lastly the tough little USS *Swearer* (DE-186) (just typhooned). To Guam and later Leyte; opening days at Iwo Jima observed tragic landings and was very close to flag raising. Later, spent 102 continuous days at Okinawa (kamikazed relentlessly). Discharged in November 1945.

Married Martha in 1946 and has three children: Robert, Rex and Nancy. He is author of book, *Standby Arnold,* telling of many humorous experiences. Currently in antique business of buying and selling.

NICHOLAS BALOGH, CMM (PA), born Dec. 25, 1920, Mahwah, NJ. He joined the USN May 15, 1940, New York City, NY. Duty stations include NTS Newport, RI; RS USS *Constellation* and USS *Niblack* (DD-424), 1940, promoted to chief aboard *Niblac* in 1944; NTS NOB Norfolk, VA, 1945; RS Bath, ME; USS *Charles P. Cecil* (DD-835), 1945; and USS *Duncan* (DD-874), 1946.

Awards include the American Defense Medal w/star, Bronze "A" European Medal w/5 stars, American Theater Medal w/2 stars, Asiatic-Pacific Theater Medal, Victory Medal and the Good Conduct Medal. He was discharged May 17, 1946, Lido Beach, Long Island, NY.

Married Eleanor Reiling, March 4, 1945, and has three children: Nicholas II, Jeffrey and Joan, nine grandchildren and one great-grandson. He was power plant chief engineer, Lederle Labs Div. American Cyanamid Co. and is currently retired.

JOSEPH R. BARBIERI, CMM, born in September 1921, Waterbury, CT, and joined the USN in February 1942. Went to boot camp at Newport, RI and naval net depot, Boston, MA. Served in USS *Matagorda* (AVP-22), USS *Wasp* (CV-18), USS *Albany* (CA-123), NTC Waterbury, CT, and USS *Kyne* (DE-744).

Made chief in October 1944, age 23 - CM3C, CM2C, CM1C and CMM (later changed to DCC). Medals include the American Campaign, European African, Asiatic-Pacific w/8 stars, WWII Victory Medal, Navy Occupation Service Medal with Asian clasp, National Defense, Korean Service, Purple Heart, Good Conduct w/3 stars, Navy Marine Corps Medal, Navy Commendation Medal, Navy Unit Commendation w/bar, Philippine Liberation ROK and the Presidential Unit Citation. After 14 months on USS *Wasp* and covering 110,000 miles, he was discharged June 1959 to Inactive Reserve, Waterbury, CT.

He owned his own plumbing and heating company until retiring in 1981. Married to wonderful woman, Helen, for 48 years and has one daughter Sandy, four grandsons and two great-grandsons. Barbieri and Helen enjoy trips to Atlantic City and Foxwoods casinos; they made some good hits and left some there. All in all, he and the wife love it.

C. JOSEPH BARRON, AOC (Ret), enlisted in the USN in 1942. Aviation ordnance training was in Norman, OK; aerial gunnery training, Purcell, OK; and Bombardier School and operational training at NAS JAX. He attended PB4Y Consolidated Aircraft Training in Chincoteague, VA, prior to transfer to VPB-107 in Brazil, South America. As WWII antisubmarine warfare (ASW) activity was reduced to the North Atlantic, he was assigned to ASW and bomb runs over Europe with other Navy squadrons based in England.

Following WWII, Chief Barron served in many activities involved in ASW and aircrew training. He retired from the Navy in 1963 at Andrews AFB while director of aircrew training. During his Navy career, Chief Barron accumulated 143 college credit hours and returned to civilian life as an education specialist for BUPERS. While there, he was one of three people responsible for establishing the Electronic Warfare (EW) rating.

He retired from federal service in 1982 as a program manager under CNET in Pensacola, FL. Chief Barron has two daughters and four grandchildren. Doreen, his wife, died in April 1993. He lives in Gulf Breeze, FL, and is a real estate and mortgage investor.

ARTHUR THORLEY BAUGH, ENC, born Nov. 1, 1921, aboard the USS *Henderson*, Indian Head, MD to CMM Arthur H. Baugh. He entered the USN on Nov. 14, 1939, at Los Angeles; sent to boot camp at San Diego, then reported aboard the *Mugford* (DD-389) in April 1940 and served until January 1942.

Went to Diesel School at NOB Norfolk, then to Class C Diesel School, Grove City, PA. Commissioned the USS *Annoy* (AM-84) at Portland, OR, then transferred to NAS Dutch Harbor. Made CMNO on Nov. 16, 1943. Next duty at NAS Whidbey Island, WA.

Re-enlisted in February 1946 and reported to USS *Corduba* (AF-32) and cruised around the world for one and a half years. Left the USN in 1947 and joined the Surface Division B-24, Billings, MT.

Retired in June 1981 with 42 years total service.

Made chief motor machinist's mate on Jan. 1, 1945. Awards include 23 service medals including Pearl Harbor w/6 Battle Stars and Good Conduct.

Married Gladys on May 17, 1944, and has three children: Ronald, Richard and Randy. Was in the profession of restoring antiques; currently retired, and enjoys traveling.

NORMAN HENRY BELL, (Ret), SDC, USN, USS *Zane*, was born at Perry, GA. He joined the USN in 1934 and received boot training at Newport News, VA. Assigned to USS *Langley* then to USS *Zane* (DMS-14) for nine years. Bell was a Pearl Harbor survivor aboard the *Zane* on Dec. 7, 1941. Cited by Adm. W.F. Halsey for skillful and effective performance of duty.

Also served on USS *Boxer, Marcus Island, Princeton, Calvert* and Seaplane Squadrons VP-48 and VP-50.

Enthusiasm, dedication, initiative, with a spirit of friendly helpfulness towards his fellow shipmates—these qualities forecast for Chief Bell the many commendations he received at NAS Alameda, CA; Kingsville, TX; Whidbey Island, WA; and Iwakuni, Japan. Bell was featured in the fleet newspaper and *Jet* magazine for his work with the Aiji-no-ie Orphanage in Japan.

After retirement and graduating from Laney College, Bell continued his interest in veterans affairs and is a member of FRA Branch 87. Retired Chief Petty Officers Assoc., Naval Minewarfare Assoc., Navy League and Pearl Harbor Survivors Assoc. Bell and wife Mae have one son Norman Johni and granddaughter Denise Bell.

Bell retired in 1964. His awards and medals include the Good Conduct Medal, American Defense, American Theater, Asiatic-Pacific w/stars, WWII Victory, Korean Service and five commendations.

MYRON W. BROBST, Chief Carpenter's Mate (AA) (T) USNR, was born in Youngstown, OH, April 12, 1920. He joined the USN, May 20, 1942, attended USNCTC Great Lakes, IL, before being assigned to Casco Bay Portland, ME, aboard USS *Denebola* (AD-12). After six months was transferred to USS *Monterey* (CVL-26). Earned chief carpenter's mate rating in April 1944.

While serving aboard the USS *Monterey*, two years and eight months, he was certified at Fire Fighters' Schools in Norfolk, VA, June 19, 1943; completed a course in velocity power tools at Pearl

Harbor in October 1943; and participated in 12 major battles which included the Gilbert Islands, Marshall Islands, Kavieng (Bismarck Archipelago), Asiatic-Pacific Raids, West New Guinea, Marianas, Western Caroline Island, Leyte Operation, Philippine Liberation First Phase, Philippine Liberation Second Phase, Okinawa and Third Fleet against Japan. He also survived a disastrous typhoon on Dec. 18, 1944, while making a strike on Luzon.

Chief Brobst received an honorable discharge at Lido Beach, Long Island, NY, Nov. 5, 1945. He returned home where he, his father and two brothers, built and owned Hy Skore Bowling Lanes. He also worked as construction superintendent for several companies, retiring in 1982. An avid bowler and golfer, he has a sanctioned "300" bowling game and a "hole-in-one" in golf.

Married to a teacher, Margaret Burton, since June 8, 1946, and has two daughters, Lynn and Laraine (both teachers) and three grandchildren.

JAMES BROGAN, HMC, USN (Ret), born Jan. 18, 1931, Newburyport, MA. He joined the USN on Jan. 19, 1948, Boston, MA. Boot camp and Hospital Corps School at NTC Great Lakes, IL. Served in USS *Cadmus* (AR-14), USS *Albany* (CA-123), Medical Research Lab, Sub Base, New London, CT, Naval Hospital, Naples, Italy; USS *Pima County* (LST-1081) NAS Pensacola, FL; and Naval Shipyard, Brooklyn, NY.

Promoted to chief while attached to Radar Picket Sqdn. #2. He retired from the USN, Oct. 17, 1967, from NAS Brunswick, ME, where he was the leading chief at the dispensary.

Divorced, he has two children, Brian and Scott. After the service he was a registered sanitarian in Massachusetts and worked 18 years for the Brookline (MA) Health Dept. He retired in January 1986. Returned to the States in July 1994 after a two year contract with Pacific Architects and Engineers as a custodial supervisor at the US Embassy, Moscow, Russia.

JACK L. BROWN, CMOMM, born in Minco, OK. He enlisted in the USN, Oklahoma City, OK, 1937. Recruit training was at San Diego, CA. He served on *Salt Lake City* (CA-25), *North Carolina* (BB-55) and took exam for CMM in April 1943. Served in *Frontier* (AD-25), *Incredible* (AM-249) and *Mainstay* (AM-261).

Tours of Duty: Diesel School DE, made CMOMM in August 1943, Beloit, WI. FLT ACT

Yokosuka, Japan; Engineman C-1 School, Great Lakes, IL, 1951-56; and CB Base, Coronado, CA. Transferred to Fleet Reserve at Naval Station, San Diego, CA, 1956, and is a member of FRA Branch 135.

He is employed by Beckman Instruments, Inc., 1956-78, Fullerton, CA, as QC and production supervisor of Helipot Div., senior analyst for purchasing (corporate). He makes his home in Hemet, CA.

LEONARD M. CARDIFF, BMCM, USNR (Ret), born Aug. 25, 1934, Braddock, PA. He joined the USN on Sept. 25, 1952. Attended Fire Fighting, Instructor Training School, Assault Boat Coxswain Trainer and Instructor, Military Counseling School, Philadelphia, PA; Career Information School, New Orleans, LA; Military Sealift Command 135 and 137 in Bayonne, NJ.

He served in 16 ships, the USS: *Rolette, Des Moines, Horace Bass, Rockbridge, Pawcatuck, Intrepid, Grand Canyon, Chandeleur, Wasp, Cherboygan County, Forrestal, Lexington, Guadalcanal, Santa Barbara, Independence, Alfray, Boulder, Antrim* and *Patterson.*

Special Assignments: tug boats - tugmaster, Philadelphia, PA; port services supervisor, Charleston, SC; assault boat coxswain and salvage, Little Creek, VA; bicentennial display carrier - CPO in charge, Philadelphia, PA; and USN Cargo Handling & Port Group, Pub's writer.

Exercises in MSC include: NAV REX 82, Bayonne, NJ; Command HQ NAV REX 83, Washington, DC; Joint Deployment Agency, Alexandria, Egypt (1985); Winter Cimex 1987, Strittgart, West Germany; Special Project MSCO Eastcote, London, England (1988).

His awards include (2) Battle Efficiency Awards, (3) Letters of Commendation, (6) Letters of Appreciation, (9) Meritorious Service, (3) Armed Force Reserve Medals, Armed Forces Expeditionary Medal, Beruit, Lebanon (1957), National Defense Ribbon, Naval Reserve Sea Service Ribbon, Overseas Duty Ribbon, Expert Rifleman Ribbon, Sharp Shooter Pistol Ribbon and Navy Achievement Medal.

Cardiff was promoted to CPO, 1964; master chief in 1974; command master chief at McKeesport for 13 years, 1976-89; president of the McKeesport/Pittsburgh Chief's Assoc. since 1974. He organized the Hard Hat Chapter of NERA in 1973 which later merged into the Greater Pittsburgh Chapter.

BMCM Cardiff USNR retired Aug. 25, 1994, with 42 years service.

Married 34 years to Loretta and has three children: Marty, Larry and Christine. He was employed by the Union Railroad Co. until retiring in 1996.

JOHN A. CARLIN, RMCS(SS), USN (Ret), born at Newcastle, WY, Aug. 2, 1943, and enlisted July 19, 1961, at San Antonio, TX. Attended boot camp and Radioman School at San Diego, CA. After Submarine School, December 1962, he reported onboard USS *Ronquil* (SS-396).

He made two WESTPACS earning the National Defense, the Navy Expeditionary and the Vietnam Service Medals. In October 1965, he was stationed at COMSUBFLOT EIGHT Naples and December 1967 at NAVCOMMSTA Japan. He reported onboard USS *Thomas Jefferson* (SSBN-618) (Gold) completing five deterrent patrols and earning his first Meritorious Unit Commendation; he won his second at NCSJ.

Transferred in May 1974 to COMSUBGRU SEVEN, Yokosuka where he advanced to RMC October 1976. June 1977—November 1985 included

three consecutive tours in Hawaii. USS *Thomas Jefferson* (SSBN-618) (Blue), completing five deterrent patrols, earning two Battle Efficiency "E's" and Navy Achievement Medal. Advanced to senior chief in June 1983 while onboard NAVCAMSEASTPAC Wahiawa.

Retired November 1985 at the Submarine Memorial at Pearl Harbor, after serving two years onboard USS *Queenfish* (SSN-651) and receiving his 3rd Battle "E," 2nd NAM and the Navy Arctic Service Ribbon.

Married Kumiko Sakai in Tokyo, June 2, 1969, and they have one daughter Elizabeth Ann. Their home in Makakilo, Hawaii.

JOHN R. CARLIN, ADRC, born Dec. 8, 1922, Sparrow Point, MD. He joined the USN Dec. 6, 1940. He was sent to Norfolk, VA, to boot camp; after completing he went to Jacksonville, FL for AMM School.

He was transferred to VP74, July 1, 1941. While in VP74 he served in Iceland, Bermuda, San Juan, P.R., Trinidad, B.W.I. Natal, Bahia and Rio, Brazil.

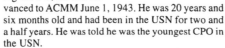

Passed exam for seaman 1/c, 3, 2, 1PO. Three months after making 1/c PO, he was advanced to ACMM June 1, 1943. He was 20 years and six months old and had been in the USN for two and a half years. He was told he was the youngest CPO in the USN.

Carlin left VP74 and went to CASU21, from there to NATTC - NAMTD in Norfolk and later in Memphis, TN. He was a leading chief of a PBM training unit, a fighter unit and later a fire fighting unit.

He left the USN in 1947 and joined the USNR at Anacosta, Washington, DC. He was leading chief of VP662 from 1949-70 when he retired. His awards include the Navy Unit Commendation w/star, Air Medal, two Good Conduct Medals and Combat Air Crew Wings.

Married first to Ruth Meekins and second to Margart Rhoades. He has four children. After the service he worked at Sparrow Point for Bethlehem Steel Co. for 30 years. Retired, he lives in Florida and enjoys it.

ANNA JEANE (BOUTWELL) COLAGROSSI, SKC, USNR, born Nov. 2, 1922, Hancock County, OH. She joined the USN April 19, 1943, Columbus, OH. Duty stations include Hunter College, Bronx, NY; Storekeepers School, Milledgeville, GA; BuSanda, Cleveland, OH (chief).

Awards include the American Campaign, WWII Victory and Good Conduct Medal. She was discharged Aug. 18, 1946, at Great Lakes Naval Training Station.

Married Alfred O. Colagrossi, May 4, 1946, and has four children: Matthew, Carol, James and Steven. After the service, she worked for Indiana Employ-

ment Security Div., Excel Corp. and quit in 1952 to be a housewife. She is member of National Chief Petty Officers' Assoc. and WAVES National. Enjoys golf, sewing, knitting, computers, babysitting and winters in Florida.

Seven of her brothers served in WWII: Clair, John, Kent, Dwight and Merlin in the USN and Wade and Wayne in the US Army. Four of her brothers and Anna attained chief rate. She kept their pictures under her desk glass and an inspecting officer advised her that one boyfriend was enough. When the officer learned they were her brothers she set a date for a Navy photographer to take pictures, and the *BuSanda Newsletter* wrote an article about "WAVE Boutwell and the fighting Boutwells."

Anna's husband and his brother were also Navy WWII vets. Two ancestors fought in Revolutionary War: USCG cutter, *Boutwell,* was named for George Sewell Boutwell, Secretary of Treasury under President Grant. Over the last 225 years many of her relatives have served in the military.

KENNETH O. COOK, OSCS, USN (Ret), born Nov. 30, 1930, Walnut Springs, TX. He joined the USN Dec. 18, 1947, and attended boot camp in San Diego, CA. Served in USS *T.E. Chandler* (DD-717), Tsingtao, China; USS *Buck* (DD-761), Tsingtao and Shanghai in China and in Yokosuka, Japan (1948-49); USS *Jason* (ARH-1), San Diego; USS EPCE(R) 857, home port at San Diego (oceanography research); USS *Fetchler* (DDR-870), Korea and world cruise.

Instructor duty, Radarman Class A School, Treasure Island, San Francisco, CA, where he was promoted to chief radarman. Attended Radarman Class B School in 1958, graduating in early 1959 at FAATC Point Loma, CA and assigned duty with Naval Beach Jumper Unit 1, Coronado, CA, team chief, BJU-1, Team 11 operating in CONUS and the Far East.

Transferred in 1963 to USS *Porterfield* (DD-682) with another cruise to the Far East and Taiwan patrol. Promoted to senior chief, 1964 and reported to USN School Cmd., Great Lakes, IL, radarman class A instructor duty. Was notified he had made senior chief in transit.

In 1966 attended NTDS and Air Control Schools, Point Loma, CA, and further transfer to pre-com and duty in USS *Sterrett* (DLG-31) upon commissioning. In 1967 at Naval Hospital, Bremerton, WA; TDY at USN and USMC Reserve Training Center, Tacoma, WA; May 8, 1968, transferred to the Fleet Reserve and into retirement.

Married in 1952 to Margie N. Horton and has four children: Kathy, Ken II, Karen and Karl. Returned with his family to Texas in 1969 where they live today. From 1978-93 in Saudi Arabia where he was maintenance and construction project manager/site manager of plant construction. Semi-retired from work at present.

JOSEPH A. CORSI, CDR, USN (Ret), born in Newark, OH, Oct. 7, 1921. He joined the USN April 14, 1941, Cleveland, OH. Attended boot camp at Great Lakes, IL; Aviation Machinist Mate Class A School, Detroit Naval Armory at Belle Isle; graduated and transferred to NAS Floyd Bennet and assigned to Aircraft Delivery Unit.

Other duty stations include Aircraft Ferry Svc. Sqdn. One (VRS-1) and reassignment to Adams Field in Little Rock, AR; HEDRON at NAS Roosevelt Roads, Puerto Rico; NAS North Island; Naval Air Transport Svc. and served in squadrons at Oakland, NAS Moffett Field, NAS Bargers Point, NAS Agana, Guam and NAS Yonabaru, Okinawa, where the war ended. He continued to serve on Okinawa until about February 1941 when sent to NAS Alameda until end of his enlistment in April 1941.

Re-enlisted in the USN and was sent to Receiving Station, Philadelphia; NAS North Island, CA; Patrol Squadron One, NAS Camp Kearney; transferred to NAS Whidbey when home station was changed there, August 1947; deployed once to Adak, AK; once to Kodiak, AK; twice to NAS Atsugi, Japan; four tours at NAS Naha and one at AFB Kadena, Okinawa. He earned five Air Medals for flights in the Formosa Straits and off the coast of Korea during the Korean War.

At the end of a tour at Naha in 1954, Patrol Sqdn. One made a flight around the world in 14 days with stops at Sangley Point, Ceylon, Saudi Arabia, Bahrein, Naples, Hutchison, KS, then back to NAS Whidbey.

In 1955 he received orders to Newport, RI, to attend indoctrination. First tour was with Fighter Sqdn. 64 (VF-64), NAS Alameda. Deployed aboard the USS *Shangri-La* for the Western Pacific; reassigned F3H Demon aircraft for training then deployed aboard USS *Midway*.

In 1959 ordered to NAS Pensacola, FL, where assigned to Sherman Field; orders to Heavy Attack Sqdn. 11, NAS Sanford, FL; deployed to USS *FDR* and USS *Independence*. After completion of two Mediterranean tours with A3D Skywarrior aircraft, he went aboard USS *Forrestal* for a trip to WESTPAC; ordered to COMFAIRWHIDBEY STAFF, NAS Whidbey Island, WA; and NAS Barbers Point.

During his 28th year of service, he received promotion to CDR (LDO) and was reassigned as AIMD officer, NAS Washington State College, Bellingham, WA. Reassigned in December 1970 to USS *Constellation* and awarded the Meritorious Medal for this tour of duty. Deployed twice on *Constellation* to WESTPAC during the Vietnam hostilities with Carrier Air Wing Two; reassigned to Commander Task Force 77; NAS Cubi Point where he served three years as department head and awarded his second Meritorious Service Medal.

Final tour of duty at COMNAVAIRPAC STAFF, NAS North Island. During this time he failed selection for captain and was retired July 1, 1982, with a total of 41 years and three months of Navy service. He was awarded a third Meritorious Service Medal; he also received numerous unit awards.

Married Delores on July 10, 1954, and has five children: Robley, Donny, Terry, Mickey and Randy. After service went to work for a corporation. Also, sec/treas, FRA Imperial Beach, Branch 287.

JAMES S. CUNNINGHAM, DKC, born in Ashland, MO, May 4, 1925. He joined the USN in 1943 at St. Louis, MO. Attended boot camp and Storekeeper School at Farragut, ID, before being assigned to USS *Uvalde* (AKA-88) and was on board when commissioned. Served in South Pacific during 1944-46 in New Guinea, Philippines and Okinawa area.

He is authorized to wear two Bronze Battle Stars, Philippine Liberation Ribbon w/star and other miscellaneous ribbons. Was released to inactive duty in 1946 and recalled in 1951 to active status with MCB 6 which was commissioned at Norfolk. Advanced to CPO in Newfoundland while serving with MCB 6 and aboard the USS *Benewah* (APB-35).

Released to inactive duty in late 1952 and was discharged from Reserve in 1958. He then served in management and executive positions with oil cooperative in Columbia, MO, until retirement in 1990.

Married Alva in 1947 and has one daughter and one granddaughter. He is member of several veteran and Masonic organization including the shrine.

ROBERT "E" LEE DAVIS, SKC, born Oct. 28, 1938, Winston Salem, NC. He joined the USN in March 1956, Winston Salem and attended boot camp at Bainbridge, MD. Served in the USS *Alamo* (LSD-33), 15ND, Panama Canal Zone, CBC Port Hueneme, CA, USN Advisory Group (made chief with the Advisory Group), RVN NROTC Unit, North Carolina Central University, Durham, NC, USS *Enterprise* (CVA(N)-65).

Awards include the Good Conduct Medal (3), Meritorious Unit Citation (NSA Da Nang), RVN Meritorious Unit Citation w/Gallantry Cross/ NAVADGRU, RVN Citation (Civil Actions Medal, 1/c Color w/Palm, National Defense, Vietnam Service Medal w/3 stars, AEF Medal (Formosa, Vietnam), Meritorious Unit Citation USS *Vega* (AF-59). He retired Aug. 2, 1976, Alameda, CA.

Married Essie June 21, 1960, and has three children: Tammy, Nancy and Melody. He worked for Williams Brothers Engineering Co. as a senior buyer for four years. Then opened up his own oil field supply business for 10 years. He retired in 1991 and is now the owner/manager of E&R Consultants, a government procurement and contracting service.

DANIEL P. DELL, ADC (AW), born April 21, 1960, Louisville, KY. He joined the USN March 21, 1979.

Duty assignments March 21, 1979, Great Lakes, IL; May 31, 1979, AD "A" School, NAS Cecil Field, FL; Aug. 23, 1979, VA-174 NAS Cecil Field, FL; July 13, 1983, HS-1 Sea Component NAS Jacksonville, FL; Dec. 2, 1986, HS-1 FRAMP instructor NAS Jacksonville, advanced to CPO Sept. 16, 1989; March

2, 1990, USS *Dwight D. Eisenhower* (CVN-69); March 10, 1993 NAVMASSO Chesapeake, VA.

Current duty - NALCOMIS system analyst. Implementation team leader for both "O" and "I" level activities. He is pursuing bachelors degree in American history for teachers certification at the middle/high school level. He intends to finish his career at RTC Great Lakes as a recruit company commander.

His awards include the Navy Commendation Medal, Navy Achievement (2), Meritorious Unit Commendation (4), Battle E (2), Good Conduct (4), National Defense, SW Asia Medal (2), Sea Service (5) and Expert Pistol Medal.

Married Sonja LeCroy and has three children: Daniel Paul Jr., Crystal Marie and Tiffany Laura.

DONALD D. DOOLITTLE, CQM, born Feb. 17, 1917, Brooklyn, NY. He joined the USN in April 1936, Brooklyn, NY. Duty stations include USS *Marblehead*, USS *Wyoming*, USN DET Worlds Fair 1940, USS *Albemarle*, USS *Bernadou* (DD-153), made chief and served in the USS *Waldron*.

Awards include the Good Conduct Medal, China Service Medal, American Area Campaign, Asiatic Theater w/ 4 stars, European and Mid-Eastern w/3 stars, American Defense w/A, Philippine Liberation Medal, Victory Medal and Presidential Unit Citation (USS *Bernadou*) invasion North Africa Safi. French Morocco. He was medically discharged (honorable) in August 1946, Sampson Naval Hospital.

Married Marguerite R. Rooney Oct. 17, 1947, and has one son, Christopher Paul. After the service he attended several schools, Paul Smiths College and Veterans Vocational School. Employed by General Electric for 26 years and retired in 1977. After retirement he moved to Melbourne Beach, FL, and did a lot of sailing—mostly to the Bahamas.

He is now an active Rotarian and interested in National Chief Petty Officers Assoc. He is president of the USS Waldron first Plankowners Club and active in the USS Waldron Alumni Assoc.

DAVIS O. EADDY, SKCM, USN (Ret), born in Myrtle Beach, SC, March 30, 1936. He worked for Floyd Repec Plumbing Co. and R.A. Harper Plumbing Co. through high school. Eaddy joined the Naval Reserves, Georgetown, SC, in 1953 and began active duty in 1954.

Duty Stations: 1954-55, USS *Conway* (DDE-507), home port Norfolk, VA; 1955-56, USS *Bennington* (CVA-20), home port San Diego, CA, and became a Horn Shellback while aboard; 1956-57, R.A. Harper Plumbing Co. (still in Naval Reserves); 1957-58, went to school and worked at Purity Market, Nick Gill owner (still in Naval Reserves).

Left the Reserves and joined the USN in 1958. Served 1958-60, USS *Vogelgesand* (DD-862), home port Norfolk, VA, made SK3 and SK2; 1960-62, Naval Supply Depot , Gitmo Bay, Cuba and made SK1; 1962, Submarine School, New London, CT; 1962-63, USS *Patrick Henry* (SSBN-599 Blue), home port New London, CT, made two patrols and operated out of Holy Lock, Scotland; 1963-64, USS *Proteus* (AS-19) home port Charleston, SC, operated out of Holy Lock, Scotland and Rota, Spain; 1964-66, Naval Radio Station, Ponce, Puerto Rico, made SKC.

Returned to the States in 1966 for Survival School at Dam Neck, VA; 1966-67, MACV-SOC (all over Vietnam, station at NAD Da Nang, Vietnam); 1967-68, JTF-8 Albuquerque, NM; 1968-69, NAS Dallas, TX, made SKCS; 1969, Chief Petty Officer Academy, Pensacola, FL; 1969-70, MACV-SOG (all over Vietnam, station in Saigon).

Served in USS *Biddle* (DLG-34), 1970-72, home port Norfolk, VA; 1972 *Biddle* spent six months off the coast of North Vietnam and shot down a number of North Vietnam MiG planes. Eaddy was advanced to SKCM while aboard *Biddle*; 1972-75 Polaris Material Office, Charleston, SC.

On July 1, 1975, he and his brother, Emmell L. Eaddy SHC, USN, transferred to the Fleet Reserve in the same retirement ceremony. From 1975 to the present, R.A. Harper Plumbing Co., Inc. Eaddy bought the business in 1980 when he got sick.

In 1985 after his first wife died, he married Sallie Tindall (also from Myrtle Beach) and they have a son Daniel (born Feb. 14, 1988) and daughter Sarah A. (born June 11, 1986). Daughter Susan was born March 18, 1968, by his first wife. Sallie has a daughter, Hannah born March 23, 1976. They live in North Myrtle Beach, SC, and are just about ready to retire from the plumbing business.

RICHARD M. ELSTER, LT(SC), USNR, SKC(SS) USNR, Sept. 16, 1987. Active duty time, June 1959—November 1966; reserve time June 1984—to date.

Born March 5, 1941, Brooklyn, NY. Attended boot camp in June 1959, SR-SA; September 1959 in Orange, TX, Reserve Fleet, SA-SN; August 1960 in USS *Waller* (DDE-446), SN-SK2; November 1962 at Sub School, Groton, SK2; January 1963, USS *Redfish* (AGSS-395), SK2; July 1964, USS *Kamehameha* (SSBN-642) (Blue), SK2(SS). In November 1966 he was discharged from active duty as SK2(SS).

SK1(SS) June 1984, NISPROD 1005 USNR, Pittsburgh, PA; SKC(SS) September 1987, COMNAVSURFLTATLANTIC, USNR, Pittsburgh, PA; ENS October 1989, COMNAVSURGRU MED 105, USNR, LDO; LT, November 1993, NR Sub Dry Docks, USNR, Columbus, OH.

Married Judy A. Byers in October 1964. They have two daughters, Bonnie (1966) and Sherri (1968). He is employed as national account manager - very fine products.

WILFRED E. EMSHOFF was born Dec. 21, 1935. He enlisted in the USNR at Bellville, TX, eight years in 1953, junior year at Bellville High School and graduated in 1954. He retired from the USNR December 1995.

Served in seven ships, naval security group activities 21 times, other naval activities five times

for two weeks ACDUTRA; six months temporary active duty tour NAVRESCEN Houston, TX; two year active duty tour at NAVCOMMSTA Morocco and advanced to CTAC in 1970.

Awards include the Naval Reserve Meritorious Service Medal (7), Armed Forces Reserve Medal (4) and National Defense Service Medal.

Retired from Pan Am World Airways in 1991. Activities include school bus driver Brenham I.S.D.; jr. vice commander of VFW Post 7104, Brenham, TX; keyboard musician in country/western dance band, Dancers Choice, of Navasota, TX; coin collector; gun collector; ammo reloader and family genealogist.

Divorced, he has one daughter and two sons.

ANNABELLE F. SUE FISCHER, YNC, USN (Ret), born in Baltimore, MD, July 1, 1926. He enlisted May 22, 1951, transferred to Fleet Reserve Aug. 24, 1971, and retired March 1, 1981. Boot RTC(W) at Great Lakes, IL.

Duty: YN-A Scol, NOB NORVA; NAVSTA Legal, NORVA with TAD NAVJUSTSCOL Newport, RI; CINCNELM London; IT Scol, SSC NORVA; RTC(W) Bainbridge, MD, as CC; YN-B Scol, SSC San Diego; NAS Legal, North Island, CA; BUPERS Briefing Board, NTC Bainbridge; Recruiting Aids Division, CRUITCOM, DC; RTC Legal, Great Lakes - last three commands as chief-in-charge.

Awards include National Defense (Korea/Vietnam), Good Conduct (6). Sports: softball, basketball, bowling, tennis, golf, pistol. Has BA degree in liberal arts, College of Notre Dame of Maryland, 1978.

Self-employed literary consultant, editor, researcher; author/editor: *Navy Women 1908-1988*, two volumes. Member of FRA, VFW, American Legion and WAVES National.

Last Fischer in direct line USN/USA: Christian and Frederick, musicians USS Constitution 1844-47; Frederick A., Sgt. USA 1882-87, MT Terr. Indian Wars; Clarence A., Pvt. USA 1918 WWI; maternal aunt Catherine Schultheis Cunzeman, Yeoman F, 1918 WWI, and uncle, William Schultheis, CPhM 1914-1927, USCG CPhM 1927-1943. Never married and resides in Westminster, MD.

J.C. GALLOP (JACK), QMC(SW), born in Monmouth, IL. He entered the USN on May 12, 1966, and received his boot training at NTC San Diego, CA. First duty station in USS *Wedderburn* (DD-684), home port in San Diego and made two WESPAC's on board.

He was transferred during the end of second tour to the USS *Wilhoite* (DER-397), home port in Pearl Harbor. Transferred in 1968 to Coronado, CA, for swift boat training and served in Vietnam with COSDIV-12 at Da Nang and Chulai. He finished his tour with Naval Advisory Group 143.

Discharged in May 1970 and re-entered the

USN in March 1976. He served aboard the USS *Paul F. Foster* (DD-964) for four years, two WESTPACs. Discharged in March 1980, he enlisted in the USCGR in 1983 and transferred to the USNR in 1989. Stationed at NRC Fresno, CA, he was promoted to chief in 1991 and currently serves as command career counselor with NRC Fresno.

Married to the former Barbara Farris and has two children, Ken is a deputy sheriff and Annjanette is married to a US Marine.

DONALD M. GLADSON, CMOMM, USS *Gustafson* (DE-182), born in 1920, Bridgeport, IN. Enlisted in USNR in October 1942 and attended boot training at Camp Green Bay, Great Lakes, IL; Diesel School Navy Pier, Chicago and General Motors School, Brooklyn, OH.

After short training period at SCTC Miami, he was assigned to DE-182 in August 1943. He was a plank owner and served 27 months aboard the "*Gus.*" The *Gus* served in the South Atlantic, Caribbean, Eastern US, Mediterranean and Eastern Pacific. At 0247 April 7, 1945, *Gustafson* sank U-857, within sight of Boston Light.

He was awarded rate of CMOMM, aboard the *Gus*, June 1, 1945. Gladson was honorably discharged in December 1945.

Married to former Mary E. Millet in 1949 and has two daughters, Vicki and Patricia, and two grandchildren. Retired from the Indianapolis Union Railway Co. after 32 1/2 years of clerical service. He now resides in Clearwater, FL, and is a member of NCPOA.

WILLIAM S. GLOVER, OSC, USN (Ret), born Jan. 14, 1926, Montgomery, AL. He enlisted in the USN in January 1943 and served aboard the USS *Kimberly* (DD-521), USS *Wm. D. Porter* (DD-579) in WWII. He served aboard USS *Springfield* (CL-66) in 1968-69, USS *Eversole* (DD-789) for two years during the Korean War, USS *McCord* (DD-534), USS *Saipan* (CVL-48), USS *Geo. Johnson* (DE-583), USS *Haverson* (DER-316) and various shore stations as supervisor and instructor in CIC Operations and served as air to air intercept instructor at Glynco NAS, Brunswick, GA, 1960-63.

Glover retired from the USN in February 1964. His awards and medals include the Purple Heart (enemy action at Okinawa in June 1945), Navy Commendation Ribbon (for meritorious service during Korean War), Good Conduct (6th award), China Service Medal, Asiatic Service Medal w/5 Battle Stars (for action in Gilbert Islands 1943, Aleutian

Islands 1944, Philippine Islands 1944-45 and Okinawa 1945), Navy Occupation Medal (Asia), National Defense Medal, Korean Service Medal w/star, ROK Presidential Unit Citation, Philippine Presidential Unit Citation and the United Nations Service Medal.

Married in December 1956, a widower in 1990, he has three children. Retired completely and mostly travels. He lives in a motor home "On the Road" again.

DAVID C. GRAHAM, SMCS, USN (Ret), born Sept. 10, 1924, Atlanta, GA. He enlisted in the USN, Dec. 13, 1941, Richmond, VA. Duty stations include USS *Idaho* (BB-42) 1942-45 COMDES DIV-42 enlisted allowance, NAVSTA Guantanamo Bay, Cuba, COMCARDIV-18 flag allowance, Service Schools Command, Norfolk, VA, COMDESRON-22 enlisted allowance, COMDESFLOT-4 flag allowance, Service Schools Command, San Diego, CA, COMPHIBRON-3 enlisted allowance, Fleet Training Center, San Diego, CA.

Retired May 1, 1971. His awards include the Navy Achievement, Good Conduct, American Theater Ribbon w/6 stars, Asiatic-Pacific w/7 stars, WWII Victory, Navy Occupation, National Defense, Vietnam Service, Philippine Liberation, RVN Campaign w/1969 clasp, Meritorious Unit Citation w/ COMPHIBRON-3 enlisted allowance.

The proudest day of his 30 year career was June 16, 1952, the day he advanced to chief petty officer. He advanced to senior chief petty officer Dec. 16, 1961, while with COMDESFLOT-4.

He was privileged to have served in the USS *Cleveland* (LPD-7) with Joe B. Havens, senior chief hospital corpsman. Let's never forget names like Peter Tomich, John Finn, Thomas Reeves—all outstanding chief petty officers. Each was awarded the Medal of Honor for heroic deeds at Pearl Harbor, Dec. 7, 1941. To those chiefs in heaven, may Almighty God watch over you.

Married Margaret J. Clarke on Jan. 12, 1963, and has six children. He was a mortician and is now fully retired.

WILLIAM J. HANNON, EMC(SS), USN, USNR (Ret), born Oct. 3, 1941, Scranton, PA. He enlisted in the USN in September 1959, completed basic training, Electricians Mate Class A and B Schools and enlisted Submarine School, New London, CT. He spent six years on Atlantic Fleet submarines, USS *Halfbeak* (SS-352) and USS *Irex* (SS-482).

Some of the memorable events during his active

duty tenure include Submarine North Atlantic Under Ice Warfare Operations; the 1962 Cuban Missile Blockade aboard *Halfbeak*; a social event as an enlisted selectee representing *Irex* at a 1964 Christmas dinner; and Palace Tour with the late Princess Grace and Prince Rainier of Monaco.

After two active duty tours, he enlisted in the USNR, January 1967. He was advanced to electrician mate chief petty officer, February 1969.

He served 22 years in the USNR, retiring Dec. 3, 1989, as command chief in a 65 member submarine support unit attached to USS *Orion* (AS-18), La Maddelena, Sardinia. He completed 30 1/4 years total naval service. Awards include the Good Conduct, National Defense, Reserve Meritorious, Armed Forces Reserve, Expert M-16 and the Presidential Unit Citation, Naval Reserve Center, Avoca, PA, 1972 Hurricane Agnes flood relief.

Married Ruth Marie Watts on Oct. 20, 1962, and has three sons: William, Sean and Kevin. After the service, he was employed as an industrial electrician for Thomson Consumer Electronics (RCA), Dunmore, PA, and is awaiting retirement.

BERNARD J. HARKINS, CPO, born Oct. 5, 1915, Erie, PA. He joined the USN on July 1, 1942, Cleveland, OH, and was one of two chiefs to open Bainbridge NTS and 1st Navy Personnel at Bainbridge training recruits.

Discharged May 17, 1946, Philadelphia Navy Yard. Received citation for training and developing best company of recruits at Bainbridge. Served in four different USN programs.

Attended Findlay (OH) College (BA degree); St. Bonaventure University (MA degree); Graduate School work toward Ph.D. at Univ. of North Carolina, Univ. of Pennsylvania, Syracuse Univ. and Allegheny College; undergraduate courses, Univ. of Pittsburgh; Honorary Doctorate, Jackson State Univ.; advanced undergraduate work, Edinboro State Univ.; and International City Managers Assoc., associate degree work in municipal government administration, assigned to city of San Diego, CA.

Married Rita A. Dill and has three daughters and two sons: Rita Jane, John C., Nora, Ann and Patrick. Active in American Legion, United Veterans Assoc. and numerous professional and civilian organizations and groups. Was a teacher, principal, coach, guidance counselor, retiring after 18 years, city of Erie, PA, city councilman and recreation director (20 years), now semi-retired.

HERMAN C. HART, EQMC, enlisted in the USN Seabees on July 26, 1967, and following Equipment Operator's School served three tours in Vietnam, 1967-68, with Naval Mobile Constr. Bn. 11; 1969-70, Naval Mobile Constr. Bn. 10; Civic Action Team 1019; and 1970-71, member of US MAG, Vietnam.

Duty assignments included: snow removal captain, NAS, Glenview, IL; transportation director, Naval Air Facility, Sigonella, Sicily; Alpha

Co. Ops., Naval Mobile Constr. Bn. 40 Det., Philippines; equipment program manager, Commander, Naval Constr. Bns., US Pacific Fleet, Port Hueneme, CA; CAT coordinator, 30th Naval Constr. Regt., Guam; active duty advisor RNMCB-18 Seattle, WA; Alpah Co. Ops. Chief, Naval Mobile Constr. Bn. 133, Gulfport, MI; director, Equipment Operator's School, Naval Constr. Trng. Ctr., Gulfport, MI; master chief of the Pacific Fleet Seabees, CO, Naval Constr. Bns. Pacific Fleet, Pearl Harbor, HI.

Navy training included Advanced Equipment

Operators School, Blaster's School, E8-E9 Management Course CECOS, Leadership Management Education Training Course, Load Planners School and Instructor School.

Awards include the Defense Meritorious Service Medal, Navy Commendation Medal, Navy Achievement w/Combat V, Navy Unit Commendation, Navy Combat Action Ribbon, Navy E (4 awards), Good Conduct Medal (5 awards), NDM, VSM, w/ Marine Corps Device (3 awards), Sea Service Deployment Ribbon, Navy Meritorious Medal, Gallantry Cross w/OLC, Vietnam Civil Actions Medal, Vietnam Campaign Ribbon, Navy Expert Rifleman Medal and Navy Expert Pistolshot Medal.

MC Hart assumed duties as MC of the Navy Seabees, CO, Naval Facilities Eng. Cmd., Washington, DC, June 29, 1990. He and his wife Donna have three children.

LESTER C. HARTLEY, Senior Chief Engineman, USN, USNR (Ret), born March 20, 1922, Beech Grove, IN. He joined the USN Sept. 6, 1940, San Francisco, CA.

Duty Stations: NTS San Diego, CA; USS *Jarvis* (DD-393); Fleet Diesel School, Norfolk; Cooper Bessemer Diesel School, Grove City, PA; Receiving Station, Philadelphia; USS *Engage* (AM-93); made CPO Dec. 1, 1944, aboard PC-1597, Palermo, Sicily; LST-838; NRTC Indianapolis; 18 different ships and shore facilities during 18 1/2 years Naval Reserve status. Was rated SCPO, Dec. 16, 1960, Indianapolis Naval Reserve Center.

Discharged from USN Nov. 3, 1946, Seattle, WA and retired from USNR July 1, 1967, Indianapolis, IN. Awards and medals include American Defense w/Fleet Clasp, Asiatic-Pacific, American, European and Middle East, Victory, Armed Forces Reserve and the Good Conduct.

Married Lucille Lutes on March 2, 1946, and has one daughter Carolyn Lynn. Was employed by Cummins Engine Co., then by Allison Div. of GM for 32 1/2 years (during this time he spent 18 1/2 years in USNR at Indianapolis). Currently retired from all paying jobs; he is on Board of Directors of Philadelphia Church Cemetery and maintains the grounds; president of Indiana Chapter #1 Pearl Harbor Survivors Assoc.; and he does upkeep of two residential properties and many other miscellaneous things.

JOHN A. HENDRICKSON, YNCS, USN (Ret) was born in White Lake, NY, Sept. 29, 1928. He joined the USN on March 16, 1946, in New York City and attended boot camp at Norfolk, VA, and Class "A" Yeoman School in Bainbridge, MD. Assigned to the USS *Mona Island* (ARG-9).

Transferred to USS *Albermarle "Able-Mable"* (AV-5) in 1947. In 1948 the ship participated in "Operation Sandstone," testing the atomic bomb in the Pacific Area. In 1949 transferred to Staff of Commander Eastern Sea Frontier, NYC, NY; assigned to VF-21 at Oceana, VA, October 1951; New York Group Atlantic Reserve Fleet, Bayonne, NJ, 1953; and HQ, LANTRESFLT, NYC, NY, 1955.

In 1957 transferred to HQ PHIBLANT, Little Creek, VA, and in 1960 was promoted to YNC and assigned to the Office of the Judge Advocate General,

Pentagon, Washington, DC. In 1962 assigned to Naval Intelligence School, Washington, DC, for course in German language; transferred to CINCUSNAVEUR REP Germany, Frankfurt, Germany; promoted to YNCS in 1964; transferred to NUCPWR TRA Unit, Idaho Fall, ID, in 1966 and retired Jan. 7, 1967. His awards include many letters of commendation and appreciation.

Married Kathryn (Walsh) on June 17, 1951, and has four children: Karen, John Jr., Teresa and Glenn, and two grandsons, John Jr. and Jeremy. After 20 years as a computer specialist with the US Government, he retired in 1986.

KENNETH J. KELLY, LTJG, born in New Albin, IA, on Oct. 9, 1917. He joined the USN July 12, 1937, and was sent to boot camp at Great Lakes, IL. After boot camp he was assigned to the USS *New Orleans* and served on that ship until August 1941, at which time he was honorably discharged, as a fire controlman second class.

Kelly re-enlisted in the USN March 1942 and was assigned to the USS *Alabama*. During his service on the Alabama, in the Secondary Battery Fire Control Div., he was promoted to fire controlman first class, chief fire controlman and ensign, USNR.

In December 1944, Kelly was transferred to the USS *Massachusetts* where he served as assistant secondary battery plotting room officer until September 1945 at which time he was honorably separated from the service with the rank of lieutenant junior grade and returned to civilian life.

During his tours of duty, his ships were involved in action in the American, European and Pacific Theaters of Operation. Today he is retired from a subsidiary of General Dynamics Corp. and lives with his wife, Fran, in Walnut Creek, CA.

JOHN E. KRUEGER JR. enlisted in the USN and was assigned to the USS *San Francisco*, 2nd Div. Highlights were Pearl Harbor attack, Guadalcanal Campaign, Battle of Esperance, Battle of Guadalcanal and Aleutians Operation. Promoted to gunner's mate second class.

He was transferred to Advanced Gunnery School, Washington, DC, 1943. Commissioned USS *Bryant* (DD-665). While aboard he was promoted to gunner's mate first class and later chief. While serving aboard the *Bryant*, he participated in the invasions of Palau, Saipan, Tinian and Leyte.

Krueger also saw action during the torpedo runs during the battles of Surigao Straits, Luzon, Mindoro,

Lingayen Gulf, Iwo Jima and culminating at Okinawa when their ship was attacked by six kamikaze planes while steaming alone on picket duty.

Awarded the Silver Star Medal for organizing firefighting operations and rescuing unconscious seaman in ammunition storage locker.

Participated in 14 major engagements. The *San Francisco* was awarded the Presidential Unit Citation and *Bryant* awarded the Navy Unit Commendation. He was discharged in 1946.

Married to Florence and has two daughters, Bonnie and Kim. He is a retired deputy chief, New Brunswick Fire Dept. and presently resides at Lavallette, NJ.

JOSEPH A. LEBENTRITT, BMC PA, USN, USNR (Ret), born April 2, 1919, Troy, NY. He joined the USN May 20, 1936, Albany, NY. Duty stations include Newport, RI Training Station; USS *King* (DD-242); USS *Saratoga* (CV-3); USS *Gherardi* (DD-637) (made chief Aug. 1, 1945, Gherrardi, D.MS.-30).

Awards include Victory Medal, American Theater Medal, European Theater Medal w/3 stars, Asiatic-Pacific Theater Medal w/3 stars and recommendation for award of Letter of Commendation and Ribbon for minesweeping operations of live mines off Japan, August to November 1945. Discharged from USN, May 27, 1946, and from USNR, Jan. 15, 1959.

Married Anna I., Nov. 17, 1946, and they have three children: Patricia, Jody and Rosemary. After the service he owned his own business and was a contractor for sheet metal and air conditioning. Currently retired and enjoys hunting, fishing, golfing and gardening.

JIM MINOR, QMC, born 1920 in Texas. He attended public school in Texas and high school and college in California. He married Anita in 1944 and has three children and three grandchildren. He was a public accountant for 35 years and business manager of a new automobile dealerships prior to retirement in 1982.

During his naval career, he graduated from the USN Signal School at the University of Illinois and Minecraft Training Center in Virginia. He was a plankowner of the USS *Fidelity* (AM-96) and USS YMS-407. Other assignments include the USS *Bombard* (AM-151) and USS YMS-355. Jim served in American, European and Pacific theaters of operation during WWII.

He is a life member of National Chief Petty Officers' Assoc., Patrol Craft Sailors Assoc., American Legion, VFW, DAV and is also a member of Assoc. of Minemen and a former president and vice-president of Naval Minewarfare Assoc.

WILLIAM B. MORGAN, LT, USNR, reported aboard USS *Chief* (AM-315) in Pearl Harbor, HI in January 1945 for duty as XO and navigator. First assignment was to become part of an escort for tankers carrying 110 octane gas for aircraft carrier in the Third Fleet. They went to Ulithi and Eniwetok, where they saw the arrival of Gen. Douglas MacArthur. Their mission terminated in Okinawa.

Spent four months in Okinawa loosely attached to the Third Fleet and worked for Adm. "Bull" Halsey. In Unten Ko anchorage they were supplying air support from Japanese submarines operating in the area and many times came under attack. They were off the coast of Ie Shima the day correspondent, Ernie Pyle, was killed by gunfire from a Japanese sniper on the island.

The A-Bomb was dropped and Japanese surrendered. They then swept minefields in and around Nagoya and Wakayama, rode out two severe typhoons, watched for any aircraft shot down during CBI operation and carried Quonset hut on their fantail from Pearl Harbor to Okinawa to be used as a communications center for Navy and Marines based there.

In November 1945 he was promoted to CO of the USS *Chief*. In December 1945, he was assigned to USS *Benevolence*, a new hospital ship for return to San Francisco, CA, and ultimate release from active duty in March 1946 after about five and a half years of active sea duty.

RUTH H. MULDOON, YNC, born in Howard Beach, Long Island, NY. Joined the USN in March 1943, NYC, and went to boot camp at Hunter College, NY. Orders to NAS Miami; FPO New York and promoted to YNC in May 1945. Transferred to Naval Barracks, NYC - ship's writer.

Was honorably discharged in August 1946. Enlisted in the USN in 1948 and assigned to NAS and Navy pier, San Diego; Naval School of Justice, Newport; transferred to NATO Intelligence Committee, Pentagon; and discharged in 1954.

Enlisted in the USNR. Upon return from Saigon, Indo-China was affiliated with USNR units at NAS Los Alamitos, CA. NAIRU, VP, VR units. Also, CPO in charge of one of first boot camps for Women's Reserve. While at Los Alamitos she was selected as Enlisted Woman of the Year.

She is widow of MGYSGT Bill Muldoon, USMC. Retired in 1982, she keeps busy with traveling, volunteer activities, bridge, etc.

HAROLD F. MULL, AFCM, USNR (Ret), born in Union County, IL, Nov. 20, 1919. Enlisted as an AS, USN, at St. Louis, MO, on Sept. 20, 1939, on a six-year enlistment. As-

signed to Co. 28, USNTS, Great Lakes, IL for recruit training. Entire company was required to sleep in hammocks during the three-month training period since some USN ships were still so equipped.

December 1939, assigned to Div. 1, USS *Portland* (CA-33). On any permanent transfer assignment, each man carried his hammock, mattress, mattress cover, blankets, pillow, pillow case, sea bag and specified clothing properly marked. A ditty bag was carried for toilet articles and small items.

In January 1940, he was promoted to SEA2/c; March 1940, transferred to V Div.; February 1941 promoted to SEA1/c; March 1941, assigned to Aviation Machinists Mate School, USNAS Jacksonville, FL; June 1941, transferred to USNAS, Anacostia, DC.

In November 1941 promoted to AMM3/c; June 1942, promoted to AMM2/c; October 1942, transferred to USNAS, St. Louis, MO; June 1943, promoted to AMM1/c; December 1943, transferred to HEDRON 5-1, Elizabeth City, NC Det; March 1944, promoted to ACMM (AA); June 1944, transferred to Hellcat Unit 602, ABATU, MTD, HQ NAS Memphis, TN.

Promoted March 1945 to ACMM(T) and Oct. 16, 1945, received honorable discharge. Re-enlisted in USNR as ADRC at USNAS Glenview, IL and assigned to Sqdn. VF-724; October 1950, transferred to Sqdn. VA-925, USNAS, St. Louis, MO; November 1955, transferred to Sqdn. VR-923 same location. Designated as aircrewman in C-54 type aircraft; March 1958, transferred to VR-881, USNAS, Olathe, KS; May 1959, transferred to VP-882, same location. Designated as aircrewman in P2F type aircraft. December 1961 promoted to ADCS; December 1963 to ADCM and in April 1968 transferred to Retired Reserve.

Married Velma Bartruff on July 25, 1943, and has three children: Sue Ann (July 27, 1946), Linda (June 27, 1950) and David (Oct. 25, 1954).

DARRELL E. NELSON was born on June 5, 1923, St. Hiliare, MN. He graduated from Hibbing High in 1941, then employed by the Learch Brothers as an iron ore chemist. He entered the USN in 1942 and did duty at Farragut, ID; Great Lakes, IL and Submarine School at New London, CT.

At New London he was on the USS 0-4(SS-65). He also served in the USS *Loggerhead* (SS-374), USS *Entemedor* (SS-340), USS *Perch* (SS-313), USS *Redfish* (SS-395), USS *Nereus* (AS-17), USS *Raton* (SS-

2170), USS *Diodon* (SS-349), USS *Oriskany* (CVA-34); and the USS *Constellation* (CVA-64).

He was stationed or staffed with or at Manitowoc Training Activity, NROTC at the University of Minnesota, Staff at SUBRON 5, Staff at SUBFLOT 1, RTC at San Diego, Submarine officer at San Diego, DATC San Diego and battalion commander RTC San Diego.

Retired on the deck of the USS *Tang* (SS-306) as a master chief gunner's mate, June 30, 1974. His awards include his Submarine Qualification, Submarine Combat Pin, eight awards of Good Conduct, Navy Medal of Achievement, Navy Unit Commendation w/2 stars and the Meritorious Unit Commendation. He also has 14 campaign medals.

He joined the Sub Vets in 1956 and was the first president of the San Diego Chapter and served a number of times as the national parliamentarian and resolutions chairman.

Nelson is married to Borgie and they live in Hudson, WI. They are making good use of their house boat, *Twin Dolphins II*, on the St. Croix River.

ALBERT C. NIELSEN, ADRC, USN (Ret), born at Maskell, NE, July 25, 1929. He took boot training at San Diego from January—April 1948; AD "A" School at Memphis, TN, then transferred to FASRON 102 at Norfolk, VA.

Received orders in 1952 to NAAS Cabaniss Field, TX (ATU-1); attended AD "B" School at Memphis in 1955; then ordered to VA-116 at Miramar.

Re-enlisted for orders to VR-8 at Hickam AFB, Oahu; ordered to VS-33 at North Island in 1961 and left the squadron and USS *Bennington* (CVS-20) at Sasabo, Japan for Instructor Training School at Memphis. Ordered to NAAS Ream Field where he received chief initiation and taught SH3A helo and GE-58 jet engine. Transferred to the Fleet Reserve on Jan. 15, 1968.

Marriage to Marilyn Johnson ended in divorce. He has four children: Debbie, Scott (deceased), Renee and Laurie. Retired from city of Wichita, KS in November 1988. He does lots of traveling and golfing.

GEORGE L. OWENBY, ADRC E7, born Feb. 21, 1928, Buckhead, GA, in Morgan County. He joined the USN March 12, 1945, NRS Atlanta, GA. Duty Stations include NATC San Diego; San Francisco; South Pacific; Adak, AK; NATTC Memphis; Pensacola; Corry and Whitting Field, FL; VX-3 Atlantic City, NJ; VR-22 Norfolk, VA; VR-24 Port

Lyautey, Morocco; VR-1 Pax River, MD; Flight Eng. School, Burbank, CA; VW-11 Pax and Argentia, Newfoundland; Pensacola, FL; Norfolk; Sicily; and Create, Greece.

His awards include the Asiatic-Pacific, American Defense, Good Conduct (5), American Theater, WWII Victory Medal, Navy Occupation Service Medal w/clasp and National Defense. Retired from Fleet Reserve Oct. 12, 1965; retired Oct. 1, 1975, Receiving Station, Charleston, SC.

Married Mittie L. Wilkerson March 23, 1951, and has three children: Lou Ann, Georgia Lynn and Janet Lee. He is retired from Delta Air Lines and currently farms and enjoys crafts in Covington, GA.

LESTER SUTTON OWENBY, LT, USN (Ret), born July 21, 1923, Buckhead, GA. He joined the USN July 9, 1941, Atlanta, GA. Duty stations include NAS Norfolk (VF-5), USS *Saratoga*, Guadalcanal September to December 1942; CASU-13 Santa Rosa, CA and South Pacific Islands (made chief September 1943); NATTC Memphis; VS-23 Norfolk; VR-22 Norfolk; COMTEN San Juan; BUWGPS, Washington, DC; USS *Shangri-La*; and NS-3 Norfolk.

His awards include the American Theater, WWII Victory Medal and the Presidential Unit Citation w/ star. He was discharged from the USN July 31, 1969, Norfolk, VA.

Married Mary Morris March 25, 1943, and has one son Ronald and three granddaughters. After the service he worked for the state of Georgia in the Natural Resources Dept. Owenby was diagnosed with Parkinson's disease in 1978 and died March 23, 1991, from complications of the disease.

GORDON DEWAYNE PEABODY, YNC, USNR (Ret), born Feb. 11, 1938, Fort Wayne, IN. He joined the USN May 31, 1956, Fort Wayne, IN. Duty stations include recruit training, Great Lakes, IL; Service School Command, San Diego, CA; USS *Genessee* (AOG-8) at Pearl Harbor, HI; HSA, Yokosuka, Japan; USN&MCRTC, TI, San Francisco, CA.

Released from active duty as YN2, May 24, 1961, San Francisco, CA. Advanced to YNC, Feb. 16, 1970, in Fort Wayne, IN, as a drilling reservist.

His awards include the Good Conduct Medal (1 awd.), National Defense Service Medal (1 awd.), Naval Reserve Meritorious Service Medal (6 awds.); Armed Forces Reserve Medal (2 awds.) and Letter of Commendation from RADM James Mantell, COMSIXTHRNCR, Gulfport, MS. Retired from the USNR on March 1, 1990, Fort Wayne, IN.

He worked for the state of Indiana for four and a half years until he was accepted for a civil service position in the federal government in October 1966. He is still with the government as a QA programs clerk in the Magnavox Plant, Fort Wayne, IN, with plans to retire in April 1988. He never married.

JAMES G. PEELER, RMC, USN (Ret), born in North Little Rock, AR. He entered the USN at Little Rock, Oct. 13, 1943, received boot training at San Diego and attended Class A Radio School at Texas A&M College, Bryan, TX, graduating as RM3 on May 8, 1944.

First duty station was Small Craft Training Center, San Pedro, CA, Radio Station NCX operating a radio drill circuit with operators on newly constructed APAs and AKAs built at Todd Shipyards on the West Coast. Next assigned to USS *Blueridge* (AGC-2) from Receiving Station, Shoemaker, CA.

Joined the *Blueridge* at Manila, P.I. as member of staff, Commander Seventh Amphibious Force, April 1945. Seventh Amphib. staff shifted to USS *Ancon* (AGC-4), June 1945. Was on radio watch and copied the first flash message advising of atomic bomb strike on Hiroshima, Japan, Aug. 6, 1945, and three days later copied message advising of atomic bomb attack on Nagasaki, Japan, Aug. 9, 1945.

Seventh Amphib. flag shifted to USS *Catoctin* (AGC-5) in Manila Bay in August 1945. USS *Catoctin* with VADM Daniel E. Barbee, Commander Seventh Amphib. Force and staff embarked and participated in transfer and convoy of GEN Hodge and the 24th Army Corps from Okinawa to Jensen, Korea to replace Japanese troops in occupation of that country.

After that operation, they participated in transfer and convoy of Marines from Okinawa to Taku Bar, China for occupation of North China. They made most of the north Chinese ports with ADM Barbee acting as mediator between the Chinese Nationalist, Chiang Kai-shek, and the Chinese Communist forces. They left Shanghai, China Nov. 22, 1945, stopped briefly in San Diego, CA on Dec. 15, 1945, and transited the Panama Canal Dec. 24, 1945, arriving Norfolk, VA, Dec. 28, 1945.

After arriving Norfolk, he was transferred to the USN Personnel Separation Center, Nashville, TN, where he was honorably discharged Jan. 6, 1946. He re-enlisted in the USN at Little Rock, AR, March 26, 1946, and was sent to Naval Receiving Station Algiers, New Orleans, LA. Next duty station was Navy Primary Relay Communication Station, 18th and Constitution Ave. N.W., Washington, DC.

He met and married his present wife in Wash-

ington, April 30, 1949. His next duty station was Argentia, Newfoundland. Discharged again Jan. 6, 1950, Farfo Building, Boston, MA. Re-enlisted Washington, DC, Jan. 23, 1950, and was assigned to Navy Receiving Station, Anacostia, Washington, DC. Next duty station was Primary Relay Communication, Port Lyautey, French Morocco. His son was born at Rabat, Morocco, May 18, 1951.

Transferred in May 1952 to duty in CNO Crypto Center, Pentagon, Arlington, VA; transferred in August 1954 to USS W*asp* (CVS-18) at North Island, San Diego, CA and was promoted to CPO and initiated aboard the *Wasp* at Yokosuka, Japan, on Nov. 16, 1954. Transferred to COM-17, Naval Communication Station, Adak, AK in November 1954. Discharged again and re-enlisted for transfer to Navy Receiving Station, Anacostia, Washington, DC, Jan. 22, 1956.

Next duty station COMSUBCOMNELM NAVSUPPACT Major Relay Station, Naples, Italy. In May 1959 he transferred to AIRTRANSPORT Sqdn. 22, NAS Norfolk, VA; transferred to USS *Midway* (CVA-41) at Alameda, CA. Left USS *Midway* in November 1963 for NAVSTA Treasure Island, San Francisco, CA to await retirement.

Retired and transferred to the Fleet Reserve Dec. 2, 1963. From 1963-65 was employed at Navy Fire Alarm HQ, Naval Base, Norfolk, VA from 1965-85, was employed as radio electronic officer aboard various ships of the US Merchant Marine. He is now enjoying retirement and attending as many conventions and reunions as he can.

GEORGE A. PETERSEN, QMC, USNR, born in Oakland, CA, Jan. 12, 1925. He joined the USN in January 1943 at San Francisco and attended boot camp and QM/Signal School at Farragut, ID. Was then sent to Amphibious Trng. Ctr. in Queensland, Australia, followed by assignment to USS LST-458 of the 7th Amphibs in New Guinea (MacArthur's Navy).

He participated in the landing of troops through the Bismarck Archipelago and witnessed GEN MacArthur wading ashore at Leyte and heard his stirring speech. Married in September 1945 in Australia; returned to San Francisco where he was discharged and immediately re-enlisted in the USNR. His wife and baby joined him in July 1946.

As a member of the Organized Naval Reserve, he went on summer cruises with the USS S*hields* (DD-596) and attended several Navy schools. Recalled to active duty in 1950 and assigned to USS *Arequipa* (AF-31) serving the entire Pacific area. Promoted to QMC in September 1951 and transferred to Naval Station San Francisco, serving as disciplinary officer. Transferred to USS *Intrepid* (CV-11) and served until his enlistment expired.

Became a firefighter and progressed through the ranks, finally becoming fire chief. His awards include WWII Victory, American Campaign, Navy Unit Commendation, Philippine PUC, Asiatic-Pacific Campaign w/6 stars, Philippine Liberation w/ 2 stars, Navy Occupation Medal w/Asia Clasp, China Service Medal, Korean PUC, Korean Service w/star, UN Service Medal and the National Defense.

He and his wife Joy have four children, nine grandchildren and four great-grandchildren.

ROBERT L. PHILLIPS, RMCM(SS), USN (Ret), born Kenosha, WI, Jan. 29, 1927. Enlisted June 1, 1944, went to boot camp at Great Lakes, Radioman Schools at Northwestern University and Sampson, NY. He served in LCI(L)-1052, PCE(C)-880, PCE-872, PGM-22, Staff COMNAVFE Yokosuka and Pusan, NAVRESTRACEN Fargo, ND, Askari (ARL-30), SUBSCHOL New London, *Pickerel* (SS-524), Staff COMSUBPAC Pearl Harbor, Staff RMSCOL San Diego, *Manatee* (AO-58), Staff COMPHIBRON-1 (CTF-76).

Transferred to FLTRES, Jan. 29, 1970, retired list on June 1, 1974. Employed by U.S. Immigration and Naturalization Service in March 1971 and served at Border Patrol, El Centro, CA. He was immigration inspector at Calexico, CA and Chicago; immigration examiner, Chicago and Charlotte, NC; supervisory immigration inspector at Laredo, TX; and supervisory immigration examiner (AOIC) Fresno, CA.

Retired November 1990 as OIC, Cincinnati, OH, Immigration Office and port director, Cincinnati International Airport. Married Shirlee Hart in 1948 (deceased in 1990) and has three children: Sue, Sara and Roger. He remarried in June 1993 to Wilma Foust and they reside in West Lafayette, IN.

JAMES PEREZ RABON, AVCM(AW) E-9, born May 31, 1945, Agana, Guam. He joined the USN Feb. 18, 1965, Naval Station, Guam and sent to NTC, San Diego, CA. He advanced to CPO Sept. 16, 1981, VS-21; AECS Jan. 16, 1986, VS-41; AVCM, July 16, 1991, HC-3 upon transfer from VS-37.

His awards and medals include the Navy Commendation, Navy Achievement w/2 Gold Stars, Navy Unit Commendation, Good Conduct Award w/Silver and Bronze Star. He earned the Enlisted Aviation Warfare Specialist, Jan. 22, 1986.

Unit awards include the Meritorious Unit Citation w/2 Bronze Stars, Navy "E" w/wreath, Navy Expeditionary Medal, National Defense w/Bronze Star, Armed Forces Expeditionary Medal w/Bronze Star, Vietnam Service Medal w/3 Bronze Stars, SW Asia Service Medal w/Bronze Star, Sea Service Deployment Ribbon w/Silver Star, RVN Gallantry Cross Unit Citation, RVN Civil Action Unit Citation and RVN Campaign Medal.

He retired from the USN, March 1, 1995, after 30 years at HC-3 NAS North Island, San Diego, CA. Married Fatima Bent Lahcen of Casa Blanca Morocco on Aug. 27, 1970, and has a daughter Madina and a step-daughter Souraya.

JERRY DON RALEY, QMC, USN (Ret), born Feb. 6, 1941, Houston, TX. He joined the USN Feb. 25, 1958, Houston, TX. Duty stations include *Gen. W.A. Mann* (TAP-112); *Pollux* (AK-54); USMCAS Quantico, VA; Da Nang, Vietnam; LCU; Vietnam Patrol Boats; *Agerholm* (DD-826); Naval Station San Diego, CA; HQ YTB-YO; advanced to QMC; Coastal River Sqdn. 1; *Dubuque*.

His awards include the National Defense, Good Conduct, Navy Expeditionary, Vietnam Campaign Medal w/OLC, Vietnam Gallantry Cross, Vietnam Civil Action Unit Citation, Vietnam Service, Navy Unit Commendation Ribbon, Bronze Star and Civil Action Unit Citation. He retired from the USN May 16, 1977, and transferred to the Fleet Reserve.

Married Karen on May 16, 1965, and has two children: Robin Lynn and Laura Jeannine. Currently a tug boat captain for Hollywood Marine Inc., Texas.

THOMAS WILSON REESE, CPO, USN (Ret), joined the USN in March 1934, Richmond, VA, went through boot training in Norfolk, VA, then assigned to the USS *Whitney* (AD-4). Commissioned USS *Hornet* (CV-8), Oct. 20, 1941, and was aboard when *Hornet* transported LTC James Doolittle and his volunteers as close as possible to where they were launched to bomb Japan on April 18, 1942. Participated in Battle of Midway June 4-6, 1942; Guadalcanal invasion; and Battle of Santa Cruz where *Hornet* was sunk Oct. 26, 1942, and he was rescued by USS *Mustin* (DD-413) then transferred to USS *Pensacola* (CA-24) for transportation to Noumea, New Caledonia.

Boarded the *Lurline* for San Diego on Nov. 18, 1942; went to Boston to fit out and commission USS *Lexington* (CV-16) Feb. 17, 1943; *Lexington* was involved in 47 air strikes during which time she survived a torpedo hit on Dec. 5, 1943, and when a Japanese plane crashed into them Nov. 4, 1944. She was decommissioned in 1947 and he was ordered to shore duty in San Diego.

Transferred in November 1949 to USS *Oglethorpe* (AKA-100); April 1950 to Underway Training Cmd. until November 1954; advanced to Electrical School, Chicago, IL; re-commissioned USS *Lexington* (CVA-16), Aug. 15, 1955, and became the only two-time plankowner for that ship.

In June 1956 went to shore duty in Saufley Field, Pensacola, FL. Retired Nov. 17, 1958, and

entered Heald Engineering College in San Francisco where he earned his BSEE degree in April 1961. Hired by Convair Div. of General Dynamics, remaining there until he retired Jan. 28, 1977.

He is now active in community affairs and serves on several Senior Citizen Committees, does volunteer and handyman work. Married Sept. 8, 1941, to Lee Fleming and has one adopted son who is married and has three daughters.

RICHARD F. REEVES, CHPHM, USN, born Aug. 22, 1917, Putnam County, IN. He joined the USNR 16th Div. in Indianapolis, IN in 1939. On Nov. 4, 1940, he was called to active duty on the USS Sacramento which was in Pearl Harbor in December and later to Hilo, HI and Palmyra Island.

Transferred to the USS *Kennison* (DD-138) as a pharmacist mate with independent duty. They didn't have a medical doctor, so he was promoted to chief pharmacist mate and was the "Doc" for the next three years. The captain of the *Kennison* was only 23 years old. Their service was mainly with search and rescue for the USS *Ranger*.

After a total of 53 months of sea duty, he was transferred to the USNH in San Diego, CA, as chief master of arms in the corpsmen mess hall. Discharged in November 1945 as a CHPHM, permanent appointment. He completed his pharmacy degree and worked as a registered pharmacist in Indiana until retirement in 1988.

His medals include the Good Conduct, American Defense, Pacific Fleet w/star, WWII Victory and Pearl Harbor Commemorative Medal.

THEODORE E. ROWAN, AOC, USN (Ret), born Sept. 15, 1924, Wheeling, WV. He enlisted in at NRS Cleveland, OH, Sept. 15, 1941, and went to boot camp at NRS Newport, RI. First duty was VP91, a PBY "Black Cat" Sqdn., South Pacific; designated combat aircrewman (Silver Wings 3 stars) AOMB-3/c - 2/c, age 18; married at NAAS Kingsville, TX; made AOM1/c in 1944, age 19.

Promoted to ACOM, March 1, 1946, while attached to VPB-71 in the Philippines/SR to Chief 4 1/2 years, age 21+; Com Commander NRTC Great Lakes, IL; recruiter in charge NR SUBSTA Ukiah, CA; Ord. Chief VF-154 - CAG-15 on USS *Princeton* (CVA-37), Korean Conflict, 1952; world tour VF-94 - CAG-9 on USS *Hornet* (CVA-12), 1954; special weapons and leading chief on staff of CAG-15, 1958-59; R-in-Charge NR SUBSTA Lorain, OH.

Transferred to Fleet Reserve NRS Cleveland, OH, May 4, 1962 (24 years). Schools attended include Air Gunner, Bombsight CL "B"; Recruiters CL "C" and Special Weapons Loading.

He is entitled to wear 12 citation, commendation, service medals, etc. including the Presidential

Citation, Navy Unit Commendation, Good Conduct w/5 stars and China Service (extended).

After Navy was lieutenant deputy sheriff, Lorain County, OH; moved to California and worked as security chief for a development company until total retirement in 1988. He keeps busy with fraternal and service clubs.

His wife Lorranie died in 1982. He has two daughters, Darlene and Sherry, five grandchildren and one great-granddaughter.

JOHN C. SANTSCHI, BMC later CSP(A), USN (Ret), born Aug. 17, 1917, Akron, OH. He joined the USN March 31, 1942, Chicago, IL; enlisted as chief petty officer. Duty Stations: Norfolk, VA; San Diego NTS; Sampson NTS; Camp Pendleton, Oceanside, CA; and Terminal Island NS.

Discharged Sept. 29, 1945, USNTS, Terminal Island, CA. After the service he was a teacher and coach at San Pedro High School, San Pedro, CA. Retired after 48 years teaching. Enjoys swimming, walking, golfing and attending football games at Ohio State and USC.

Married Mary Jayne on April 14, 1943.

FRED A. SCHLUETER, ABHC, USNR, born in Ripon, WI, and entered the USN Aug. 1, 1957, following high school graduation. Attended boot camp at Great Lakes, IL; graduated from Music A School and was stationed with bands at Washington, DC; NAS North Island, CA; USS *Ranger* (CVA-61); USS *Shangri-La* (CVA-38) and NAS Memphis where discharged Aug. 1, 1961.

Changed rates on entering USNR and served as ABH at Reno, NV, NAS Fallon, NV and NAS Alameda, CA. Currently in USNR in IRR waiting for retired pay at age 60.

Married Kathy Parsons and has two children, Daniel and Heidi, and one grandson. He is self-employed and lives in Chester, CA.

RONALD E. SECHRIST, FCC(SW), USN/USNR, born in Connellsville, PA, Nov. 9, 1948. Enlisted in the USN's advanced electronics field program April 15, 1968, Pittsburgh, PA. Graduated boot camp (Color Company 271) and FT "A" and "C" Schools at Great Lakes, IL.

Assigned to first ship, USS *Decatur* (DDG-31) homeported at Long Beach, CA; deployed to WESTPAC during the Vietnam Campaign. Next assigned to the pre-commissioning crew of USS *Downes* (DE-1070) in Seattle, WA.

Discharged from USN July 12, 1974, and joined the USNR at NRC McKeesport, PA, accepting a TEMAC assignment as a canvasser/recruiter for NRD Pittsburgh. Units assigned as a reservist include: AR-405, USS *Vulcan* (AR-5) Det 205, Phib CB2 Det 105, and is presently assigned to NR Wepsta Earle 306. Qualified ESWS and made chief in 1979 while assigned to USS *Vulcan*.

Accepted a CWO commission in March 1991 and is presently assigned as XO of his unit at N&MCRRC Pittsburgh as a CWO3. He met his wife Sue Ann (Lyon) while stationed at Great Lakes and has been happily married over 24 years. They now reside in Vanderbilt, PA.

ROBERT VERNON SIMMONS, BMCS, USN (Ret), born Sept. 14, 1926, Nashville, TN. He joined the USN Aug. 12, 1944, Nashville, TN, and Jan. 21, 1962, Jacksonville, FL, NAS.

Duty Stations: Camp Perry, VA; Camp Bradford, VA; NAB Fort Pierce, FL; Solomons, MD; USS LCS(L)(3)77, NAS Glenview, IL; NAS Twin Cities, MN; USS *Tutuila* (ARG-4); USS *Bradley* (DE-1041); Phib CBI Coronado, CA; NTC Orlando, FL; (DD-822) *McGard*; PERSUPDET, Washington Navy Yard. Was CPO at NAS Twin Cities, MN and senior chief while aboard USS *R.H.M. McGard*.

Personal awards include the American Campaign Medal, Asiatic-Pacific Campaign Medal, WWII Victory Medal, Navy Occupation Medal, China Service Medal, Good Conduct Awards (4), Vietnam Service Medal, Vietnam Campaign Medal w/device, National Defense Service Medal and Enlisted Surface Warfare Specialist. Unit Award includes Meritorious Unit Commendation. Retired Nov. 2, 1979, PERSUPDET Anacostia, Washington, DC.

Married Augusta L. Oct. 22, 1955, and has two children, Robert Jr. and Frank. After service attended two years Community College and worked one year with a fastener company, security for SAV & PAC, corrections officer; eight months Orange City Jail and five and a half years with Seminole Cty. Jail. Retired completely April 25, 1993.

JOHN CLARK SIMON, HMC, USNR (Ret), born Feb. 18, 1946, Meadville, PA. He joined the USN Aug. 2, 1964, Meadville, PA. Duty stations

include NTC San Diego, CA; USNH San Diego, CA; USS *Suribachi* (AE-21); USNH Da Nang, RVN; USS *Krishna* (ARL-38); RVN; NAS Dispensary, Lemoore, CA (Desert Storm); Nine Naval Reserve Centers.

Awards and decorations include Good Conduct, Navy Reserve Meritorious Service Medal (4), Vietnam Ser-

vice (6), Vietnam Campaign, Armed Forces Reserve Medal (2), Expert Pistol and Rifle Medals, National Defense Service Medal (2), Navy Unit Commenda-

tion, Presidential Unit Citation-USNH Da Nang, RVN Navy Unit Commendation; Vietnam Civic Action Medal and Combat Action Ribbon-USS *Krishna*.

Retired from USNR April 1, 1994, N&MCRC Las Vegas, NV. After release from active duty: drilling reservist-9 Navy Reserve Centers; Penn State Univ. 1970-74, BS degree in health care admin.; Peace Corps-Thailand, 1975-76, malaria control; Penn State Univ., 1976 Graduate School Public Admin.; National Univ., 1979 Graduate School Health Admin.

Presently, assistant shift manager, Fremont Hotel/Casino, Las Vegas, NV. He was initiated as CPO at NAS Lemoore, CA, Sept. 16, 1990. Married Alice A. in April 1985.

CARLTON L. STEWART SR., PMC, born Aug. 22, 1927. Enlisted USNR, June 15, 1945—Jan. 7, 1952; USAF Jan. 8, 1952—Dec. 19, 1955; USNR

Aug. 3, 1965—Jan. 1, 1986; retired with pay in 1987. His last assignment was with NR AS-39 Land Det 407. Appointed to CPO in 1969.

Married Aug. 28, 1948, to Peggy, and they have two sons. He was employed at R.J. Reynolds Tobacco Co. as a patternmaker for 30 years. He retired in 1987.

RICHARD STOIBER, CHIEF SHIFITTER (AA), born Aug. 31, 1921, Kenosha, WI. He joined the USN Oct. 22, 1940, Chicago, IL. Duty stations include USNTS Great Lakes, IL; Navy Service School Ford Motor Co.; USNAS San Pedro, CA; USS *California*; and USS ARD-16.

He was promoted to chief shipfitter while aboard the USS Ard-16 (auxiliary repair dock) on Dec. 1, 1945, in the Philippines.

Stoiber was aboard the USS *California* during the attack on Pearl Harbor, Dec. 7, 1941, and was commended by commanding officer for his performance of duty in a highly meritorious manner during the Pearl Harbor attack. He was recommended to Commander Base Force for advancement to shipfitter second class. He also received citation for his courage, judgment and performance of duty.

Other awards include the American Campaign Medal, Asiatic-Pacific Campaign Medal w/2 Bronze Stars, Philippine Liberation Medal w/2 Bronze Stars, WWII Victory Medal, Philippine Presidential Unit Citation, Good Conduct Medal and the Pearl Harbor Commemorative Medal issued by Congress for the 50th anniversary of Pearl Harbor attack.

Married Virginia Anna Braem on Oct. 2, 1943, and has two children, Nancy and Marsha. After the service he worked in the profession of plumbing, heating and air conditioning. He is currently retired.

ROGER D. STORMES, was a lifelong resident of Syracuse, NY. He went to Hospital Corps School at Bainbridge, MD and was stationed at the Naval Hospital there. After discharge, he married and has two fine sons, Scott and Sean. Unfortunately, six

years later the marriage ended in divorce, but the relationship with his sons remained strong and still does. They both secured excellent jobs after high school and each has a child of their own.

He has written many letters, dissertations and articles for various newspapers namely: *The Freeman's Home Journal*, Cooperstown, NY; *Winter's Eve*, *The Baldwinsville Messenger* and *The Syracuse Herald Journal*. His 1980 *Dissertation on Regulation* was read on the floor of the 96th Congress.

Fifty Years, Fore, Aft and Midships is an autobiography filled with dynamic experiences, adventures and reminiscences, as the reader is transported through the years. This is a very personal story about a boy who was reared in a medium size city in Upstate New York; a neighborhood of poor to middle class families. His existence was a proud one that experienced continual uprooting and indecision to poor decision.

The story is marked with sadness, glory, death of a close brother and adventure. Only in later years, as a young divorced adult, did the writer find complete harmony. The subsequent success of he and his two children, along with the overwhelming support from his wife, culminated the personal satisfaction that he always wanted. *Fifty Years, Fore, Aft and Midships*, will relate to Middle America at its finest.

He and present wife Donna have two children, both college graduates with communication degrees.

JERRY L. SWEENEY, AZC, USN (Ret), born at Arnold, NE, April 21, 1938. He joined the USN Feb. 24, 1959, and attended boot camp with

an all Nebraska Company at San Diego. His first duty assignment was aboard the USS *Pine Island* (AV-12), followed by tours with VF-91, VX-5, NATTC Memphis, VAH-4, VAH-10, NAS Whidbey Island and VA-196.

He was selected as a plankowner of the AZ rate in 1964, advanced to chief on April 16,

1966. During assignment at NATTC Memphis, he was a student in the AME(B) and data analysis (C) schools and an instructor in the AME(A) and AZ(A) Schools.

Married Sandra Yocum, July 24, 1960, Dunning, NE. They have four children: Gynon Nash, Daylon Sweeney, Tamara Schultz and Tonja Dent, and three grandchildren. He is presently employed by Greyhound Lines as a bus driver.

ROBERT M. TAYLOR, CWO4, USN (Ret), born Oct. 10, 1920, Worcester, MA. He joined the USN in November 1939, Springfield, MA, with recruit training, Newport, RI; Hospital Corps School, Naval Hospital Corps School, Portsmouth, VA; Naval Hospital, Chelsea, MA; 1st Mar. Div., FMF (served consecutively at Quantico, VA; GTMO; MCB, Parris Island, SC; MCB, Camp Lejeune, NC; Staging Area,

Apia, Upolu, Western Samoa; Guadalcanal Campaign, Solomon Islands; R&R, Camp Mount Martha, Melbourne, Australia; Staging Area, Oro Bay, New Guinea; Cape Gloucester, New Britain Campaign).

Naval Hospital (made CPO Aug. 23, 1944), St. Albans, Long Island, NY; Naval Med. Research Lab., Camp Lejeune, NC; USS Sims (APD-50); USS ATA-201; USS LCI(L)985; Patrol Sqdn. 8, NAS, Quonset Point, RI; Naval School of Hospital Admin., Bethesda, MD; HQ 1st Naval Dist., Bost, MA; Naval and Marine Corps Res. Trng. Ctr., Worcester, MA; USS Van Valkenburgh (DD-656).

On Station Korean War; COMSERVPAC, Honolulu, HI; Marine Officer Procurement Office, Boston, MA (selected for WO1, May 1, 1958); Naval Hospital, Pensacola, FL; Naval Aerospace Medical Ctr., Pensacola; 2nd Mar. Div., Camp Lejeune, NC (afloat during Cuban Crisis as CO, Field Hospital attached to 2nd Mar. Div.) later deployed with Marine Corps Landing Force Mediterranean-62 as CO, Field Hosp., embarked in USS Rockbridge (APA-228) from January—July 1962; Naval Aerospace Medical Institute, Pensacola; Naval Auxiliary Air Station, Saufley Field, Pensacola, FL; Naval Aerospace Med. Ctr., Pensacola. Retired May 31, 1970.

Received two Letters of Commendation (Commander, Service Force, US Pacfic Fleet, 1956 and Surgeon General, USN, 1965), Five Good Conduct Medals and the Presidential Unit Citation.

After the service was in sporting goods sales; executive director, Multiple Sclerosis Society Chapter; district director, state of Florida, Development Services and retired after 12 years in February 1983. He has served on board of directors, Multiple Sclerosis Society, United Cerebral Palsy; Mental Health Assoc.; local Assoc. for Retarded Citizens; regional vice president of Florida Assoc. for Retarded Citizens; local Special Olympics Steering Committee and served two interim periods as executive director, Epilepsy Society of NW Florida. He served two years as developer for corporation building and managing retirement centers.

He is life member of ROA, American Legion, VFW, DAV; KC's, Deluna Assembly, Kiwanian and former member of Lions International. Charter inductee of Guadalcanal Campaign Veterans, FRA; plankowner, USN Memorial and American Association of Navy Hospital Corpsmen.

Married Constance Shea, Feb. 10, 1945, and has eight children: Maureen Anne, Robert M. Jr., Mary Lee, Constance P., Timothy O., John J., Ann Elizabeth and Michael S. Currently, board member of Pensacola Care, Inc. in Florida.

ANTHONY TESTEVERDE, CBM, USNR, born in Boston, MA, Sept. 5, 1920, and joined the USN, Nov. 6, 1942, at the 1st Naval Dist. in Boston as a third class petty officer. Went from fishing boat to LST without boot camp or Navy training. Commercial fishermen and yachtsmen were asked to join the USN M-2 (the Inshore Patrol of Coastal New England) and within a few months he was sailing off the coast of New Guinea.

Plankowner on LST-459 which was built in Vancouver, WA. He made BOSEN2/c and participated in D-Day and landings in SW Pacific, earning three Battle Stars while the 459 earned six stars.

Sent home to pick up LST-746 at Pittsburgh, PA (plankowner). Within six months and a day after LST-746 was commissioned, he was promoted to chief BOSEN mate. He participated in the Philippine Liberation: Leyte, Mindano, Lingayen Gulf,

Zambales, Subic Bay, Mindanao and Palawan Island landings. LST-746 shot down three kamikaze planes and received four Battle Stars for WWII service. He was recommended for CWO.

While operating a smoke machine during enemy attack, the smoke machine exploded, causing him first degree burns and some hearing loss.

His awards include the Philippine Campaign, Victory Medal, Asiatic-Pacific, Philippine Campaign, American Theater and seven Battle Stars. He was discharged Sept. 25, 1945.

He is a member of the Navy League, plankowner-life member USN Memorial, plankowner-life member National CPO Assoc.; member National U.S. LST Assoc.; plankowner Massachussets, U.S. LST Assoc.; life member AMVETS, 50 year member of American Legion.

Today he is retired and he and his wife enjoy traveling and having fun together.

WILBUR M. THORPE, CRM, USN, born in Houston, TX, Oct. 17, 1922. He joined the USN Oct. 17, 1940, and after attending boot camp at San Diego, CA, he was sent to Pearl Harbor to attend the first Radio School held there. After graduation, May 1941, he was transferred to Naval Radio Honolulu, Wailupe, Oahu. Promoted to RM3/c November 1941 and was at the radio station during the Japanese attack on Pearl Harbor.

The radio station was moved to Wahiawa, Dec. 17, 1941, and he remained there until May 1942 when he was transferred to Palmyra Island for a six month tour of duty. In August 1942, he was promoted to RM2/c; November 1942, transferred to NAS Ford Island until March 1943 when he volunteered to return to Palmyra Island. He returned to the Naval Radio Station at Wahiawa in March 1944 as RM1/c.

He taught classes at one session of the Radio School, and returned to the watch list. In January 1945 he was transferred, along with 14 other RM1/c's to Joint Communication Activities in Guam for four months of temporary duty to install and operate high speed tape equipment between Guam and Honolulu. Was promoted to chief radioman, May 1, 1945, after four years, six months and 15 days USN service, at the age of 22 years, 6 months and 15 days.

Assigned to NAAS Santa Rosa, CA. In January 1946, he received orders to EE&RM School at Great Lakes, NTC. On May 2, 1946, he went aboard USS Rockbridge (APA-228) to participate in "Operation Crossroads" atomic bomb tests at Bikini Atoll.

Returned to CONUS in September 1946, then

transferred to Treasure Island for separation as term of enlistment was about up. Upon discharge, he remained in the Inactive Reserve until November 1954.

Divorced, he has two daughters, one son and six grandchildren. He is retired and writing a book on his USN experiences.

LESTER B. TUCKER, CWO4, born Feb. 22, 1921, El Paso, TX. Enlisted Nov. 9, 1939, San Francisco, CA with recruit training from Nov. 10, 1939—Feb. 10, 1940. He retired as CWO4 on Feb. 1, 1967.

Duty Stations/Squadrons: USS Memphis (CL-13), USS North Carolina (BB-55), Aviation Unit (VO-6), USS Oriskany (CVA-34), USS Salisbury Sound (AV-13), VPB-26, VPHL-5, VP-42 and VP-46.

Shore duty includes USN Recruiting Station, Modesto, CA; NAS Atsugi, Japan; Bureau of Ordnance, Washington, DC; COMFAIR, Hawaii; Nuclear Weapons Trng. Ctr. Pacific, San Diego, CA (twice); USN Missile Facility-Pt. Arguello, CA. Promoted to ACOM at NAAS Oceana, VA on Aug. 15, 1945.

Awards include the Distinguished Flying Cross, Air Medal (4), Good Conduct (7), American Defense w/A, American Theater, Asiatic-Pacific Theater w/3 stars, WWII Victory Medal (2), China Service Medal, Korean Service Medal w/star, UN Service Medal and Korean Presidential Unit Citation.

Married former Helen Yates and has children from previous marriage: Lavone Anita, Juanita Ann, Jeannette Gail and Kenneth Lane. He resides in Oak Harbor, WA.

He holds a commercial pilot license (now inactive). In 19th year of writing a multi-volume book covering USN ratings from Revolutionary War to date. He is the author of the History of the Chief Petty Officer Grade and several articles on navy ratings.

VINCENT J. VLACH, LCDR, USN (Ret), born June 19, 1917, St. Michael, NE and enlisted Jan. 12, 1937, in Omaha, NE. Sent to NTS San Diego; he is survivor of USS Arizona (BB-39) and remained until March 1942 closing out ship.

Promoted chief yeoman (AA) at COMSERVFORPAC and (PA) at COMDESPAC. Appointed ensign and assigned to USS Gambier Bay; CINCPAC Advanced HQ, Guam; Office Chief of Naval Operations; XO of HQ, COMNAVFORCES, Marianas; COMSURASDEVDET, Key West; flag secretary and aide, COMPHIBGRU-3 in USS Eldorado and USS Estes; chief, administration (J2) on Staff CINCPAC.

Retired May 31, 1960. He married Jeanne in March 1940. Employed at Rockwell 23 years and retired 1982; was one of three original incorporators for the NCPOA and holds membership 0003-CL. Resides at Continuing Care Retirement Community and his hobby is genealogy.

ROBERT J. WATSON JR., YNC, USN (Ret), NCPOA member 0005, born Dec. 19, 1921, Trenton, MI and enlisted in USN July 2, 1940. He retired June 29, 1959.

Reported to Commander Patrol Wing Two at NAS Pearl Harbor in October 1940. Was so stationed during the Japanese attack on Dec. 7, 1941.

Left COMPATWING TWO in September 1942 for NAS Alameda, CA, for duty in Patrol Sqdn. 63 when commissioned Sept. 19, 1942. With VP-63 and saw duty in Quonset Point, RI; Iceland; RAFB, Pembroke Dock Wales; and Port Lyautey, North Africa.

Authorization for CPO received in Wales and went to London for uniforms. Got necessary authorizations from naval attache for uniforms, shoes, etc. and had them made by London tailor.

At Marseille he set up communications center three days after the invasion of Southern France. Other duties were Palermo, Sicily; recruiting duty, Detroit, MI; USS *Badoeng Strait*; Fleet Air Electronic Training Center, San Diego; Fleet All Weather Training Unit, Barbers Point, HI; recruiting duty, Salt Lake City, UT; Kwajalein, Marshal Islands; USS *Polaris*; USS *Blue* and USS *Mansfield*.

He and his wife Betty have five children, six grandchildren and a great-grandson.

VAN WATTS, Naval Philosopher and Sea Power Advocate Supply Corps, CWO W4, USN (Ret), born Aug. 26, 1920, Mooers, NY. He joined the USN Dec. 15, 1937. Promoted to chief disbursing clerk at 22 years, 2 months and 5 days on Nov. 1, 1942, while serving in USS *Mackinac* (AVP-13) during the Battle of Guadalcanal.

Promoted to SC warrant officer at age 23 years, 3 months and 19 days, Dec. 15, 1943, while serving as deputy paymaster at Milne Bay during the Battle of New Guinea. Retired Oct. 1, 1962, as SC CWO4. He declined higher promotion but CO entered recommendation for LCDR in his official record.

Originator/producer of TV and radio Navy-slanted shows from Norfolk in early 1950s. Founded Navy's Sailor of the Week, Month, Quarter and Year Programs which implement his ideas and philosophy acquired coming up through the ranks throughout Navy in 1952.

Created Norfolk's big ship welcoming ceremonies called by CHINFO "The Centerpiece of Norfolk-Navy relations," 1954. Enrollment of all TV and radio media in area to promote Norfolk-Navy goodwill inspired formation of Navy League's Hollywood Council, or "media division" of which he is a life member, in 1954.

Receipt of numerous honors including city of Norfolk official thanks 1954 and Royal Mace Pin in 1988, Hollywood Council Navy League's Gold Plaque for the founder of programs that spread around the globe with three of his nautical celebrities returned from Persian Gulf duty for ceremonies at Bob Hope Hollywood USO 1988.

Inspired by Maury, Mahan and Dana, Watts came into the Navy well-prepared to continue their work and is today honored for his own contributions to the sea service throughout the Navy. "Much of the positive civilian awareness of the role and value of the Navy can be attributed to Van Watts." Naval Historian Kit Bonner in *Treasure Island Museum Association Newsletter*, Fall 1993.

Decorations include the National Defense Service, American Defense Service, American Campaign, Asiatic-Pacific Campaign, WWII Victory Medal, Navy Occupation, Armed Forces Expeditionary, Guadalcanal and New Guinea Battle Stars.

Member of USN Inst., Naval Hist. Found., Navy Sup. Corps. Assoc., Guadalcanal Campaign Vets Assoc., VFW, FRA, Hollywood Council Navy League, Botsford Family Historical Assoc. and the New Hampshire Hist. Soc.

Married Lilie Remoreras in 1971 and has daughter Michelle Remie. His children from previous marriage are Philip, Charlotte, Britt, Lance and Douglas.

ANNABELLE D. WEGELE nee MENGE, CSKD, USNR, born Dec. 31, 1919, Fond Du Lac, WI. She was sworn in March 20, 1943, in Milwaukee, WI and started active duty May 17, 1943. Duty stations include USNTS, (WR) Bronx, NY; USNTS, Milledgeville, GA, Storekeeper School; USNTC San Diego, CA; and USNRB San Diego.

Was platoon leader at USNTS, Bronx; USNTC, SD, discharged illiterates; USNRB, SD, worked on pay lines on base and aboard ships; worked officers per diem, transfers and decommissioned ships to be put in mothballs; and received chief rating at USNRB San Diego.

Awards include the WWII Victory Medal, American Area Campaign Medal and Good Conduct Medal. Received honorable discharge July 2, 1946, USNTC San Diego, CA.

Charter and field rep. of WISMA Memorial Foundation; 2nd vice commander, Trier-Puddy Post; member of WAVES National; charter member and former treasurer of Badger Unit 39; member of FDL Cty. WWII Commemorative Comm. and 50 year member of the Eastern Star; staff of Senior Center.

Married Robert A. Wegele Sr. June 8, 1946 and has one son Robert Jr.

W.A. "BILL" WILLIAMS, EMC, USN, born March 30, 1925, Rogers, TX. He received boot training, followed by Class A and C Schools at NTC San Diego, CA. First duty station was USS *Cheleb* (AK-138) which he commissioned Jan. 1, 1943, San Francisco.

Attached to SERPAC, landings were made in the Admiralties, Marshall and Gilbert Islands. Later the *Cheleb* was part of landings at Ulithi, Iwo Jima, Okinawa and Leyte Gulf.

Transferred to the USS *Benevolance* (AH-13) and released to inactive duty in 1945. Received BSC degree in civil engineering, recalled to active duty in August 1950 and placed aboard USS *Manchester* (CL-83), transferred to USS (LST-611) in time for landing at Inchon, Korea.

Later transferred to the US for recommissioning USS *Formoe* (DE-509) and released to inactive duty

at Newport, RI, in 1953. Recalled again in 1955 and served aboard USS *Keywadin* (ATA-213) in Boston, MA, transferred to Great Lakes, IL, for EM Class B School, then went aboard USS *Diamond Head* (AE-19).

Released from naval service in 1957 as EMC, USN. He is now retired in Texas Hill Country and is a widower with one son, Gary Alfred.

He is one of the founders of the National Chief Petty Officers Association and served as its third president.

ANNA MARIE WILBER (see Anna Marie Wilber-Anderson)

RONALD L. WILSON, HMC, USN (Ret), born Dec. 9, 1940, St. Louis, MO. He enlisted in the USN July 10, 1959; attended recruit training at NRTC San Diego, CA; Hospital Corps School, San Diego, then stationed at Naval Hospital Camp Pendleton, CA. Attended Urology Technician Training at Naval Hospital San Diego then again stationed at Naval Hospital Camp Pendleton.

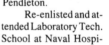

Re-enlisted and attended Laboratory Tech. School at Naval Hospital San Diego, then assigned to Naval Hospital Sangley Point, RP. Transferred to USS *Piedmont* (AD-17); Naval Hospital Great Lakes, IL; attended Medical Services Tech. School, Naval Hospital Portsmouth, VA.

Assigned to recommission crew of USS *Diamond Head* (AE-19), then commissioned crew of USS *Thomas C. Hart* (FF-1092) and earned (HM-8425). Final duty stations was with I&I Staff 3/24/4th Marines at Lambert Field, St. Louis, MO. Retired July 14, 1978.

Married Susan P. Nachtsheim on July 14, 1978, and has three children: Susan, Becky and Christopher, and one grandchild James. One daughter, Theresa, died in September 1975.

He is now retired and coaches softball and baseball; he is on the board of directors of the St. Louis Junior Football League; an active member of the Lions in St. Louis and a charter member of the American Association of Navy Hospital Corpsmen.

JACK E. WINGERTER, ABHC(AW), USN, born in Quincy, IL. He entered the USN on June 24, 1971. After recruit training, he attended Fleet 'Prep' School at NATTC Memphis, TN, before receiving orders to crash and salvage responsibilities at NAS Guam. Transferred to Helicopter Combat Support Sqdn. 3 at NAS North Island, CA, and was honorably discharged in June 1975.

Re-enlisted Nov. 15, 1979, and reported to USS *Belleau Wood* (LHA-3) homeported in San Diego, CA. Transferred to NATTC Lakehurst, NJ, for ABH "A" School instructor duty. Re-enlisted to USS *Nimitz* (CVN-68) on Nov. 7, 1986. Received chief's initiation in 1989. Transferred to NAS Corpus Christi, TX, where he again re-enlisted for orders to USS *Tripoli* (LPH-10), where he is currently serving in 1994.

His first marriage ended in divorce and they had one child, Jeremy, who is presently a student at Iowa State. His marriage to Ann Francis (Summers) has produced two children, Jason, a freshman in high school, and Jacob, a 7th grader.

ANDREW "ANDY" WELCH YANCYE,

AVCM, USN (Ret), born April 29, 1925, and was raised and educated in Savannah, TN. He joined the USN Sept. 17, 1943;
went to boot camp at Great Lakes, IL; AO School, Millington, TN; Aerial Gunnery School, Yellowwater, FL; and CAC School, Banana River, FL. Further assignments included VH-5, CASU-13, AROU-2 and AO"B" School.

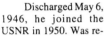

Discharged May 6, 1946, he joined the USNR in 1950. Was recalled to active duty with VP731 Sept. 29, 1950; September 1951—July 1952 FASRON 110; released to inactive duty July 11, 1952; recalled to active duty Aug. 16, 1955; August 1955—October 1957 VA196, AO"B" School, VA155, VF91; 1957-58 change rate program AOU2 to AQ2; 1958-60 VF141; 1960-62 VA44; 1962-63 AQI"B" School; April 1963—May 1966 USS *Constellation* (CVA-64).

Promoted to AQCA on Sept. 16, 1963; September 1966—November 1967 NAMTRADET Whidbey Island; promoted to senior chief on Nov. 16, 1966; November 1967—May 1970 A6 tech. coordinator NAMTRAGRU Memphis; promoted to master chief

Feb. 16, 1970; May 1970—May 1973 VF-213; May 1973—June 1976 F14 tech. coordinator NAMTRAGRU Memphis; June 1976—October 1978 USS *America* (CVA-66); October 1978—Jan. 1, 1980 F14 tech coordinator NAMTRAGRU Memphis; retired Jan. 1, 1980.

Andy and wife of 49 years, Kay, reside in Memphis, TN. They have four children and 10 grandchildren.

LEWIS W. YORE, BMC, USN (Ret), born in

Scotland County, MO, Nov. 28, 1924. He joined the USN Oct. 26, 1942, Minneapolis, MN; attended boot camp and Torpedo School at Great Lakes, IL, before being assigned to the pre-commissioning detail for USS *Mobile* (CL-63).

This was followed by assignments to USS *Santa Fe* (CL-60) TACGRPACREFLT, BREMGRPACRESFLT, and Minesweeping Division (MSB DIV) One. At this last duty station, he served as a "boat captain" and was assigned TAD to eight different ships. While with MSB DIV One, he was authorized to wear the Navy Unit Commendation Ribbon for operations against enemy aggressor forces in Korea. He was also personally commended by COMSEVENTHFLT and authorized the Commendation Ribbon w/Combat Distinguishing Device and by SENAV who presented him with the Navy and Marine Corps Medal. He was earlier commended by COMFIFTHFLT and authorized the Commendation Ribbon while serving aboard the USS *Santa Fe*.

Other assignments include USS *Hummer* (AMS-20), USS *Ruddy* (AMS-380), USS *Loyalty* (MSO-457), NAVSTA Subic Bay PI YON-144 as a "craft master" and NTC GLAKES. Units he was assigned to received 16 Battle Stars. He also received, in addition to previously mentioned awards, the Philippine Campaign, Victory Medal, American Theater, Asiatic-Pacific Philippine Campaign, Good Conduct w/5

stars, Korean Presidential Unit Citation, US Service, National Defense and Korean Service Medal w/3 stars. He made chief while at NTC GLAKES and retired from the USN March 22, 1963.

He and his wife, Betty, have one daughter and two grandchildren. Today, he is "retired and doing nothing."

RAYMOND D. YOUNG, TMC, USN (Ret),

born Dec. 15, 1948, Keokuk, IA. He entered the USN at Chicago, IL, received boot training at Great Lakes, IL, followed by Torpedomansmate Class "A" School at AUW School, Key West, FL.

First real duty station was USS *Fulton* (AS-11), homeported New London, CT. While abroad the *Fulton*, he attended MK45 (ASTOR) Torpedo School at AUW School, Key West, FL.

Subsequent duty stations were MK28 (SUBROC) Missile School at AUW School, Key West, FL; Naval Weapons Station, Seal Beach, CA; MK45 (ASTOR) Torpedo Test Equipment School at AUW School, SCC Orlando, FL; USS *Fulton* (AS-11), New London, CT for a second tour; Naval Inactive Ships Maintenance Facility, Valleyjo, CA; Instructor School, SCC, Great Lakes, IL; Nuclear Weapons Training Group, Pacific (instructor duty), NAS North Island, San Diego, CA (Received chief's initiation); USS *Sperry* (AS-12), Naval Submarine Base, San Diego, CA; USS *McKee* (AS-41), Naval Submarine Base, San Diego, CA; MK48 Torpedo Shop, Naval Submarine Base, San Diego, CA; transferred to the Fleet Reserve on Oct. 31, 1985.

Chief Young owns and operates a small computer repair business in the city of Coronado, CA. He married Georgetta (Georgi) in 1982 and has a stepdaughter, Victoria (Vicki). From a previous marriage, he has three children: Raymond C. is on duty with the US Army, Donetta K. is a registered nurses aid in Illinois and Michele R. lives with her sister and works in Illinois.

C Division, USS Pennsylvania, (BB-38) in Bremerton, Washington, March 21, 1946. (Courtesy of William F. Moreland)

Sideboys mann the rail to pipe Senior Chief Petty Officer Joe B. Havens ashore.

Medical Department, USS Estes, AGC-12.

Joe B. Havens and graduation class, HM "A" School 1950, San Diego.

Boot Training Ship, U.S. Naval Training Center, San Diego, CA.

Index

The biographies were not included in this index since they appear in alphabetical order in the biography section.